B
U
P

PRODUCTION, OPERATIONS AND SUPPLY CHAIN MANAGEMENT

Grando · Belvedere
Secchi · Stabilini

Translation: Andrew Spannaus for Language Solutions for Business – London
Typesetting: Corpo4 Team, Milano

EGEA S.p.A.
Via Salasco, 5 - 20136 Milan
Tel. 02/5836.5751 - Fax 02/5836.5753
egea.edizioni@unibocconi.it – www.egeaeditore.it

First edition: September 2021

ISBN Domestic Edition 978-88-99902-57-5
ISBN Digital Domestic Edition 978-88-238-8318-5
ISBN International Edition 978-88-31322-41-6
ISBN Digital International Edition 978-88-31322-42-3

Table of Contents

This book is dedicated to all doctors, nurses,
and workers of factories and distribution chains
who have never stopped their Operations, saving lives
and sustaining all of us during one of the worst
and unexpected crises experienced by the planet.

Preface

Industrial production has always been a propulsion element in any economy, be it from a perspective of value-creation, employment potential, or its impact on sustainability and technological innovation; the ability to manage operations and supply chain is therefore a key skill for a company's competitiveness.

If this has always been true, today as never before, in a world hit hard by a medical emergency first, and then by economic and social ones, the reflection on the most appropriate management practices in manufacturing has become a central element for individual companies, entire sectors and countries.

New manufacturing strategies more oriented toward flexibility, agility and resilience have replaced traditional paradigms and choices oriented toward efficiency, pursued through outsourcing and offshoring in countries far from the assembly plants and their markets, sourcing policies based on very few suppliers, lean approaches and reduction of the working capital invested in inventories, planning systems based on optimization criteria and sell-in in channels crowded by the proliferation of product ranges. In this new landscape shaped by the recent medical and economic crisis the most appropriate manufacturing foot-prints and supply chains are those redesigned on the basis of local-for-local principles, quickly reconfigurable, characterized by multiple suppliers, compensation flows among warehouses and distribution centers to offer product availability where needed, insourcing decisions aimed at saturating production capacities and re-internalized skills, choices favoring agile practices, revision of end-to-end planning systems fueled by sell-out dynamics, focalization on products, channels and clients based on priorities linked to the necessity of insuring business continuity, cash-to-cash cycles control, creation of buffers justified by the volatility and uncertainty of the markets.

In summary, from an approach that, at best, protected itself from probabilistic events with risk management measures, to one aware of the need to put crisis management processes in place, aimed at preventing and reducing the impact of unforeseeable events.

In this context, deep reflection aimed at analyzing which production models to choose and how to redesign the value-creation chains has begun inside single production units and entire supply chains. The ability to analyze the need of change in the operations and, consequently, to select and implement appropriate choices, can only be based on robust approaches of analysis and diagnosis of manufacturing systems, the measurement of the current and prospective performance, and the knowledge of the foundations and principles of Operations and Supply Chain Management.

This work has therefore the objective of collecting and systematizing what has been developed through research and didactic production by the group of scholars who for more than 20 years have been collaborating on the topic of Operations and Supply Chain Management. In addition to integrating into a single perspective what has been developed and, at times, published by the authors of this volume, it is intended to collect here the scientific production traceable to three streams of contributions in the field of Operations and Supply Chain Management:

- the main works published at international level that represented and continue to form the foundational reference for this field;
- the works of Italian Authors who placed the topic of Operations Management in the studies conducted in our country;
- the most recent productions by a few colleagues with whom we share the commitment to focus on and teach Operations and Supply Chain Management.

In trying to provide the reader with a useful key to understanding its content, the work is written from the perspective of those who – be it a young student, a master participant, a manager or entrepreneur – are interested in approaching the main methods and practices of Operations and Supply Chain Management in order to start processes of analysis, diagnosis and improvement of company's performance. In accordance with the objectives set, the text is characterized by a managerial point of view, accompanied by multiple quantitative insights and examples of application of the main tools and algorithms illustrated.

This volume offers an extensive coverage of the issues discussed, selected based on the programs of Operations and Supply Chain Management taught in Management and Engineering faculties, ranging from the basics of the field to the most recent approaches and solutions, among which it is worthwhile mentioning the following:

- Analysis of production systems
- Operations and supply chain strategy decision making
- Performance measurement
- Problem setting and solving in the operations management
- Production planning and inventory control
- Production Scheduling, Just in Time and Theory of Constraints
- Lean management
- Procurement and vendor management
- Supply chain management
- Operations data management and Information systems
- Digital manufacturing and Industry 4.0.

In publishing this work, we wish to express our gratitude to all colleagues who inspired us, to Nadeem Abbas for his support in reviewing the English translation, to Cinzia Facchi and Cristina Casati for their patience in dealing with the authors during the editing process and to Claudia Valerii, who read most of the chapters catching mistakes we would have surely missed without her help.

Finally, thanks to all the university students, master candidates and participants to the executive courses met along our teaching path, whom we have taught something, and from whom we surely learned much.

Alberto Grando, Valeria Belvedere, Raffaele Secchi and Giuseppe Stabilini

1 Strategic Alignment in Operations and Supply Chain Management

by *Alberto Grando*

1.1 Introduction

The complexity and competitive dynamics that characterize manufacturing companies today have gradually led to the search for multiple organizational, management, and technological arrangements of Production and Supply Chains. Within the same sector, we see the co-existence of production structures that are different in terms of size, or technology used, with equal opportunity of success. Some are oriented towards targets of efficiency, others to reaching high levels of performance in terms of services offered or flexibility, and still others towards the search for absolute quality. The borders of the Supply Chains themselves become elastic; dense communication webs are woven, physical and informational flows with suppliers and clients are created, and alliances are composed to in order to compete, while it is necessary to be highly competitive to access the most coveted alliances. The goals change more rapidly over time and it becomes harder to reach them. Clients, in both B2C and B2B contexts, thanks to increasingly powerful and sophisticated connection technology, can access a virtually unlimited supply that heightens competition and imposes the ability to adapt supply to evolving demand, and the search for increasingly extreme levels of performance. A significant component of the ability to prepare an offer consistent with the needs of the market lies in each organization's system of Operations and in the ability to effectively coordinate with other company systems, regarding: Human Resources, Sales, Finance, etc. That "**set of systems**" de facto represents a true "organism" that, like any other living organism, consumes resources to produce others, in an incessant process of transformation of which Production processes, whether regarding goods or services, are undoubtedly the element of propulsion. Therefore, production-logistics systems must be seen as components of that organism, and interpreted as such, i.e. their physiological and pathological behavior observed, together with the context in which they are situated.

1.2 The scope of this book

With the goal of correctly framing the issues treated below, it seems appropriate to provide some initial remarks allowing us to define the perimeter of analysis of this book (Grando et al., 2007).

Operations

Each organization has an Operations function, because each business produces goods and/or services through processes, consuming resources to produce its outputs. In reality, operations exist within each company function. Creating a marketing campaign requires operations activities to select, coordinate, and manage media partners, prepare appropriate means of communication, conduct promotional activities, etc. Likewise, starting a selection of personnel entails performing operational activities linked to processes of receipt of resumes, their selection, preparation of interview calendars, etc.; just as closing a loan agreement with a bank or with outside lenders will require conducting operations. It is therefore useful to distinguish between Operations as an activity and Operations as a function, and their different meaning.

- "**Operations as an activity** means the management of the process within any of the organization's functions.
- **Operations as a function** means the part of the organization which produces the products and the services for the organization's external customers" (Slack et al., 2007, p. 15).

Below we will focus in particular on the second meaning, that is, the "business function that plans, organizes, coordinates, and controls the resources needed to produce a company's goods and services" (Reid and Sanders, 2005, p. 3).

Operations System

That function monitors an articulated system of means, people, resources, and knowledge, defined the **Operations System**, which exchanges information, physical, and financial flows, connecting with other company systems: financial, informational, commercial, etc.

Operations Management

The term **Operations Management (OM)** therefore refers to the management of that function, which is to monitor that system. More in particular, "Operations management is about the management of the processes that produce or deliver good and services" (Greasley, 2005, p. 5). In other words, the main task of Operations Management is the most appropriate management of the Operations System with respect to the company's goals, i.e. the organization and management of the processes that guarantee the **transformation of inputs**, consisting of materials, information, and in the case of services, of the clients themselves (**to be transformed inputs**) and machines, personnel, and methods (**transforming inputs**) into **outputs expressed by goods, services, and performance**. Transforming inputs into outputs does not only mean transforming leather into shoe uppers or bags, cotton into shirts, sheet metal into pots, or in the case of a service, a sick patient into a healthy person, or a hungry customer into a full and satisfied one, but it means doing all of that while offering the market those goods and services by guaranteeing performance that allow for effective and efficient acquisition and consumption, or in the case of services, use. This consists of the **performances in terms of time, cost, quality, flexibility, and sustainability** that accompanies the production of a good or the provision of a service and that are an integral part of the value transferred to the customer (see Chapters 4 and 5). The issue of sustainability, in its environmental and social forms, in which we place also the area of safety, should be considered a sort of constraint function. Indeed, in facing the possible trade-offs between the goals of a production system, company management should never search for compromises with respect to the issues of **environmental protection** and **social sustainability**; it's not only about good business practices, but in particular an ethical imperative.

A company able to produce a perfectly functioning good, with high quality and at low costs, will not be able to compete if its Operations System ends up systematically making late deliveries. Think of the case of businesses that operate in markets subject to seasonality, such as clothing or some sports articles: starting to deliver beautiful bikinis in October or skis with amazing performance in April will not open up good prospects for profitability and success. Similarly, a company offering tourist services with cut-rate prices, but cramped rooms, poor service, and unprepared personnel, will rapidly lose its reputation and market, just as an excellent chef, who systematically makes his customers wait an hour between one dish and another, will be destined to fail. Those specific "outputs" are accompanied by other collateral outputs generated by a production-logistics system, that have positive or negative value, and are often given scarce importance or defined as "**externalities**": employment, wealth creation, reutilization of recyclable materials, and reduction of environmental impact, among the positive ones; unemployment, pollution, social exploitation, and poverty among those that are negative (Grando and Vicari, 2019).

Operations and Operations Management are thus central elements in any business activity, equal to those of other functions such as marketing and finance. Whether one is producing tractors, lipstick, motorboats, ballpoint pens, or books, and whether services are provided by banks, restaurants, hotel chains, machine shops, universities, or consultants, in the absence of an adequate Operations System and appropriate Operations Management, the probability of generating value is very limited.

1.3 The sustainability imperative[1]

To speak of sustainability, even from the specific perspective of Operations and Supply Chain Management that are the subject of this book, means confronting the question of how to pursue the objectives of the present – whether they are those of a com-

[1] The analysis of Sustainability and Sustainable Operations Management is a synthesis of Belvedere and Grando, 2017 (Chapters 1 and 3).

pany, an institution or a broader socio-economic system – while ensuring that adequate standards of development are also guaranteed in the future.

The subject of **Sustainability** is therefore intrinsically connected to a vision of the future that postulates intergenerational equity. The socio-economic context in which future generations will live and companies and institutions will operate, will largely be the result of the decisions made today, and in particular, of the actions to be taken due to these decisions. Recognizing this great responsibility, it therefore seems appropriate to rethink management choices, placing attention on a more extensive, complex, objective function than the one that traditionally characterizes company operations, often summarized in the objective of value creation for shareholders. This involves devising decision-making processes based on the values of responsibility, ethics, and sustainability, within a timeframe that is consistent with the ability of the system to generate and regenerate adequate resources for sustaining its development. The development of awareness in this important field has mainly led to the different origins of the concepts of **Sustainable Development**, which for the most part refers to environmental impact and attention to the ecosystem, and **Corporate Social Responsibility** (CSR), which concerns guidelines on social matters for business managers. The two tendencies, however, have found elements of convergence and integration over time, in the sense that CSR is becoming an important tool that public actors and companies can use to pursue sustainable development objectives.

The most widespread definition of Sustainable Development is, in fact, the one that refers to the possibility of promoting development by looking not only at current needs, but also future ones. More precisely, **Sustainable Development** is defined as: "development that meets the needs of the present without compromising the ability of future generations to meet their own needs" (Brundtland, 1987, p. 54).

According to the European Commission's work, this definition "contains within it two key concepts:

- the concept of needs, in particular the essential needs of the world's poor, to which overriding priority should be given; and
- the idea of limitations imposed by the state of technology and social organization on the environment's ability to meet present and future needs." (Brundtland, 1987, p. 54).

With reference to the **needs** component, apart from increasing awareness of the topic of waste, consumer choices, and methods for the use of goods, which companies can certainly contribute to in terms of awareness and communication, little can presumably be done from a corporate perspective. On the other hand, with regard to the subject of **limitations**, namely the development of technology, processes, products, and practices that enable current limitations to be overcome, the responsibility of companies and institutions seems paramount. The role that a company ought to promote responsibly is that of developing business strategies, technological innovations, and practices that enable business and sustainability objectives to be pursued together from a dual social and environmental perspective, in the interest of all stakeholders, in a broad sense and in the long term.

The increasing attention of society to the subjects of sustainability and sustainable development has led to companies gradually paying more attention to the integration of CSR objectives and practices in the formulation of their development strategies and their pursuit of objectives. Several definitions of CSR have been coined in the literature and there are numerous studies that have analyzed the origins and evolution of the concept of sustainability over time and its implications on a corporate and public policy level (Perrini et al., 2006). The European Commission (2011) produced a new definition of CSR, as "the responsibility of enterprises for their impacts on society." In the same document, the Commission specifies that "Respect for applicable legislation, and for collective agreements between social partners, is a prerequisite for meeting that responsibility. To fully meet their corporate social responsibility, enterprises should have in place a process to integrate social, environmental, ethical, human rights, and consumer concerns into their business operations and core strategy in close collaboration with their stakeholders, with the aim of:

- maximizing the creation of shared value for their owners/shareholders and for their other stakeholders and society at large;
- identifying, preventing and mitigating their possible adverse impacts." (EC, 2011)

It is within the framework of this definition that the interaction between business, environmental, and social objectives is discussed below, with specific reference to the role played by Operations and Supply Chain Management.

Sustainability objectives therefore refer to three performance levels, defined as the **Triple Bottom Line** (Elkington, 1994; 1997) or the 3P, illustrated in **Figure 1.1**, which each company must try to jointly maximize.

- **Profit**: expression of the performance that leads to economic and financial sustainability and its development prospects in the medium to long-term.
- **Planet:** refers to the performance that guarantees environmental sustainability, in terms of environmental protection and the overall impact of the business on the environment.
- **People:** connected to the performance that measures the social impact of the business, in terms of social equity and cohesion, economic prosperity, and the protection and promotion of fundamental rights.

Unfortunately, a deep-rooted vision in several business contexts may conflict with these objectives, presuming that the maximization of profit may justify paying less attention to the subjects of environmental and social sustainability, as well as to conduct that does not respect these imperatives. The trade-off between these objectives must however be rejected, first of all on ethical grounds, and second of all, by extending the scope of the analysis of the results attainable by each company, with

Figure 1.1 **3Ps and Triple Bottom Line**

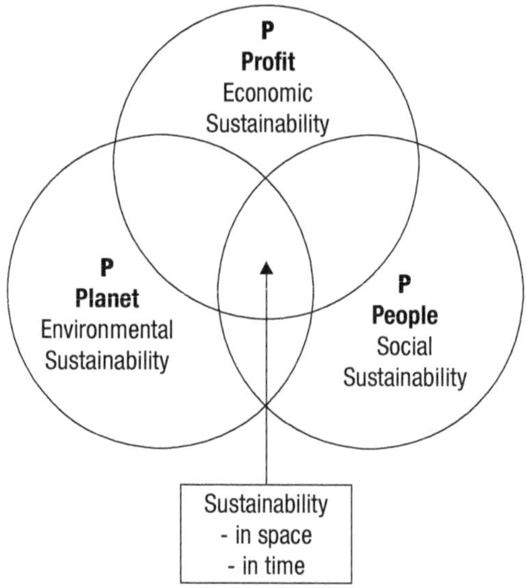

reference to both the timeframe of the aforesaid results subsequent to managerial or public policy decisions and the type of stakeholders that the achieved results are measured against.

If economic and financial sustainability is the main objective of value creation and is mainly connected to the interest of shareholders, the support of the other two kinds of performance, linked to environmental and social sustainability, extends the ultimate purpose of the company to all stakeholders, be they current or future, with the objective of guaranteeing an overall quality of life on the planet and fairness in standards of living in terms of time and space:

- **in space**, in terms of a better distribution of the value created by more fortunate populations, located in the areas that, although representing a minority of the world population (20%), absorb far greater wealth (80%), and the remaining part that in many cases find it hard to reach the threshold of survival;
- **in time**, with reference to the need to guarantee intergenerational equity, offering future generations the same opportunities that are offered to those of today's generations and gradually encouraging balance in all forms of growth and development.

In this regard, economic and financial sustainability must not only refer to the economic entity capable of producing it, i.e. the company, but to society as a whole, which by hosting and interacting with it, contributes to its creation and must therefore benefit from it. The concept of the triple bottom line therefore refers to broad objectives, which must nevertheless be summarized using performance, metrics, and specific indicators (Belvedere and Grando, 2017).

1.4 The new paradigm of the Circular Economy

The need to provide clear answers to the dilemma linked to the trade-off between population growth and people's aspirations to prosper, and the requirement to reduce the pressure on the environment and preserve the planet's resources, defined as **absolute coupling** at both the macro-level (in the drafting of consistent policies) and the micro-level (influencing the conduct of companies and supply chains) has led to the development of the approach known as the Circular Economy (Ellen MacArthur Foundation, 2012; 2013).

This approach sets forth the need to base the development of an economy and its industrial system on the reuse of products and raw materials, and to the extent possible, the need to replenish natural resources. The **Circular Economy** is essentially based on the possibility of intentionally contemplating an industrial and economic system in which all waste streams and emissions can be a source of value creation, limiting environmental impact and moving towards the search for forms of decoupling that allow for sustainable development and the restorative capacity of natural resources.

As shown in **Figure 1.2**, the traditional development of an industrial cycle, shown in the center of the diagram, proceeds according to a **linear logic** that starts with the extraction of raw materials and moves towards production, consumption, and the destruction of value though waste disposal. On the contrary, sustainable development must be based on **circular cycles**, shown along the sides, which differ due to the nature of the reusable materials and products involved – that is, the **biological and technical nutrients**, respectively; the former are non-toxic and can be composted, hence guaranteeing the restorative capacity of natural resources, and the latter must be designed to have multiple life cycles from a **cradle-to-cradle** perspective, achieved through their reuse with the aim of minimizing the impact on environmental resources. In particular:

- As far as **Biological Nutrients** are concerned, the cycles shown in the diagram demonstrate how biomasses and biotic waste streams ought to return to the soil as nutrients, once their capacity to create value has been exhausted. Prior to this, however, the value of these nutrients may be increased through cycles based on the **Cascade of Processes** in which valuable feedstocks can be extracted and put back into circulation through composting or biorefining processes, aimed at extracting high quality materials such as biofuel and biogases from the biomasses, which are useful for the production of energy, such as methane, or even fertilizer for farming.
- As far as **Technical Nutrients** are concerned, the **feedback loops** shown in the diagram are based on the possible lengthening of the life cycle of products through design choices and the improvement of maintenance and repair methods or through recovery options that range from the reuse and reselling of products on secondary markets to the refurbishing and remanufacturing of parts and components that can re-enter industrial processes and the recycling of reusable raw materials.

Figure 1.2 The Circular Economy

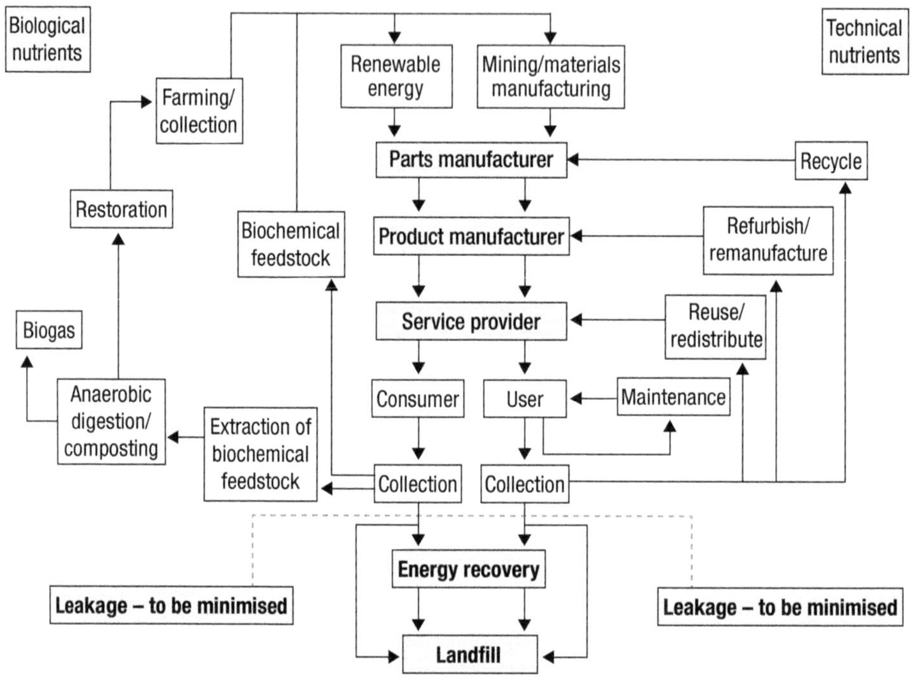

The study promoted by the Ellen MacArthur Foundation (2012) offers considerable insight, as well as operating instructions that are useful from both a macro and micro perspective. With reference to **Figure 1.2**, it appears evident that the different feedback loops shown in the diagram offer possibilities to create value that are characterized by varying intensity. **Box 1.1** summarizes the main ways of creating the value implied by circular economy logics.

1.5 Sustainability and operations management

Operations Management choices must therefore be adopted from the binding perspective of Sustainability. Too often, the goals of efficiency and business are placed ahead of those of environmental and social sustainability.

With the goal of providing a synthetic presentation of how sustainability goals cut across decisions regarding Operations and Supply Chain Management, we refer here to the seminal work of Corbett (2009), which puts forward an effective approach for outlining the areas of possible interaction between the business processes developed during the lifetime of a product and their implications in terms of sustainability. For our purposes, the original model has been modified and integrated with the goal of emphasizing the link between the stages of the **Life Cycle** of the product and the players

Box 1.1 The Power of Circular Economy

The power of the inner circle refers to minimizing comparative material usage vis-à-vis the linear production system. The tighter the circle, i.e. the less a product has to be changed in reuse, refurbishment, and remanufacturing and the faster it returns to use, the higher the potential savings on the shares of material, labor, energy, and capital embedded in the product and on the associated rucksack of externalities (such as greenhouse gas (GHG) emissions, water, and toxicity).
The power of circling longer refers to maximizing the number of consecutive cycles (be it reuse, remanufacturing, or recycling) and/or the time in each cycle.
The power of cascaded use refers to diversifying reuse across the value chain, as when cotton clothing is reused first as second-hand apparel, then crosses to the furniture industry as fiber-fill in upholstery, and the fiber-fill is later reused in stone wool insulation for construction—in each case substituting an inflow of virgin materials into the economy—before the cotton fibers are safely returned to the biosphere.
The power of pure circles, finally, lies in the fact that uncontaminated material streams increase collection and redistribution efficiency while maintaining quality, particularly of technical materials, which, in turn, extends product longevity and thus increases material productivity.

Source: Ellen MacArthur Foundation, 2012.

along an ideal supply chain involved in their management. As illustrated in Figure 1.3, and as will be stated in the next paragraphs, the goals assigned to the departments and company managers of the different phases of a product life cycle must necessarily derive from broader strategic goals, typically linked to positioning choices of the focal company that governs the supply chain, and shared, although through mediation processes, with other players. The consistency between higher-ranking corporate goals and sub-goals pursued on a functional level or in a sub-set of the supply chain must emerge in the medium to long-term, guaranteeing **strategic alignment** (also defined as strategic fit), and a focus on defined priorities.

The central part of the diagram compares, on the one hand, the three components of sustainability, previously summarized in the **Triple Bottom Line concept** and on the other hand, the typical phases along which the product life cycle is developed, according to the **cradle-to-cradle approach**, assuming a Circular Economy Perspective. This perspective is based on the development of products and processes designed so as to generate (or save) value in **several life cycles** through the reuse of their parts (or at least some of them) at the end of their lifetime.

In order to integrate Corbett's proposal (2009) based on product life cycle with the need to highlight all the main players in the supply chain and accountable for its management, the Figure also illustrates the sequence of **internal and external players** and departments involved in each stage of the product lifecycle. In this regard, the main phases can be summarized as follows:

- **design and pre-production**, which, based on principles of Design for Environment, are crucial in determining the possibility of recovering value along the life cycle of a product, and in particular, at the end of its life-time;

Figure 1.3 A reference framework for the development of sustainable Operations and Supply Chain Management

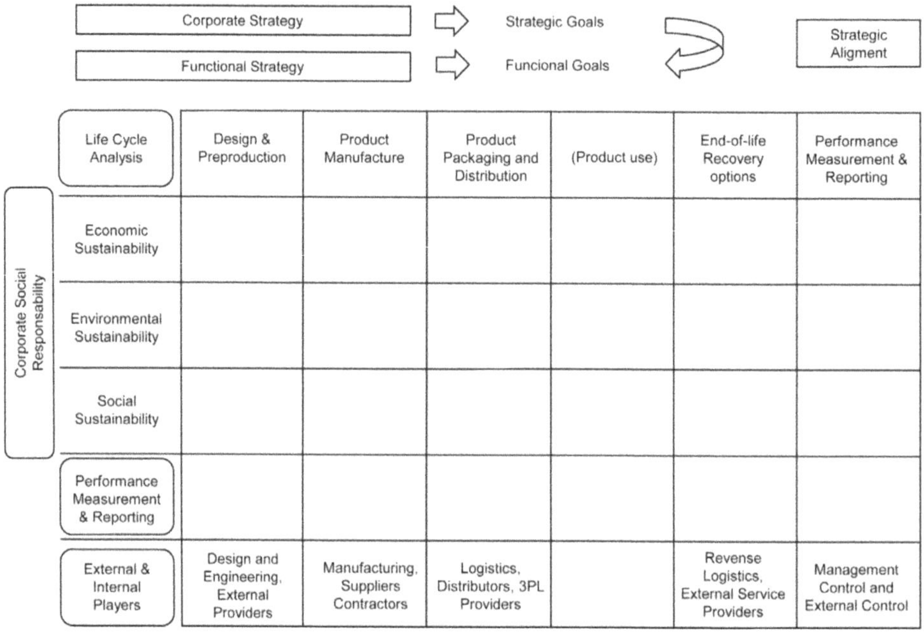

Source: adapted from Corbett, 2009.

- **production** of the product, which, in terms of the organization of the players involved along the supply chain, can be divided into two stages, involving the supply system, which is increasingly relevant in an environment dominated by outsourcing decisions, and the production activity carried out by the focal company;
- **packaging and distribution** of the product, that are crucial in terms of sustainability due to the impact of primary and secondary transportation, and the possibility of minimizing and reusing packaging; in this case as well, the focal company can use third party logistics providers for the activities necessary to reach the final customer, especially for consumer goods in B2C systems;
- **use of the product** by customers, who play an increasingly important role in directing corporate choices through their preference for products that respect environmental and social sustainability principles, and in the development and spread of a culture of accountability that is implemented through specific actions, such as the tendency to save energy, focus on the reduction of waste and separate recyclable waste, and so on.
- **end-of-life management** of a product, through Reverse Supply Chains and the adoption of different recovery options, which, if planned and managed properly, can offer the possibility to extract residual value, through processes that range from high-impact choices, such as reuse, re-manufacture, and recycling processes,

to low-potential processes like the decision to dispose of the exhausted products or send them to incineration plants.

The proposed model is completed with the analysis of **Performance measurement and reporting systems**. In all systems that are geared towards ongoing improvement, it is in fact necessary to monitor the performance of the processes under analysis, designing Key Performance Indicators (KPIs) suitable for measuring targets and achievements. In this regard it is worth highlighting how these measures should have a double purpose: on the one hand, in accordance with **feedback control** logics, to highlight any variance between target values and actual ones; and on the other hand, following the principles of **feed-forward control**, to be a diagnostic tool, able not only to express the size of the gap, but also, if possible, to highlight the main causes behind it, and thus to lead to the implementation of appropriate improvement measures. For example, take a company that has consumed more energy than its target values or has used lower volumes of recycled materials than expected. The usefulness of a deviation analysis as an end in itself appears to be limited to an ex-post assessment of the increased cost, or lower savings, obtained through a comparison with the expected values. A **reporting system** that enables an in-depth analysis of these deviations, such as, for example, a breakdown of the energy losses or of the materials wasted in the various phases of the production process, will make it possible to reconsider previous choices and to implement appropriate corrective actions.

It is therefore evident that Operations and Supply Chain Management will have to change and will also become more and more important in the future. For example, according to a survey, the production of internationally traded goods accounts for approximately 30% of global CO_2, while agriculture is responsible for the consumption of about 70% of worldwide freshwater resources. The globalization of trade increases the need for companies to also pay attention to the environmental and social impact of their choices, especially for those that act as the focal companies of their supply chains. To exemplify the effort made by several companies in the direction of being more sustainable, it is worth drawing attention to several actions undertaken by leading global industrial groups, which have been described in a report Coca Cola has spent almost 2 billion dollars since 2003 in order to reduce water consumption in its 863 production plants, and more than 1 billion dollars for the development of water treatment systems and recycling processes. Nestlé, another food and beverage giant, spent 31 million euros on the introduction of new water treatment technology in 2013. In its Spanish factory, it managed to reduce the consumption of water per ton of product by 60% with an investment of 1 million euros. Google is experimenting with the use of seawater to cool its digital archive in Finland and is testing the use of rainwater for another factory in South Carolina, whereas it is using water from the sewer network in a U.S. factory in Georgia. In 2014, Barilla, the Italian leader in the production of pasta, reduced the consumption of water for each ton of finished product by 20% compared to 2010, in addition to cutting CO_2 emissions by 20%, and has set the goal of reducing both by 30% by 2020 (Belvedere and Grando, 2017).

1.6 Operations Strategy development

Operations Management decisions must be coherent with the **corporate's strategic choices**. "Strategy is the direction and the scope of an organization over the long term: ideally, which matches its resources to its changing environment, and, in particular, its markets, customers or clients, so as to meet stakeholder expectations." (Johnson et al., 2005) The process of strategic formulation, generally, but not exclusively, follows a top-down path and develops along three levels – corporate, business, and function – in a manifestation of increasingly specific goals.

- At the **corporate** level, the long-term goals and guidelines are developed for the entire organization, deriving from the **company's *vision*** (what we want to become) and the ***mission*** (why we exist).
- At the **business** level, distinct plans are developed for each Strategic Business Unit (SBU), in order to define the elements at the basis of the creation of competitive advantage, for the products and services offered to specific markets, or for each business area defined at the corporate level.
- At the **function** level, long-term plans are drawn up to assign to the functions – Operations, Marketing, Finance, etc. – or to any other organizational structure that is chosen (by division, project, etc.), such that they can sustain the creation of competitive advantage, defined at the corporate or business level.

The hierarchical approach described, despite being guided by **top-down** indications, imposes two-way interaction, such that the function-level choices can contribute to the definition of business-level goals and so forth, in a **bottom-up** process. The decision-making weight of the choices coming "from the bottom" can be decisive, as argued by **emergent strategies** theories (Mintzberg and Waters, 1995), according to which the implementation of a strategy, rather than being based on a long-term hierarchical planning structure, can emerge from daily operational experience.

Therefore, **Operations Strategy** is located on the third level of the sequence described, and like other functional strategies, such as Marketing, Sales, and so on, is implemented through the preparation of the governance model (through the formulation of policies and plans) of decisions and strategic actions aimed at defining the role, objectives, and activities of operations in the long-term (Slack et al., 2016; Jacobs and Chase, 2018).

The term Operations Strategy may seem like an oxymoron, as operations are principally linked to daily routine activities, apparently far from any strategic content. However, as Slack et. al. (2016) state, we must **not confuse "operational" with "Operations"**, where the former certainly refers to routine activities, while the latter, as said, regards the governance of the processes useful for the realization of products and services, and as such, can take on a strategic nature, due to the contribution they are able to provide to the competitive success of a business. In this regard, it is customary to distinguish between three distinct levels, the expression of the different quality of the contribution offered by Operations to competitive success (Slack et al., 2016).

- **Operations implementing business strategy**, in which Operations perform a basic function, effectively and efficiently implementing the company's business strategy. These are execution activities necessary to give substance to the strategy. Thus a bank, in order to issue new financial instruments, will need to design and manage some processes useful for their presentation on its website, the preparation of the necessary documentation, the contact with the potentially interested customer, and so on.
- **Operations supporting business strategy**, in which Operations go beyond the role of mere execution, developing capabilities that allow an organization to achieve its strategic goals, perfecting and improving them. The role of Operations, from being merely passive, takes on more pro-active characteristics. A company that competes on the ability to personalize its own products will leverage the flexibility of its Operations System and the reconfigurability and versatility of its production resources, that will demonstrate their capacity to suggest improvements, solutions, and innovations able to support the business strategy.
- **Operations driving business strategy**, in which Operations assume a role in guiding business strategy. This is the case of companies in which over time, Operations have developed competences allowing to continuously provide unique products or services, that are the basis of the company's competitive success. In this case, the role goes from pro-active to fully propulsive. It is the case of some e-commerce providers, that base their success on a perfect operations and logistics system.

The development of an **Operations Strategy** can also be implemented through two different perspectives (Greasley, 2006).

- The first is defined as the **Market-driven Operations Strategy**, according to which the company's choices derive from the positioning objectives on the target markets, and within those, from the customer satisfaction objectives. From these, derives the level of performance necessary to compete successfully, and thus the methods of organization, management, and control of processes and the resources suitable for their efficient and effective generation. By way of example, think of the case of some fast food chains, that aim to penetrate broad markets with a positioning founded on a business formula tending to offer their customers quality products, at low costs, ready in just a few minutes, in a clean environment, etc. They pursue quality performance, rapid service, and efficiency, based on the design and management of strongly standardized processes, the use of specifically trained personnel, technological applications developed *ad hoc*, and infrastructure designed with ergonomic and quasi-industrial criteria.
- The second is defined as the **Resource-based Operations Strategy**, that formulates the business choices starting from the awareness of its tangible and intangible resources (**competences**), accumulated and exploited over time, based on which it is able to structure a set of **operations capabilities** that are functional to the pursuit of competitive success in its target markets. To provide an example, this is the case of some web application design companies, that thanks to the competences of their

programmers and web designers and the availability of technology, develop production and personalization capacities able to find success on the constantly evolving market for applicative software.

The two approaches can be appropriately integrated and provide mutually support; and over time, due to the dynamism of the markets and the need to develop new competences, one or the other may prevail in different periods, in moments of relative calm and moments of rapid acceleration.

1.7 Order qualifier and order winner

From the resource-based perspective, the accumulation of skills and core competences of a company's operations system in the long term will lead to successful performance, appropriately oriented towards the target markets; while in the market-driven perspective, operations must constantly adjust their strategy and management as a function of the needs expressed by the market. With the goal of satisfying customers' requests, it is thus necessary for the Operations System to identify and select the **key elements and performance levels** on which to invest, aligning with – or better, anticipating – the expectations of the target customers. The competitive factors and performance goals can be divided into two categories, defined as **order qualifier** and **order winner** (Hill, 2005). A third category is added, defined as secondary or **less important** factors, that do not directly or significantly impact the possibility to beat out the competition and convince the customer to buy (**Figure 1.4**).

Order qualifier

The first category consists of all of the characteristics and performance profiles for which clients take into consideration the offer of a company's products and services, on the same level as those of other competitors. These are factors and levels of performance, sometimes defined as "**points of parity**," in the absence of which (or below which) a company loses any possibility to compete successfully. In fact, potential customers consider these factors to be indispensable, as "givens", a necessary condition in order to take the company's offer into consideration, although not sufficient to convince them to buy.

Order winner

The second category, often defined as "**points of difference**," are the specific factors and elements of performance for which a given company, among the others with which it competes, acquires the customer's trust and ultimately wins the order. These are factors that contribute directly and significantly to creating a successful business.

Figure 1.4 Order winner, order qualifier and less important features

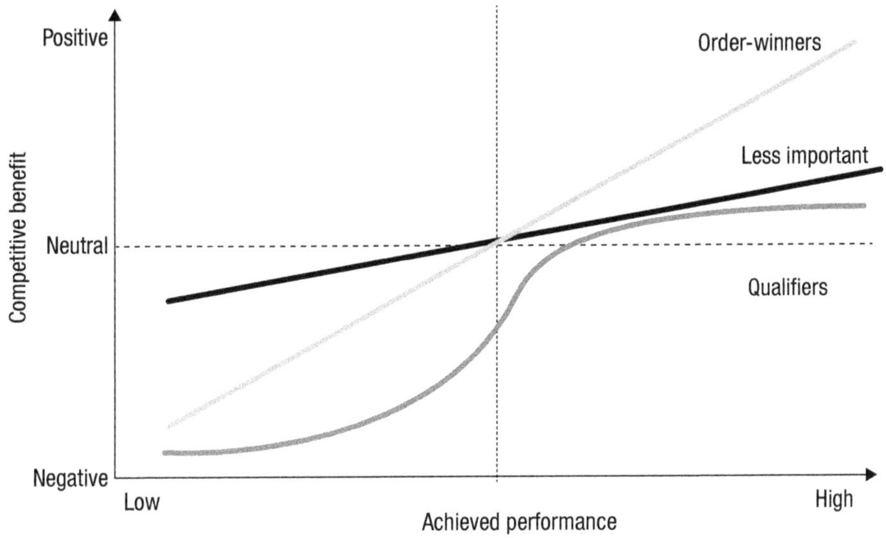

Source: adapted from Slack et al., 2016.

Less important

The third category, named **Less important factors**, does not influence the customer in their choice, but may be relevant in other parts of the operations processes.

Both order qualifiers and order winners are important. The former are necessary to participate in the competitive game, the latter to win it. Thus, a company that has reached an adequate level of performance in order qualifiers to compete, should not invest further in those elements, because it would not increase its chances of victory. It should rather concentrate on the elements and performance levels that constitute **order winners**, for which each improvement translates into a greater probability to beat out the competition.

Within a defined class of cars, many automotive companies offer related products and services, with levels of comfort, safety, power, costs, and delivery that are often comparable, but ultimately, the customer makes a choice and the order is obtained by only one company. So for a middle class car, with certain levels of power, fuel consumption, and accessories guaranteed, investing further on those elements may not produce particular advantages. To the contrary, offering a lower price, innovative payment formulas, and prompt delivery may represent the key to success.

A winning factor for a specific market segment can be a non-important one for another, such as price for a top, fully customized product. In addition, a non-important factor can be irrelevant from a customer's perspective, but crucial for some operations processes. It is the case of car diagnostic tools, utilized to optimize maintenance and repair activities.

Therefore, in order to compete, every business must understand what type of performance allows it to be considered as a potential supplier (or, as the saying goes, to be on their customers' short list). The fact of being qualified in the eyes of the customer is not enough. It is necessary to do better than others in developing (or fostering the perception) of the types of performance that allow the customer to make his/her choice, allowing one of the qualified companies to "win." However, since the majority of companies produce products and services with different positioning, aimed at different segments of customers, it will be necessary to define the respective **order qualifiers, winners, and less important for each cluster of customers**. Lastly, it is known that characteristics that are considered innovative at the start of the life cycle of a product/service tend to be consolidated over time, moving from elements considered to be order winners to qualifiers, and in some cases, to the category of less important.

In automobiles from a few decades ago, characteristics such as air conditioning, sound systems, or electric windows or electronic doors were considered important innovations, while today the consumer considers them standard.

1.8 Strategic choices and design levers

The use of Design Levers must start from a clear identification of the production system's primary goals, in turn derived from the company's more general strategic goals, according to the approach defined as **Strategic Alignment** or **Strategic Fit**.

Those goals, often linked to a company's positioning on its market, must in fact pervade all of the company's activities; and while indicating the trajectory of the competitive efforts, must define the sub-units' goals (in terms of process, project, etc.), with a cascading effect, including those in the area under investigation here, Operations. As illustrated in **Figure 1.5**, and limited to the area considered in this book, the sub-units' goals can be numerous and often antithetical. For example, think of a logistics system that attempts to jointly pursue limited investments in stock and high levels of service, or a production line which seeks maximum saturation and versatility, or of a company that pursues low costs and a high quality of resources used. The traditional approach to production choices is thus based on the concept of **focalization**: in the presence of divergent objectives, it is necessary to prefer one (or some that are coherent with each other) and consider the others residually, like dependent variables. Competing on cost cannot translate into sacrifices of quality and service. Searching for personalization of production, all other conditions being equal, necessarily leads to an increase in management complexity.

Having defined the set of objectives assigned to the logistics-production system, it is now necessary to shape a situation able to adequately respond to the priorities identified, through the use and coherent integration of all of its components, or design and management levers.

The **design and management Levers** available to management can be distinguished depending on the degree of irreversibility of the choices that characterize them. We can thus define:

Figure 1.5 **The strategic approach to production choices**

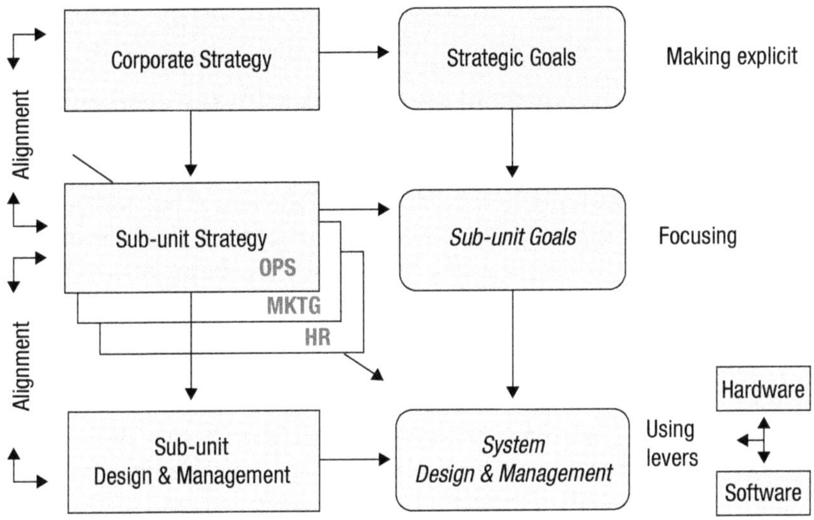

- **hardware levers** (also called configuration or structure levers): the choices concerning the type of systems and technologies adopted, the production capacity installed, the degree of vertical integration, and potential decentralization of Production Units. The decisions concerning these issues are adopted for medium to long intervals of time, and define the *pro-tempore* permanent character of the production structure or manufacturing footprint;
- **software levers** (also called management levers): those relating to the planning, execution, and control systems of typical Operations core processes. For example, this includes processes of Production Scheduling and Control, Materials Management, Quality Control, Maintenance, organization of work and factory personnel, etc. These are susceptible to modifications at closer rhythms and lend themselves more to short- to medium-term management.

Software levers are generally linked to a defined hardware, driving it and guaranteeing its optimal behavior. This means stressing the need for intimate coherence between the different choices made, of a unitary organizational and management vision that makes the configuration and management choices compatible. Frequent pathological behavior in fact originates from inconsistencies between the production structure and management logics. This is the case in those companies where production structures, designed for specific aims, are managed with operational mechanisms modified over time, with the intent of pursuing the variability of competitive pressure. The room for discretion guaranteed by the software management of a production structure encounters a limit in the principles of its hardware design, such that every forcing risks degenerating into inconsistency. We see this with plants characterized by limited levels of versatility that have difficulty responding effectively to potential requests for flexibility coming from the market through only planning levers, except when faced with high costs and impact on overall efficiency.

1.9 Operations Management and its link with the economic-financial perspective

Among the many perspectives from which it is possible to look at business performance, the economic-financial approach undoubtedly plays a key role, in terms of immediacy, synthesis, and interpretative capacity, in particular for the top management. Most of the decisions and choices are in fact authorized and implemented evaluating **economic return**. This often constrains those functions, such as Operations and Supply Chain, for which performance improvements are usually measured through **operational indicators** – efficiency, utilizations, lead times, defects – that are harder to trace to economic and financial measures. However, we should stress that all business decisions have an impact on economic-financial performance, in addition to on operations. So this is where we will start: although this book is dedicated to Operations and Supply Chain Management, it is crucial to quickly examine the link existing between the areas and decisions placed under the control of the processes and functions analyzed in this book and the synthetic economic-financial results.

To put it very briefly, businesses tend to increase their **Economic Value Creation (EVC)**, that is to ensure that their decision-making processes generate a **Return On Invested Capital (ROIC)**, also called Return On Capital Employed (ROCE), higher than the **Weighted Average Cost of Capital (WACC)**, according to the following equation (Cachon and Terwiesch, 2013):

$$\text{EVC} = \text{Invested capital} \times (\text{ROIC} - \text{WACC}) \qquad [1.1]$$

Considering that **Cost of Capital** is unlikely to be modified in the short-term, below we will examine the second component of the equation, the Return on Invested Capital (ROIC). To understand that equation we should go back to other measures, typical of economic-financial analysis.

On this point, see **Figure 1.6**, which illustrates the deployment of **ROIC** into its principal components, in turn deployed in increasing detail until reaching the elements impacted by the most common Operations Management decisions.

As shown in Figure 1.6, **ROIC** is the ratio between **Earnings Before Interest and Taxes (EBIT)** and **Net Invested Capital (NIC)**. In turn, in order to investigate the dynamics of ROIC, that measure can be split into its two components, **Return On Sales (ROS)** and **Net Invested Capital Turnover (NICT)**.

- The first **(ROS)**, given by the ratio between **EBIT** and **Revenues**, is an expression of a merely economic dimension, being the result of prices, volumes, mix and costs, and expresses the company's operational efficiency, i.e. the quantity of operating (gross) margin generated by each unit of sale.

Figure 1.6 **Return On Invested Capital Equity and its determinants**

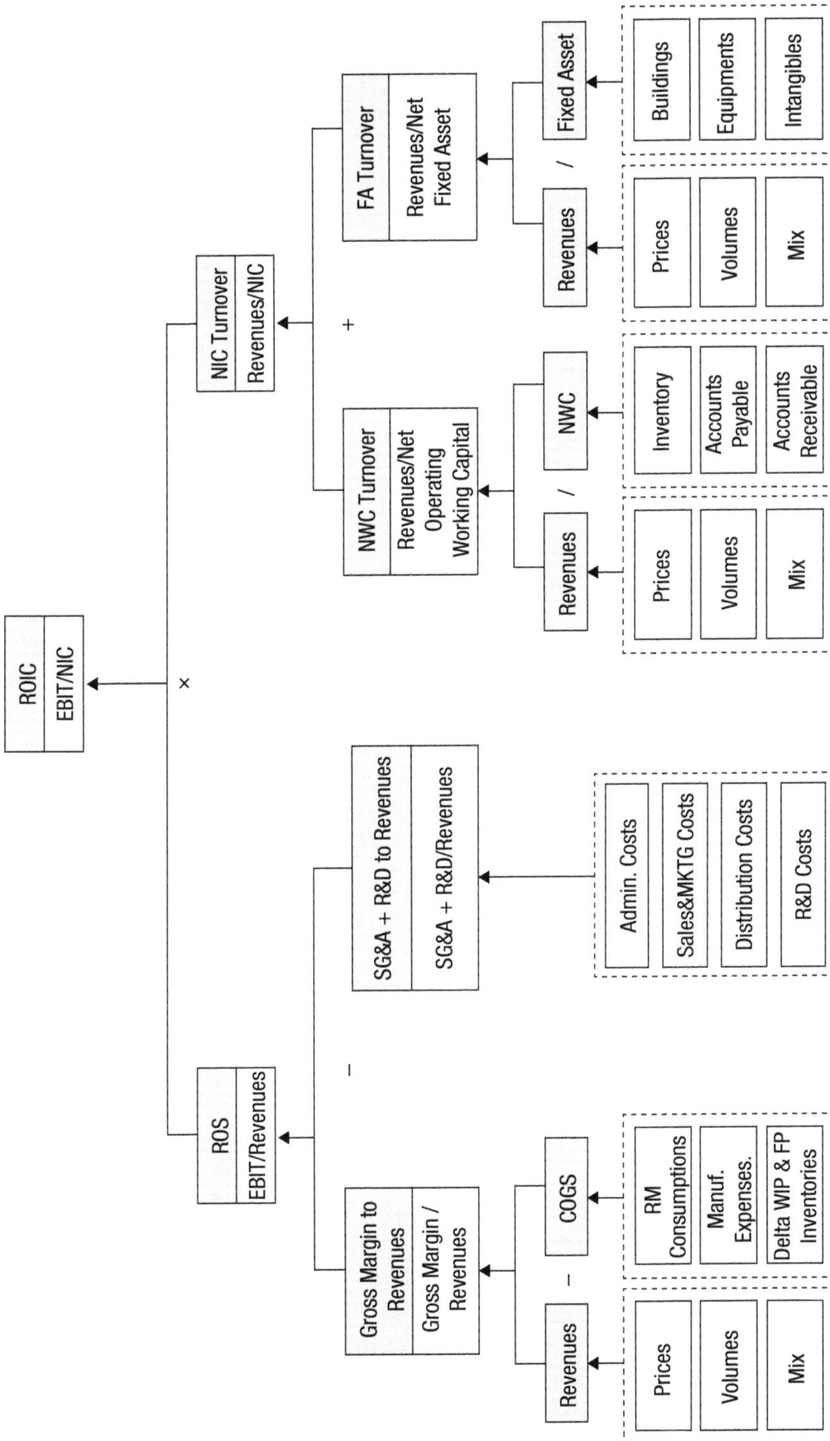

- The second **(NICT)**, given by the ration between **Revenues** and **NIC**, represents the annual income generated by a unit of Net Invested Capital, i.e. the structural efficiency of a company, and also encompasses a financial perspective, deriving from the rotation of its two components: **Net Operating Working Capital** and **Net Fixed Assets**.

The incidence of the two components can vary significantly in different sectors and lead to very different ROIC, in terms of entity and composition. By way of example, think of the following cases:

- in the case of asset-intensive sectors, such as steel or food, on the one hand NICT is low and significantly influenced by Fixed Assets, and on the other, ROS is linked to often limited margins;
- in other asset-intensive cases, such as pharmaceuticals, NICT is also low, but ROS can be significant;
- in still other cases, such as the production of luxury or premium fashion goods, NICT rotates rapidly and ROS has high leverage.

Continuing in the deployment, it is evident that the economic profile expressed by **ROS** is influenced by choices regarding **Revenues**, and thus on the one hand by **Prices, Volumes** and the **Mix** of production and sales; and on the other, regarding **Costs**, whether variable or fixed, as well as the degree of **efficiency** of use of resources, all variables that are directly or indirectly influenced also by Operations Management choices. As regards the elements that directly impact the relationship under examination, think of the costs linked to production factors, such as raw materials and components purchased, labor, the machineries amortization costs, and energy consumption, as well as efficiencies and yields that can be achieved, all of which are significant elements in an industrial company. In addition, the possibility to increase (or reduce) volumes and sustain prices (or grant discounts), is also indirectly linked to the quality of products, the level of service offered in terms of speed, dependability, completeness of deliveries and after sales services. In this case as well, the performance is influenced by Operations and Supply Chain Management. The considerations presented above thus make it evident that a potential reduction of ROIC can be ascribed to the need to support demand – and thus production volumes – in periods of crisis, through reductions of price, that will impact ROS. This is particularly true in Asset Intensive plants, for which the need to saturate Fixed Assets and reach the break-even is crucial. In other cases, the reduction can be motivated by an increase in variable costs, such as the cost of labor, energy, or materials not entirely transferable to the market in terms of price increases.

As concerns **NICT**, both components – **Net Operating Working Capital Turnover** and **Net Fixed Asset Turnover** – are strongly influenced by production-logistics choices. Once again, think of the effects generated by **Inventory Management** policies on **Working Capital** or **investments in plants, machinery, and equipment** on **Fixed Assets**; or, indirectly, of the impact of product quality and customer service on the amount of **Accounts Receivable** or the volumes purchased and the methods of interaction with suppliers as regards **Accounts Payable** policies. Analogously, on the Fixed Asset front, saturation choices of production capacity, the life cycle of machines and the related maintenance policies, as well as the rationalization policies of the spaces occupied, can affect investment choices for renewal or expansion of company assets.

Continuing with the deployment of the system of indicators, **ROS**, in turn, is equal to the ratio **Gross Margins** to **Revenues** after deducting the ratio of **Sales, General and Administration plus R&D Cost (SG&A+R&D)** to **Revenues**, where as we have said, revenues are influenced by Prices, Volumes, and the Mix offered, while Gross Margin is equal to Revenues after deducting the **Cost Of Goods Sold (COGS)**. The split between the two components of ROS aims at distinguishing the major elements under the direct control of the Operations and the other costs related to Sales, Marketing and Administration processes.

- **COGS** is equal to the sum of Raw Materials consumption, Manufacturing Expenses, and the variation of Work in Process (WIP) and Finished Product inventories.
- **SG&A+R&D**, on the other hand, are equal to the sum of Administrative, Sales and Marketing, and Distribution costs plus R&D costs.

Analogously, **NICT** is equal to the sum of **Net Operating Working Capital Turnover** and **Net Fixed Asset Turnover**.

Both ratios have Revenues as the numerator, as already stated. As for the denominator:

- the first has Net Working Capital, which is known to be equal mainly to the algebraic balance between Inventory, Accounts Receivable, and Accounts Payable;
- the second considers **Fixed Assets**, in an industrial company principally linked to Tangible Assets, such as Buildings, Machineries, and Equipment and Intangible Asset, such as acquired patents.[2]

The breakdown of ROIC into the two indexes, ROS and NICT, shows two different strategic orientations of management and increase of EVC of a business, to which Operations Management contributes significantly.

- The **"Value" strategy**, that bases the income of a business on a Profitability premium, i.e. a differential of ROS with respect to competitors determined by a supply differentition.
- The **"Volume" strategy**, that bases business profitability on high Capital Productivity, i.e. on a differential of NICT with respect to competitors determined by a differential of volumes and of consequent economies of scale.

[2] Current assets generally consider Tangible Assets (Building, plant, equipment and properties), Intangible Assets (acquired brands, patents, licences and goodwill, etc.). In the text we provide an example with reference only to the assets generally influenced by Operations Management decisions.

Three reflections emerge from these brief points.

- Each choice, even if apparently confined within an organizational unit or a function, always has a more or less significant impact on the other units that make up the business system, given the principle of **unitariness** that characterizes the business processes.
- The decisions adopted, that are often oriented towards the search for improvement in operational performance, have – and must have – importance also from an **economic-financial standpoint**, and management must be aware of these links.
- The possibility to use financial performance measures alongside non-financial measures, enriches the tools available to management in the incessant **process of analysis, diagnosis, and improvement** (Chapter 3) of each production system oriented to development.

From the standpoint of the choices of Operations Management and their interpretation through the indicators highlighted above, ROIC and its deployment appears the most widespread. In Figure 1.7, an example of ROIC Tree is reported. Considering that the determination of the Price (P) is in general outside of the sphere of action of Operations, Figure 1.7 shows an example of explosion of ROIC in which the costs of materials are considered variable (cv), the costs of direct and indirect labor are considered fixed (CF) and the Production Volume (Q) is assumed to be equal to the minimum between Demand and Production Capacity in the bottleneck phase.

Knowing that:

- Revenues = P × Q
- Return = [P × Q – (FC + vc × Q)]

in **Figure 1.7**, the common language of Operations Management – namely fixed and variable costs, volumes, scraps, etc. – is used, in a cascaded of detailed information that can be observed in two directions:[3]

- from left toward right side with the aim of understanding how the ROIC can be deployed until its main levers directly managed by Operations;
- from right toward left side to estimate the impact on ROIC generated by interventions in the Operations processes, such as a set-up reduction, improved materials or labor efficiency or throughput time, etc.

[3] In the reported example ROS is fully deployed, while NICT could be further deployed as well.

Figure 1.7 An example of ROIC Tree

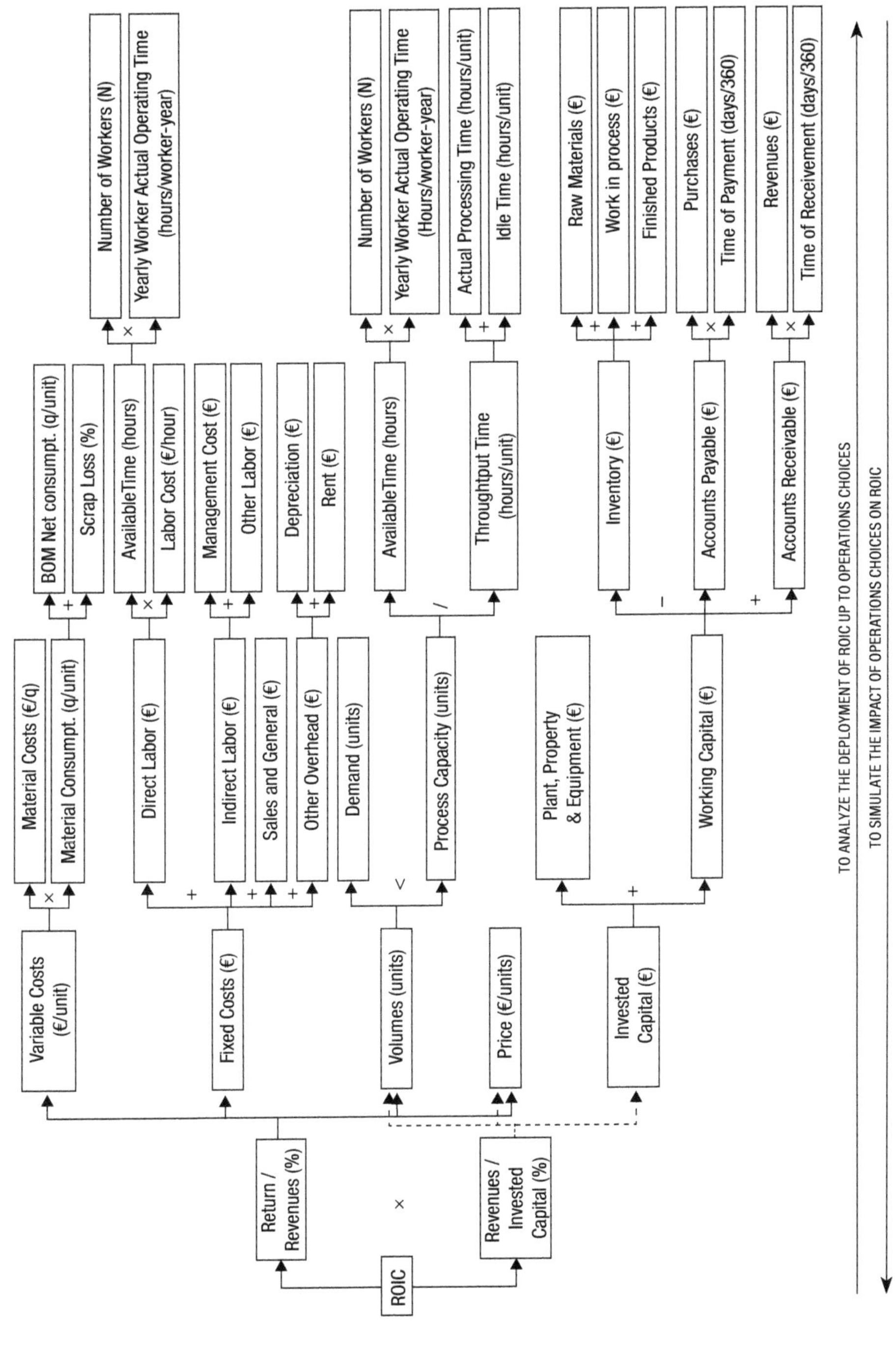

Source: adapted from Cachon and Terwiesch, 2013.

2 Design of Production Systems

by *Alberto Grando*

2.1 Typological analysis of production processes

Before addressing the relevant issues in Operations Management, it is necessary to at least briefly examine the characteristics of the different production processes that we will consider. Here, we limit ourselves to suggesting two models of **typological analysis of production processes**, that in our view lend themselves more than others to highlighting the distinctive characteristics and to supplying the framework in which to situate the problems dealt with in this book.

The two classifications, despite dealing with the same theme, are not to be considered alternatives, due to the different intentions behind them. The first is particularly useful for descriptive purposes, despite offering an essentially static perspective. The second, in introducing elements regarding management and due to greater intrinsic dynamism, comes closer to the areas explored in the book. As we proceed we will thus constantly refer to both, with the aim of maintaining a single conceptual frame of reference regarding the issues observed, that unites the proper terminological rigor with the technical-management approach adopted as the leitmotiv of the work.

Figure 2.1 shows the first classification, that identifies **three essential profiles** for the analysis of production systems: the methods of responding to demand, the criteria for managing the production volumes, and the characteristics of realizing the product (Brandolese et al., 1991).

On the vertical axis of the three-dimensional representation, we see the distinction relating to the **Methods of response to demand**: **Production** aimed at **replenishing a stock** level, also defined as **MTS – Make To Stock** (or Assemble to Stock or Delivery to Order) is highlighted – which generally entails working based on forecasts, anticipating the production phase with respect to that of acquiring orders, and **Production to order**, in which the transformation takes place following the acquisition of the client's order. We can also distinguish between **a differentiated (or single) order**, also known as **ETO – Engineering To Order** – for which it is necessary to await the order to launch the product's design and execution phases, and a **characterized (or repetitive) order**, for which the general design is prior to the time of acquisition of the order, except for detailed interventions and customization, and it becomes necessary to proceed only with its execution. In terms of the method of managing **Production to order** flows, this is possible through systems such as **ATO – Assemble To Order, MTO – Make To Order and PTO – Purchase To Order**, depending on the section of the production cycle influenced by the specifications of

Figure 2.1 Essential profiles for the analysis of production systems

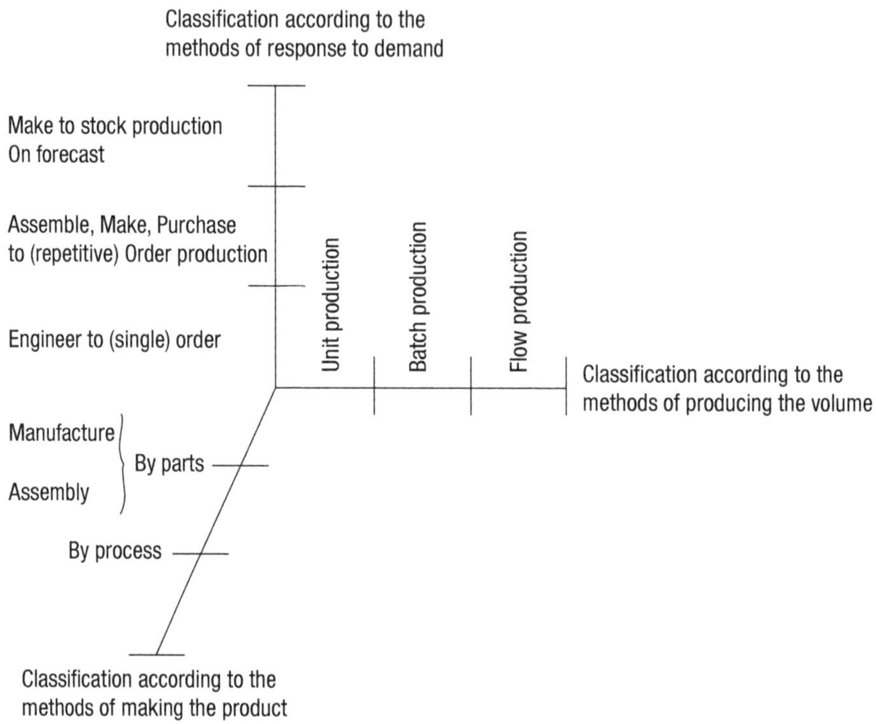

Source: Brandolese et al., 1991.

the client order. In cases where both the design and the definition of product specifications completely precede the order and the physical production, we generally speak of catalogue production to order.

On the horizontal axis are the **Methods of producing the volume** of the product, i.e. **unit production** (or one-piece flow), **batch production** (or intermittent flow), and **flow production** (serial or continuous flow). The first is characterized by an extreme variability of work cycles, that results in the creation of unique articles, variability that is reduced in intermittent production, marked by appropriate sequences of batches (or lots), managed according to interwoven or alternated flows, and is cancelled for extended time intervals in series or continuous flow productions, carried out on substantially undifferentiated products. This is called continuous production flow due to the substantial absence of interruptions between the entry of the raw materials and the components and the production of the finished product.

Along the third dimension, there is the aspect represented by the **intrinsic nature of the product and its Modes of production**. This involves the distinction between **production by process** and **production by parts** (also called manufacturing). The former types of production, which include steel, petrochemical, paper, food and other such processes are also called **irreversible production processes**, because the raw materials used to create the finished product are no longer distinguishable and identifiable in the finished product. The transformation takes place, thanks to exchanges of energy, due to alterations in the chemical and physical characteristics of the product, such that it seems virtually impossible to trace the finished product back to its component materials, just as it is not possible to do so for products such as cracking of crude oil, from the steel to the iron and carbon contained in it. In turn, those types of production can be divided into **integration processes**, in which there is a synthesis of materials not found in nature (such as in the case of steel and synthetic fibers) or **disintegration**, which is used to separate the components (as in the case of petrochemical production). These are processes generally carried out in large volumes, such that it is crucial to control the relevant variables (temperature, pressure, etc.) and the rapidity of corrective interventions of retroaction, in order to avoid significant issues in terms of quality losses and even financial impact. **Manufacturing production processes**, on the other hand, are obtained through the production and/or assembly of components or parts, entailing changes in the morphological characteristics of the parts used. So these are **reversible production processes**, in which it is possible to go backwards from a hypothetical whole (an engine, a car, a piece of furniture, or a shoe) to all of its elementary components. Production by process generally consists of production with a fixed technological cycle, i.e. in which the sequence of operations is technologically constrained, thus leaving relatively limited room for management, while presenting considerable technological and plant constraints. This is different than in manufacturing (by parts) – often a non-fixed technological cycle – that by virtue of the broad variety of possible solutions, acts as fertile ground for interventions for optimization from a management standpoint as well.

As stated repeatedly, every classification sets the goal of identifying the process archetypes, that are useful for descriptive purposes. In factory settings, **mixed systems** in any event prevail. By way of example, think of some food or pharmaceutical productions in which the product is produced in continuous flows (a dough, a drink, or a drug in liquid solution), and production by process in a first phase, to then be packaged in different formats or boxed, through systems that operate separately, in flows organized by batches, at times with customization interventions (for example labeling or marking) using customer specifications.

The model in question, despite clearly describing the boundaries of the possible types of production system, offers a **static vision** of the issue and does not set the goal of illustrating the **different organizational solutions** that, each time, best adapt to the management of the factors used in the transformation, the goals assigned to them, or the critical tasks to be carried out. Those limits are overcome by the model of the **Product-process matrix** (Hayes and Wheelwright, 1979; Schmenner, 1984), described in Figure 2.2. The representation is built by comparing, respectively, the **characteristics of the products processed**, in terms of mix, volumes and customization, with the **specificity of the production flows** suitable for their transformation. Therefore, as concerns the first aspect, we observe the evolution that, from the creation of the unique and unrepeatable specimen, also called a **specialty**, leads, through a process of gradual standardization and an increase of volumes, to

Figure 2.2 The product-process matrix

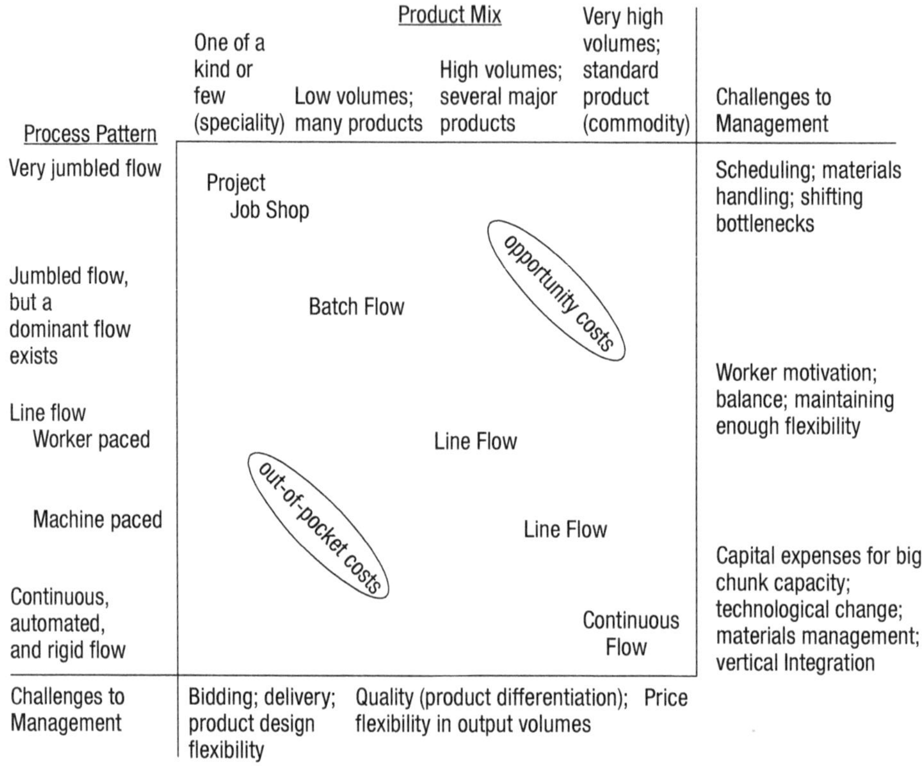

Source: Schmenner, 1981.

the offer of a **commodity**. Regarding the second aspect, we stress the parallel evolution of the production flow that, from being **extremely fragmentary**, takes on connotations of **growing continuity**. The points of intersection between the degrees of composition of the production mix and the moments of discontinuity in the production flow highlight the **physiological and pathological states** of the Production Units placed there.

The cases that lie along the diagonal that joins the vertex on the top left (specialty product and extremely fragmentary flows) with the opposite vertex (commodity product and continuous flow) are generally physiological. With reference to the **organizational and management implications** of those points of intersection, there is generally a distinction between:

- **job shop** or fixed production solutions, characterized by general-purpose equipment, highly qualified and multi-skilled labor. In general, this regards production processes of work with a high level of artisanal or artistic content, typically to order, or unique items, even if they are very complex, such as special plants or a yacht;

- **batch (or lot) flow**, characterized by the production of uniform quantities of equal or similar products, staggered by product change operations, defined change over or set-up. Batch production, as we will see below concerning layout alternatives, can be organized by functional departments or in flexible (not dedicated) lines able to produce different products. The first case includes, for example, upstream shoe production (cutting, seaming, sawing, etc.), furniture production (cutting, squaring, edging tape, finishing), and mechanical production (drilling, welding, assembling), etc. Examples of the second case are conveyor systems along which different batches of footwear are assembled (shoes, boots, and sandals);
- **line flow**, characterized by linear flows that follow a sequence of operations and machines, generally dedicated and arranged in line, sequenced according to needs dictated by the specific technological cycle (routing) of a product or a product family:
  - a line is defined as loosely connected**, or decoupled**, when the transfer from one workstation to another takes place non-automatically, through operators, fork lifts, and other handling systems, generally in the presence of accumulations of stock appropriately placed in inter-operational inventory (buffers), positioned at the workstations. In that case, the production rhythm, not imposed (asynchronous lines with non-constrained transfer), is dictated by the labor;
  - this is to distinguish from a **continuous transfer line**, in which the movement, at a given rhythm (paced line) takes place thanks to the use of automated transfer lines or similar methods such as ceiling conveyors, floor chains, conveyor belts, etc., such that the frequency of progress is dictated by the system moving the pieces and is equal for all of the workstations (synchronous lines).

  By way of example, the first case includes some industrial machine production, and the second the automotive, white goods and appliance industries;
- **continuous flow**, which we have already mentioned with reference to process production, in which the flow follows a technologically fixed cycle in processes conditioned by continuous supply, as in the case of oil and gas pipelines, and some types of food, paper, or chemical production.

Each case pursues its **own goals**, described on the sides of **Figure 2.2**, in terms of differentiation of the offer, service to the client, production efficiency, and allocation of resources. The opposed vertices of the diagonal contain incoherent placements that represent **pathological states**, whose most evident symptoms reside in the emergence of unjustified **out-of-pocket costs**, whether actual costs or more simply **opportunity costs**. The first is the case of a flow production process (line or continuous) for which a broad range of products is required, in low unitary volumes with consequent machine stoppages, retooling, greater discards linked to the rebooting of the line and lack of utilization, all phenomena that rapidly lead to growing costs, that are often unsustainable from a financial standpoint. The second is the opposite case, in which there is an attempt to create uniform production in large volumes with insufficient and generic tools, thus giving up the opportunity to satisfy the market with a more coherent offer, characterized by specific dedicated structures, oriented to volume and efficiency goals. Both of these examples show strong incoherence between product policy and the process focus, clearly indicated by the incompatibility between the different critical tasks which management should face.

The product-process matrix also lends itself to a **dynamic reading**, as the different types of production process are the archetype of the process of development of a Production Unit, both in their structural and functional characteristics, and in the performance pursued. A company characterized by a high level of flexibility and a broad range of products, that due to the success of some product families decides to gradually increase volumes, eliminating those less valued by the market, will sooner or later have to adjust its production structure moving towards organization, methods and flow management that are gradually more rigid. Lastly, it should be stressed that the advent of flexible automation and the most recent developments underlying the paradigm of Industry 4.0 (or digital manufacturing) have partially modified the profile of the area of coherence between processes and products, increasing the occasions for compatibility between needs of flexibility connected to customized production and plants traditionally oriented towards efficiency.

The classifications proposed can be accompanied by others, and lend themselves to further investigations. For our purposes, it seems advisable to only provide some considerations and further discussion regarding:

- the main **manufacturing processes**, with the goal of providing a merely didactic form of support to the following analyses;
- the **fundamental layout** choices adoptable in the production units, for the crucial implications induced from an operations management standpoint.

The table in appendix to this chapter shows the characteristics of the different types of production organization (Schmenner, 1981).

2.2 An overview of manufacturing processes

Manufactures regard the work and transformation processes performed on materials and components, aimed at altering their properties and geometries; they are distinguished from assembly, understood as the creation of sets of multiple components, through processes of unification or mixing.

Manufacturing processes can be divided into processes:

- that **preserve mass**, for which the mass of outgoing materials (outputs) is essentially equal to the mass of incoming materials (inputs), and thus entail alterations of form and appearance. Think of the case of drawing and pressing processes or surface treatments carried out through galvanic baths or tempera processes, obtained through heating or cooling cycles of articles;
- that **reduce mass**, for which the geometry of the final product (output) is the result of the removal of material from the incoming semi-finished product (input).

This is the case of processing of cutting leather to obtain shoe uppers, that will generate scrap or waste, or the case of roasting of coffee that will entail an outgoing mass lower than the incoming mass due to the drop linked to the reduction of humidity;
- **of union**, for which the final geometry of the article (output) corresponds to the union of multiple components (inputs) and the mass to their sum. Assemblies can essentially be considered processes of union, characterized by dismantling or reversibility, i. e. the possibility to restore the original separation of the components. Mixtures are obviously an exception.

The three types of processes can be represented by flows of materials, respectively defined as (a) **monoline**, (b) **diverging**, and (c) **converging**, as illustrated in **Figure 2.3**.

Every manufacturing process can in fact be observed as the combination of flows of **materials, energy, and information**. The first of these – materials – which we will discuss later, requires the supply of energy to give rise to a transformation, according to the indications contained in appropriately structured information. For example, in a process for the shavings removal or plastic deformation, it is necessary to provide the material, in the solid state, with energy in the form of rotating movement of a tool or the pressure of a mould, according to trajectories, forces, and geometries contained in an organized system of information (in the mould, the work cycle, and the design and instructions by the technical office).

Figure 2.3 Materials flows in manufacturing processes

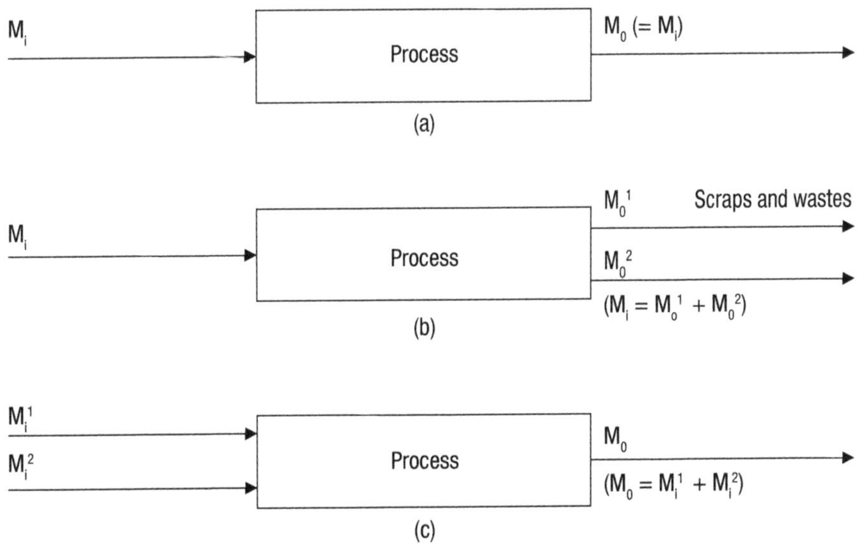

Key: T principal types of flows: a) conservation of mass; b) reduction of mass; c) union or increase of mass. M indicates mass of material, i is for input, and o is for output.

The analysis of the materials flows entails the definition of two elements.

- The **states of the material** on which the intervention is performed to alter geometries (forms and volumes) and properties (thermal, chemical, mechanical, electrical, etc.).
- The **base processes** of transformation, usable to carry out those transformations.

There are various **states** in which materials can be processed: liquid, fluid, granular (i.e. loose solid material), and gaseous. Unlike the classifications of the state of matter in physics, granular material (loose) is distinguished from solid material, since the technologies generally used in the two cases are very different.

The **base processes** of transformation, as mentioned, produce alterations of the geometries and/or properties of the materials, and when appropriately sequenced, define a manufacturing process. The typical sequence is generally divided into three adjacent phases.

a. The first consists of **preprocesses**, also known as secondary or preparatory, that have the aim of bringing the material into the state suitable for the subsequent production of the desired change. This phase includes processes of heating, melting, cutting, etc.
b. The second, central phase consists of **primary processes**, since they produce the changes that are the aim of the transformation, such as the moulding or alteration of the properties.
c. The third consists of **postprocesses**, which are also secondary or finishing phases, intended to bring the product to the final state desired. This is the case of solidification, cooling, shaving, etc.

The nature of manufacturing processes can be seen by observing together the state of the materials and the interaction between them and the **base processes** applied to them. In this regard, we can distinguish between:

- **mechanical manufacturing processes**, such as plastic deformation, elastic deformation, removal of shavings, fragile or ductile fracturing, mixing, separation, etc.;
- **thermal manufacturing processes**, such as heating, cooling, melting, solidification, condensation, vaporization, etc.;
- **chemical manufacturing processes**, such as solution and dissolution (electrolytic and chemical), combustion, surface hardening, precipitation, etc.

By way of example, **Table 2.1** shows the possible primary processes utilized for changing a geometry, while **Table 2.2** shows some cases, in summary form, specifying the type of flow, the state of the material, the category of the base process, and the primary base process.

For our purposes, it is useful to stress here the strict connection existing between technological profile of production processes and the related management choices.

Table 2.1 Primary processes used to change a geometry

Category of base process	Primary process
Mechanical	Plastic deformation (strain)
	Fragile or ductile fracturing
	Elastic Deformation
	Flow (filling)
Thermal	Melting
	Evaporation
Chemical	Solution and dissolution (electrolytic and chemical)
	Deposition (electrolytic and chemical)
	Combustion

Table 2.2 Classification of technological processes in relation to type of flow

Type of flow	State of the material	Category of the base process	Primary process	Examples
Conservation of mass	Solid	Mechanical	Plastic deformation	Forging and laminating
	Granular	Mechanical	Flow with plastic deformation	Powders compactation
	Fluid	Mechanical	Flow	Casting
Reduction of mass	Solid	Mechanical	Fragile or ductile fracturing	Turning, Milling, etc.
		Thermal	Melting and Vaporization	Arc cutting Electrical Discharge machining
		Chemical	Dissolution	Electro-chemical machining
		Chemical	Combustion	Cutting
Increase of mass (Processes of union)				
Union at atomic level	Solid	Mechanical	Plastic deformation	Friction welding
	Fluid (at the union point)	Mechanical	Flow	Fusion welding
Adhesion	Solid (Fluid filling material)	Mechanical	Flow	Brasing

The summary examination of manufacturing processes has only the goal of demonstrating that many of the solutions identifiable from the standpoint of production management cannot be divorced from the knowledge and full understanding of the specific features of the underlying technological processes. This is also due to the significant economic and financial implications of process choices, such as the opportunities to exploit economies of scale or convertibility options linked to certain plant engineering choices. Technology and management are in fact two inseparable and interrelated elements, that mutually condition each other, as the former often poses constraints on the latter; as illustrated below, only with awareness of these links is it possible to identify useful areas of design and improvement.

2.3 Summary of types of layout

The study of the links between process and product has led to the identification of different types of organizational structures of production, according to a distinction that develops from methods of managing extremely fragmentary and heterogeneous flows, until the definition of an orderly and continuous flow. From an organizational and management standpoint, an additional element of differentiation of production systems lies in the characteristics of the **layout**, i.e. in the floor plan of areas, machines, and equipment, around which the materials and components used in the transformation and placed at the operators' workstations are moved.

Although in reality mixed layouts are prevalent, it is possible to identify some **principal types**, represented in Figure 2.4, such as:

Figure 2.4 **The principal types of layouts**

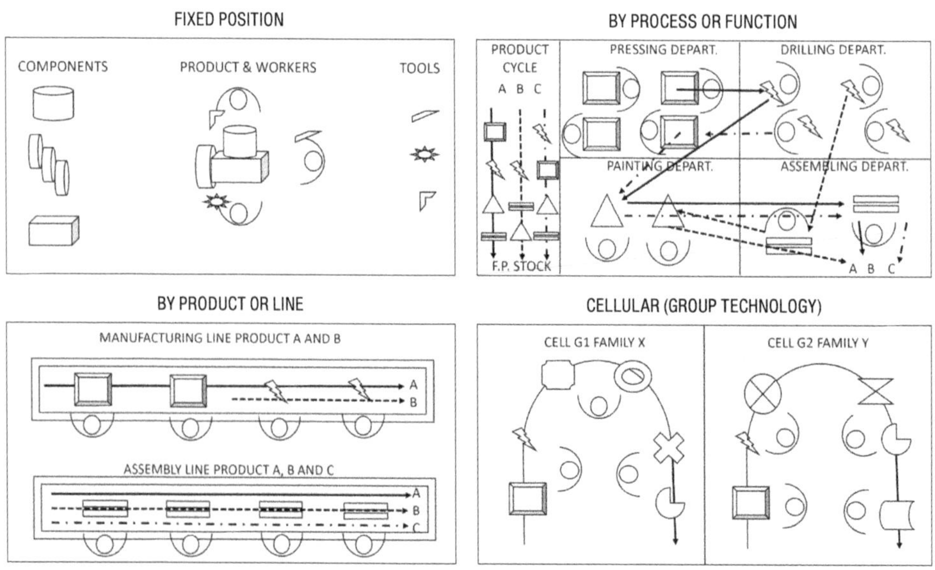

- fixed position or fixed product layout;
- layout by process or function;
- layout by product or line;
- layout by group technology or cell.

A **fixed position** layout (also known as worksite or fixed product layout) entails the stationariness of the article in a set location, around which the equipment, labor, and materials and components for its production revolve and are moved. This solution, essentially linked to processes of assembly, is dictated by objective **logistical reasons**, as in the case of civil engineering works or large plants, that necessary must be located at the construction site, or by considerations of **economic convenience**, as in the case of naval and aviation worksites, in which the product is removed at the end of the work cycle, and the equipment is not, as in the previous example.

A layout **by process or function** is characterized by areas (functional departments) in which machines are grouped that are uniform in terms of technology, or the function performed and the operations carried out. This is the case for most of the mechanical or furniture industry, where we see a transfer of semi-finished products from one department to another, such as departments for cutting, lathing, milling, assembly, etc. The main justification for that choice lies in the **concentration of specific skills** on the part of workers, the efficiency linked to the **optimal exploitation of machines**, according to sequences dictated by the uniformity of operations, and to the considerable flexibility of the work processes that allow for the production of very broad product ranges and production for single orders. On the other hand, we generally see a considerable accumulation of stock of materials and semi-finished materials in **interoperational inventory** and in the departments themselves, and very complex flows to optimize.

The solution **by product or line**, on the other hand, entails an arrangement of the machines and production resources that is coherent with the technological cycle of a given product or a limited family of products. Along the line (manufacturing or assembly, connected or not connected) the product undergoes successive phases of processing until its completion. In line solutions there can be a flow by batches or that tends to be continuous, depending on the degree of flexibility of the line itself. As in the case of the home appliance industry, arrangement by product is justified in the need of high levels of saturation, and significant volumes of uniform products, produced in long and uninterrupted series. The undoubted advantages that can be obtained in terms of **productivity, efficiency, and containment of work in process**, are accompanied by significant limits in terms of investments, possible inelasticity of production, and consequent costly reconfiguration and conversion.

Layout **by Group Technology or cell** developed with the aim of combining the advantage of versatility, typical of functional departments, with that of efficiency, typical of a line arrangement. This is a grouping of machines and equipment necessary for the production of a very limited range of products belonging to the same family. It should be stressed that, during the planning of industrial processes, belonging to a **family** is not due to elements of a commercial nature, but can be recognized by the geometrical uniformity of the pieces or the commonality of the used technologies. For example, there can be mechanical processing on prismatic pieces, that are very different in terms of final destination, but similar in terms of geometries and sizes, or that require the same work cycle. That option is characterized by the **containment of**

Table **Comparing Processes of Different Types**

FEATURES	(1) JOB SHOP	(2) BATCH FLOW
Product Features		
Product mix	Generally custom products	Lots of generally own designed products
Products compete largely on:	Speed of delivery, product customization new product introduction	Product performance, product reliabili and workmanship, delivery reliability, new product introduction, flexibility to produce either low or high volumes
Products unlikely to compete on:	Price	
New product introduction	All the time; easy	Frequent; routine
Process Features		
Process pattern	No rigid pattern; product can be routed anywhere; sometimes a dominant flow	Not all procedures performed on all products; product can be routed many ways; often a dominant flow
Type of equipment	General purpose	Mostly general purpose
Balance of equipment	Balance of speed and time done in only the grossest, long-run terms. At any one time, an imbalance is likely to exist	Balance likely to be imperfect betwee segments of process but better coordinated than typical job shop
Capital utilization	Labor intensive; machines frequently idle	Labor intensive, although less machin idleness
Typical size of operation	Generally small	Generally medium-size
Economies of scale	None	Few, if any
Notion of capacity	Very fuzzy; definable vaguely in dollar terms; useful only in long run	Fuzzy; product mix implies a dollar definition only
Product Features (continued)		
Additions to capacit	Incremental over full range of possible capacity	Larger increments possible, but over range of possible capacity

work-in-process, satisfactory saturation indexes, versatility – within the limits of the range of products that can be processed – and generally limited series.

The majority of manufacturing production processes are characterized by a layout by process or product.

(3) WORKER-PACED ASSEMBLY LINE	(4) MACHINE-PACED ASSEMBLY LINE	(5) CONTINUOUS FLOW	BATCH/ CONTINUOUS FLOW HYBRID
ostly standard products; me opportunities for selected tions	Same as for 3	Standard products with little or no customization possible	Same as for 3 and 4
oduct performance, ice, product reliability d workmanship, delivery liability	Product performance, price, product reliability and workmanship, delivery reliability	Price	Product reliability and workmanship, price
		New product introduction, product customization	
metimes	Sometimes; generally expensive	Hardly at all; very costly	Infrequent; expensive
ear pattern, though special eatment of some products metimes permitted	Clear, rigid pattern, though some off-line work possible	Clear, very rigid pattern	Same as for 5
ecialized	Same as for 3	Same as for 3	Same as for 3, 4, and 5
achinery speed and size in od balance with peak needs	Speed and size of equipment very well balanced. Capable of being adjusted together over small changes in line speed	Good balance of speed and size. Any excess capacity often placed in latter portion of process to provide insurance against breakdowns or unusual order requests	Same as for 5
though equipment is ecialized, it is fairly cheap; bor still a big item	Capital intensive	Very capital intensive; machines nearly always utilized	Same as for 5
riable	Large	Mammoth	Same as for 4
me as for 2	Some, perhaps	Yes	Same as for 5
creasingly clear; some ysical unit measures possible	Clear; physical unit measures	Same as for 4	Same as for 4 and 5
me as for 2	Changes can be made, but costly to do so because capacity comes in large chunks; otherwise, mild fluctuations possible within relatively narrow limits (e.g., line rebalance)	Only modest range for incremental change; otherwise, huge chunks of capacity required	Comparatively modest range for incremental change; then huge chunks of capacity required

Table (*continued*)

FEATURES	(1) JOB SHOP	(2) BATCH FLOW
Speed of process (dollars of output/ unit of time or dollars of output/dollars of input/unit of time)	Slow	Reasonably slow
Pacing	Worker discretion key	Worker-paced
Bottlenecks	Movable; frequent	Movable, but often predictable
Nature of process change	Incremental	Mostly incremental; some significant radical changes possible
Place of technological change in process itself	Little impact; unlikely to be revolutionary	Important once in a while, usually incremental though
Setups	Many; varied expense	Some setups needed, but generally easy to
Run lengths	Short	Medium
Materials-Oriented Features		
Materials requirements	Uncertain	Can often be placed statistically withir reasonably narrow bounds
Vertical integration	None	Sometimes backward, sometimes forward
Inventories		
Raw materials	Small; most raw materials purchased to coincide with orders	Moderate; some purchased to coincide with orders and some purchased to provide buffer stock
Work-in-process	Large	Moderate
Finished goods	Low, if any	Varies
Control over suppliers	Low	Moderate
Control over customers	Utile or none	Same as for 1
Supplier ties	Informal; spot buys	Some spot buys, some longer term contracts

(3) WORKER-PACED ASSEMBLY LINE	(4) MACHINE-PACED ASSEMBLY LINE	(5) CONTINUOUS FLOW	BATCH/ CONTINUOUS FLOW HYBRID
creasingly fast	Fast	Fast, sometimes astounding	Same as for 5
orker-paced but set with me bouinds by management tion (e.g., line balance)	Machine-paced, but can be a management goal as well (e.g., line balancing)	Determined technologically; built into equipment	Generally determined technologically but some leeway available
ovable, but often predictable	Generally known and stationary	Known, stationary	Same as for 5
ostly routine (rebalance); metimes radical (equipment)	Same as for 3	Most change incremental but radical change possible; means big bucks and sweeping conversions	Sometimes radical; sometimes incremental
creasingly important; nbodied in equipment	Important; embodied in equipment	Far-reaching; surprisingly regular	Same as for 4
me as for 2	No setup required; line already set up	Few and expensive, if any; process organized to simplify most kinds of setups	Little or no setup required; process simplifies need for most kinds of setups
me long; some short	Long	Very long	Same as for 5
own statistically within fairly ose limits	Certain once production plan established	Same as for 4	Some aspects certain, others known statistically
me as for 2	Sometimes backward, often forward	Often backward and forward	Same as for 2 and 3
ries; often steadily purchased nce material needs are nerally known within asonably narrow bounds	Varies; often steadily purchased to coincide with production plan	Often large, but can vary and be steadily purchased to coincide with production plan	Varies; some purchased for buffer stock and some purchased to coincide with production plan
tle	Same as for 3	Very little	Inventory placed between batch and continuous flow segments of process; moderate batch WIP, little continuous flow WIP
me as for 2	Can vary; often thrust down distribution channels	Varies; often thrust down distribution channels	Same as for 5
eat	Same as for 3	Same as for 3	Same as for 3, 4, and 5
me as for 1	Some, as to delivery	Same as for 4	Same as for 4 and 5
ntracts increasingly long m	Formal; long term	Formal; generally long term	Same as for 5

Table *(continued)*

FEATURES	(1) JOB SHOP	(2) BATCH FLOW
Customer ties	Informal; repeat business encouraged, however	Some informal, some formal distributi
Information-Oriented Features		
Order handling and sales	Often bid for; sales are to order	Varies; some to order, and some from stock with lagged adjustments to that stock
Degree of information coordination outside factory	Needed only for bids, receipt of any supplied materials, and to initiate supplies	Needed only to monitor sales and to initiate supplies
Information-Oriented Features (continued)		
Information systems within factory	Elaborate, viewed as central; lots of flow between factory and workers and management	Less elaborate, but still considerable; less feedback required
Trigger for production	Order itself	Could be order or level of finished goods inventory
Scheduling	Uncertain, flexible; always subject to change	Flexible but not as uncertain; less subject to change
Quality control	Informal, by each worker; spot checks	Can be more formal, though often is n
Response to cyclicality in demand	Overtime, shift work, some subcontract, hire/fire	Overtime; adjustments can be made b building and depleting inventory, shift work, and hire/fire
Labor-Oriented Features		
Labor content per $1 of product value	Very high	High
Job content	Large	Medium

(3) WORKER-PACED ASSEMBLY LINE	(4) MACHINE-PACED ASSEMBLY LINE	(5) CONTINUOUS FLOW	BATCH/ CONTINUOUS FLOW HYBRID
Can be informal or formal	Formal distribution channels	Same as for 4	Same as for 4 and 5
Same as for 2	Sales well established ahead of time or from finished goods inventory; lagged adjustments to stock	Sales well established ahead of time or from finished goods inventory	Same as for 5
Needed to monitor sales and to place orders for supplies; sometimes for order scheduling	Elaborate order scheduling, materials tracking; various levels of forecasts; great deal of corporate communication	Elaborate order scheduling, materials tracking, forecasting; great deal of corporate communication	Monitoring sales, placing supply order; little corporate communication on day-to-day operations
Little information needed, basically just to communicate order; flow from management to workers. Opposite flow used primarily to signal breakdowns; more informal	Same as for 3	Little information needed, basically just to communicate product change; flow from management to workers. Opposite flow used primarily to signal breakdowns; generally more informal	Same as for 5
Level of finished goods inventory, longer term forecasts, or "frozen" orders	Same as for 3	Same as for 3	First portion of process triggered by forecasts; second portion triggered by "frozen" orders
Process designed around fixed schedule	Same as for 3	Easy to group similar jobs or orders; fixed schedule often set technologically and well in advance	Similar jobs or orders grouped together; fixed schedule set in advance, at least for second portion of process
Some on-line checks, some postassembly checks; brame traceable	Same as for 3	Essentially done by process control people; designed into process; periodic sampling often done	Much of quality designed into process; periodic sampling often done
Overtime work or closing line early; Some adjustment: can be made by building and depleting inventory, rebalance, and dose down shift or plant	Same as for 3	Inventory adjustments; otherwise, shift or plant is shut down	Same as for 5
Medium	Low	Very low	Same as for 4
Small	Same as for 3	Often small "push button" staff; but can be "art" as well	Same as for 5

Table (*continued*)

FEATURES	(1) JOB SHOP	(2) BATCH FLOW
Importance and development of standards for labor	Key for scheduling and planning growth; big purpose of information system; sometimes key for wage payment	Often key for wage payment; needed also for scheduling product mix
Worker payment	Hourly; piece rate or other incentive wage	Often piece rate or other incentive wage
Advancement for worker	Knowledge of more and more machinery and machine capabilities. With more skills and/or seniority acquired, greater pay and responsibilities given, such as "lead worker." Seniority can lead to change in shift assignment	Knowledge of more and more machinery and machine capabilities. With more skills and/or seniority acquired, greater responsibilities given, such as "lead worker." Seniority can lead to change in shift assignment or department
End-of-month syndrome (i.e., more produced at end of month for billing than at beginning)	Inevitable	Often happens
Management Features		
Staff-line needs	Small staff (information, quotes, and new product development). Great line supervision needed	Small staff (generally methods related). Line supervision more critical
Degree of corporale influence on operations, if plant within a multiplant company	Modest	Same as for 1
Means of control	Usually a profit center	Can be either a profit cent[illegible]
Challenges to management	Scheduling, bidding, information flows, expediting, product innovation, shifting bottlenecks	Order processing, labor issues and pay, handling cyclicality

Source: Schmenner, 1981.

(3) WORKER-PACED ASSEMBLY LINE	(4) MACHINE-PACED ASSEMBLY LINE	(5) CONTINUOUS FLOW	BATCH/ CONTINUOUS FLOW HYBRID
Crucial for process design	Same as for 3	Important largely for assigning workforce to equipment	Same as for 5
Hourly base; sometimes also tied to percentage of standard that cremo achieves	Hourly	Same as for 4	Same as for 4 and 5
With more skills and/or seniority acquired, greater responsibilities given, such as "lead worker" or "group leader." Seniority can lead to change in department or shift assignment	Seniority can lead to change in department or shift assignment	Seniority can lead to change in shift or, sometimes, department assignment	Same as for 5
Close to nonexistent	Same as for 3	Nonexistent	Same as for 5
Large staff for process redesign, methods, forecasting, capacity planning, and scheduling. Line supervision and troubleshooting still critical	Same as for 3	Large staff for technology and capacity planning, and scheduling. Line supervision less crucial	Same as for 5
Same as for 1	Great, for both operations and capital expenditures	Same as for 4	Variable; more likely that corporale influence is with capital expenditures than with day-to-day operations
Same as for 2	Usually a cost center	Same as for 4	Same as for 4 and 5
Balance (process design), product design, managing workforce, technological advances, capacity planning, materials management	Same as for 3	Capital needs, site, technological change, materials management, vertical integration	Same as for 5

3 Problem Setting and Problem Solving in Operations Management

by *Alberto Grando*

3.1 Physiology and pathology in Operations Management processes

Design interventions regarding production systems can be divided **into *ex novo* interventions** (green field), the original action carried out at the time a Production Unit is created, or **adjustment** (or reorientation) **actions**, if they involve revisions in the processes of management of operations induced by the natural evolution of company activities. Such adjustments, that are possible from the standpoint of organization, management, or technology, can assume the character of gradual evolutions or true breakthroughs. The intensity and breadth of the change can entail more or less radical modifications in the management policies and industrial structures in order to adapt them to the needs of competition. In the context of adjustment actions, we will thus speak of **evolution**, to describe the continuous adjustment of the production system driven by the search for superior performance, **reorganization** for the changes to the organizational and management structures of production, and **restructuring** (or foot-print redesign) in cases in which the assets (plants, structures, technologies, etc.) are affected by the redesign process.

Leaving aside newly-created companies, that by their nature deal with problems of assessing investments that are very complex and linked to much more general considerations, the others – evolution, reorganization, and restructuring of the production system – become unavoidable in the presence of inadequate operations responses to the needs of competition, whether obvious or only perceived. Those interventions can start with changes to business strategies, manifest technological obsolescence, the opening to innovative management solutions that do not fit well with the existing processes, the competitive pressure produced by competitors or new entrants into the markets, and so on. Over time, such phenomena tend to translate into unsatisfactory performances with respect to market expectations or corporate governance, and tensions among the components of the production structure; **multiple signals** that the management must be able to recognize promptly and process, in order to prepare the most appropriate solutions. Delays or superficiality in the perception of signs of inadequacy that emanate from businesses or the preparation of appropriate responses can translate into pathologies that at times are much worse and can become irreversible (Grando, 1993).

The company management is thus pushed, on the one hand, to govern increasingly complex phenomena, and on the other, to search for timely and reliable solutions faced with the greater rapidity of change. In that sense, a need arises to grasp the warning signs of such changes in advance and to fully evaluate their repercussions, not setting limits of a functional nature or regarding the time horizon of the choices to be made.

3.2 Problem Setting and Problem Solving

Considering the above, it follows that there is a need for the company management, and in particular the operations managers, to have models of investigation, conceptual schemes, and competences aimed at **Analysis and Diagnosis** (i.e. Problem Setting) of the company's structure functioning and its systems processes.

It is necessary to develop **Problem Setting** skills, even before those of **Problem Solving**. People who work in the business world are in fact aware of the difficulty of identifying the priorities on which to intervene. Faced with the numerous small and large problems and plans for improvement that every company and every manager must constantly deal with, in contexts of resources availability that by definition are limited – not only the **budget**, but more frequently **time** – the need arises to define the **priorities for intervention**, due to the relevance of the expected results, to be expressed in terms of coherence with strategic goals, and the probability of their pursuit. On the issue of relevance, and of the connected order of priority among possible improvement interventions, particular attention is required to the phase of *Problem Setting*, which is necessarily prior to that of *Problem Solving*. It is in fact evident that perfectly solving an irrelevant problem translates into an unjustified waste of resources, while developing a **"good enough" and timely solution** to a priority problem can lead to significant advantages. Within a business an optimal solution does not exist, because in the time used for its identification, the variables observed change, new constraints can emerge, or goals can shift. Management never operates in conditions of certainty, and in any event must search for solutions that by their nature are imperfect because they are always influenced by the human factor. The search for perfection is sometimes an indicator of poor managerial skills or the inability to assume the inevitable risks of every decision-making activity. This often leads management to request continuous studies and analyses, pushing the company into the trap known as "**Paralysis by Analysis**."

To assume a controlled risk in processes of decision-making, management must develop a structured approach based on rational analysis and on measurements that are as objective as possible. This involves first of all **identifying the improvement areas** or the problems to be dealt with, understanding how to assign priority to each of them, and as a consequence deciding how to best allocate the resources available to improvement projects aimed at resolving them or at least mitigating their negative effects.

With this, we intend to stress that in the context of the classical phases of an intervention, represented in **Figure 3.1**:

- **Analysis** of the picture of symptomatology;
- **Diagnosis** of pathologies;
- **Therapy** or implementation of solutions.

Figure 3.1 **Analysis and diagnosis of operations systems**

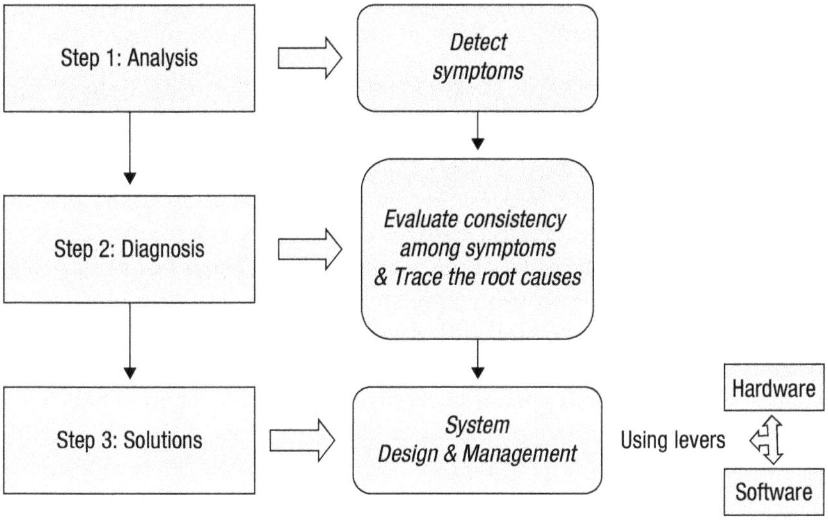

The first two phases are particularly critical, involving analysis and diagnosis, i.e. the Problem Setting phases. If the specter of therapeutic interventions (and solutions) is at times known and often corroborated by sufficient cases, significant difficulties are often found in the explanation of the problems requiring an adequate response and in the methods of adapting the theoretical solution to the specific context in which the intervention takes place.

The Operations System thus continuously issues **weak and strong signals** concerning its state of functioning. These can be evident, such as a sudden increase of costs, an increase of defective products or delays, the emergence of missing components during planning, or the reduction of the production efficiency, or almost imperceptible, inextricably bound up in a broader accumulation of indications. Generally, we are used to considering only the good or service produced as the output of a production system; for our purposes, on other hand, it is necessary to stress that a fundamental component of the output is also represented by the sum of those signals, and particularly by their informational content. The "weakness" of the signal is not to be understood as an indication of poor predictive relevance, but as an expression of a general difficulty to grasp phenomena of change in the embryonic state that can also have significant importance from a diagnostic standpoint, in foreshadowing changes in trends.

The phase of **Analysis** thus consists of monitoring the functioning of operations systems in order to grasp the information content of those signals, i.e. to **identify the symptoms** of particular physiological or pathological states, that deviate from expectations. Those symptoms clearly represent only the exterior manifestation of causes often located elsewhere with respect to the emission of the signal, which the process of diagnosis must aim to identify.

The **Diagnosis** is in fact the phase in which the coherence between the symptoms detected is analyzed, and the **root causes are traced**, through a backwards search that by passing through numerous links of cause-effect, aims to reach the original cause.

Think of the case of a mechanical company worried about increasing levels of stock. The symptoms, expressed by slow inventory turnover, induced management to launch an intervention aimed at identifying the most appropriate techniques for stock management. A careful process of analysis and diagnosis found that the unexpected level of stock was due to excessive production batches with respect to the fragmentary nature of requests from the market, despite being justified by the high setup costs. In turn, those costs were correlated with significant retooling times, due to obsolete equipment and methods of changing dies. The simple example provided here stresses that an intervention on the inventory would only regard the effects – the symptoms – without impacting the true causes of the trouble, the set-up technologies and methods, producing either no benefits or benefits that would be dissipated quickly, because they are the result of symptomatic, ineffective, and palliative therapies. To the contrary, the removal of the causes, launched through the revision of the methods and die changing equipment and the adoption of organizational interventions inspired by SMED (Single Minute Exchange of Die) methodologies in the example, automatically limited the cascading effects, producing the hoped-for benefits on the inventory level. Similar cases can be found starting with the detection of an excess of defects, that despite being identified downstream, can be the results of phases placed further upstream in the production process or even be originated by the suppliers of materials or components. Similar considerations can be carried out with reference, for example, to reduced productivity of labor, motivated by erroneous layouts, irrational production flows, and wastes that ends up affecting the working time of the operators.

The **Diagnosis** phase thus takes place through the construction of a picture of symptoms, that is, through the analysis of the logical correlations between them that, only if observed jointly, contribute to highlighting the origin of certain inefficiencies. A defect generated by a manufacturing process, for example, can be the symptom of inadequacy of the incoming materials or the machines that processed them, of poorly-trained labor or the procedures and methods used. Only by **cross-checking multiple levels of observation** is it possible to isolate the elements of non-conformity. This is the case of a department, for example, in which the source of defects found during inspection, through a series of investigations and checks on the processed pieces, was identified as the labor employed in the Assembly phase, with particular reference to specific shift, that included recently-hired workers and for which it became necessary to intensify training activities and supervision by expert operators.

Once an accurate Diagnosis has been completed, we move towards the third phase, the identification and preparation of the most appropriate **Solutions**, through the maneuvering of specific Design Levers. The choice between multiple options responds to criteria of efficiency and efficacy, respecting three constraints: **time constraints**, **technical feasibility** and **economic-financial convenience**. For example, in the case of a distribution network afflicted by a service level considered insufficient due to the frequent unavailability of requested products, or the reduced deliveries dependability, the hypothesis of introducing shipments by air transport or multiplying peripheral safety stock would be a solution that, despite being technically possible, would probably be anti-economical. On this point, we should observe that it is not rare for the range of theoretically plausible solutions to be reduced by **constraints** that are unmovable in the short to medium-term, that lead to the adoption of second best or compromise choices amply justified in practice. Those constraints, either inside or outside the production system, can produce two effects:

- reducing the spectrum of conveniently applicable solutions;
- reducing their efficacy.

To provide some examples, among the most frequent **external constraints** we find the pressures dictated by social, legal, regulatory, or market imperatives. These are cases of recurring needs regarding labor, precaution towards the ecosystem, constraints imposed by regulators, or policies implemented by competitors. An additional element that slows investments is linked to the uncertainty of the markets served or the more general economic context. The mitigation of some constraints can be stimulated from the outside: for example, think of fiscal incentives to stimulate investments in technologies or employment.

Internal constraints refer to limits on exploitation of technologies, the scarcity of certain resources, in particular financial resources or knowledge, bargaining power with respect to counterparties, or the factory environment or culture.

It is very rare, in any event, to have constraints that do not allow for any solutions; at the most, the breadth of the Design Levers used, the intensity of their use, and the entity of the expected results change.

3.3 A structured approach to Problem Setting and Problem Solving

The ability to successfully face a process of problem setting and problem solving depends on at least two elements.

1. A first element has to do with some **managerial skills and abilities**. In particular, we refer to:
   - the **mental attitude to face problems or projects based on priority**, using approaches that allow for identifying the principal criticalities or bottlenecks of the processes observed on which it appears reasonable to launch justified drill down actions (considering resources and times that are limited by definition). Indeed, very often, especially in technical areas, there is a tendency to prefer detailed analysis and the search for perfection, leading to unjustified times and costs with respect to the relevance of the phenomenon investigated and to the possibility to implement heuristic approaches that lead to "good

enough" solutions. This often leads to an unjustified use of resources that is disproportionate to the value generated;
- the **ability to make decisions in conditions of uncertainty**, typical of every managerial problem, avoiding ambitions of determinism and the already-defined syndrome of "paralysis by analysis," in which there is a tendency to request incessant, and at times useless, investigations. Sometimes these generate very little in terms of added informational value, but dangerously extend times and costs of diagnosis, at times leading to solutions that are already obsolete with respect to the evolution of the competitive context.

2. The use of **structured methods of analysis and diagnosis**, based on principles of **Focused Management** and **Complexity Reduction**, able to guide action in non-deterministic contexts (Ronen and Pass, 2008). In fact, the more the issues to be addressed appear to have a limited structure and uncertain borders, the more, paradoxically, it is necessary to rely on tools able to give rationality to very structured processes of analysis and evaluation. We refer to managerial tools that allow to:
   - **express the goals and expected times** linked to a project for improvement, focusing the managerial action;
   - **reduce complexity of analysis**, offering the ability to monitor over time the achievement of goals or gradual improvements;
   - **facilitate the involvement of resources** and communication in the project team and with other members of the company;
   - **suggest for which cases it is advisable to initiate drill down actions**, or in-depth examination, aimed at identifying the root causes at the origin of the problem or the base of the project implemented.

Below, using a well-known framework originally proposed by General Electric in the context of the Six Sigma approach and then adopted as a guide in business improvement projects, we illustrate a set of approaches and tools applicable to the processes of problem setting and problem solving.

The framework proposed here – **DMAIC** – is an evolution of the well-known approach named the **Deming Wheel** (Deming, 1990), amply used in kaizen or continuous improvement approaches. The latter evokes the circular nature of the process that moves from the definition of an improvement goal, to its potential achievement and the advisability of extending the solution identified to other similar processes and areas. In a kaizen perspective, there is always a priority project to be dealt with, until every area of inefficiency or waste is completely eliminated. The Deming Wheel or Cycle, also called **PDCA**, that is illustrated in **Figure 3.2**, is divided into 4 phases.

1. **Plan,** in which an improvement intervention is planned.
2. **Do**, in which what is planned is done.
3. **Check**, in which it is checked to see if the planned goal has been reached. If the outcome is negative, it is necessary to review the previous phases to investigate the causes behind the failure. If positive, move on to the next phase.
4. **Act**, in which the solution used is spread to other areas of the system.

Figure 3.2 The Deming Wheel and continuous improvement

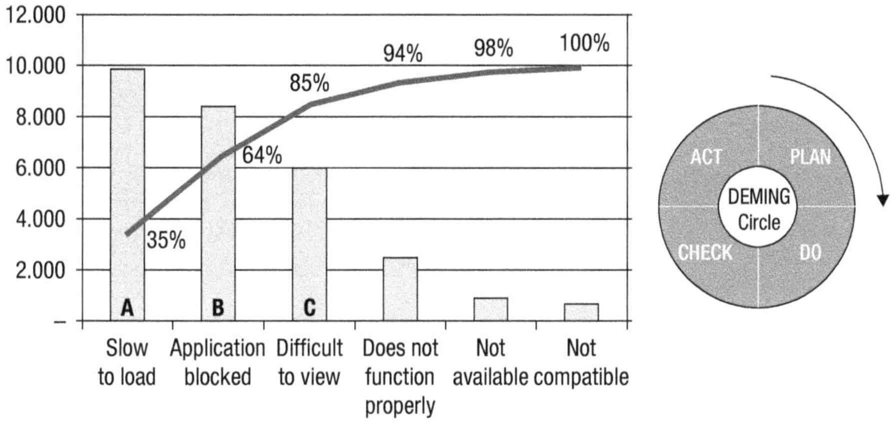

Once a priority problem has been faced and resolved – class A in the ABC curve shown in **Figure 3.2** – the next problem takes on priority, and thus will be faced with the same determination as the first. Regardless of the gradual resolution of problems and elimination of defects and waste, according to the principle of **continuous improvement**, there will always be a class A problem to address, even if it is relatively less important than the previous ones faced and resolved, that is still a priority at that moment for that organizational system. **Figure 3.2** shows the example of a malfunction of a software application for which the principal problem have been detected, classified by their frequency. The management, applying the Deming cycle, will thus define the goals of the action (Plan) to address the principal problem – the slow rate of loading the software – then will then carry out the actions (Do) and verify the results in terms of loading speed (Check). If the actions are considered to be decisive, they will be standardized and extended to all applications that show the same problems (Act). At that point, the new class A, or priority problem, will be the frequency with which the application freezes up, that will be dealt with following the same approach.

The Deming Wheel has an evocative value, as we have said. The approach shown here follows the same path but adds more structure and tools to the process of Problem Setting and Solving.

3.4 The DMAIC method

The acronym **DMAIC** refers to the principal phases of Problem Setting and Solving; **Define, Measure, Analyze, Improve, and Control**. Below we provide a summary illustration of the single phases, using an example that will facilitate comprehension.

Imagine the case of a manufacturing company operating in B2C of mechanical and electronic products, in which the sales functions complain of continuous prob-

Figure 3.3 **The Project Charter and Project Rough Schedule**

Problem Statement	**Business Case and benefits**
• *Problem description*: – delivery delays grew in the last 3 quarters – Customers started complaining for delays and uncompleted delivered orders – Salesforce started increasing extra-discount and terms of payment – Risk of order cancellations • *KPIs As-Is* – OTIF 76% – Average Delivery Delay: 6 days – Order-lines completeness: 95%	• Customers ask for a sharp improvement in the delivery performance • Bridging the gap with the best competitors in terms of service means gaining Market share by 2%, revenues by 10% and EBIT by 3% • Reducing the length of the terms of payment will improve our financial performances • Need to recover market positioning, customers loyalty and reputation
Project Goals • Planned improvement KPIs To-be – OTIF: 90% – Average Delivery Delays: max 2 days – Order-lines Completeness: 98%	**Project Scope** • *What is in scope* – Mechanical articles produced by the plant of Lyon • *What is out of scope* – All the other articles produced in the other plants – Electronic components produced in Lyon – Articles manufactured by outsourcing partners

Project Team (Sponsor: Mr. S. Mattioli - Mechanical Components Division Director, Project Leader: Mr. L. Maré, Lyon Plant Operations Manager, Team members, Mr. A. Marzagalli, Plant Manager, Mr. P. Stringa, Quality Manager, etc.)

• **Project Rough Schedule**: Project Progress at: dd/mm/yyyy

Phase	START	END	Progress
Define	01/04/20XX	31/05/20XX	100%
Measure	01/06/20XX	30/06/20XX	100%
Analyze	01/07/20XX	15/09/20XX	100%
Improve	18/09/20XX	28/02/20X+1	70%
Control	01/03/20X+1	31/03/20X+1	00%

lems linked to delays in delivery, attributable to production. Those delays translated into increasing problems with customer satisfaction and the need to grant extra-discounts and deferments of payment, with consequent impacts in economic-financial terms and growing reputational risk. To deal with these criticalities in a non-episodic way, the management decided to launch a DMAIC project to identify the causes and design an improvement action.

The first phase – **Define** – is characterized by the use of a **Project Charter** able to:

- **define the improvement project** through statements, identifying one or more performance measures that represent the starting point, i.e. the As-Is KPIs;

- **estimate the expected benefits**, through the development of a business case of credible guesses;
- **represent the project goals** in descriptive and quantified terms through the same performance measures used for the definition of the starting point, To-Be KPIs;
- **identify the project scope**, thus defining what is in scope and what is out of scope.

In general, the Project Charter, of which an example is provided in **Figure 3.3**, shows the list of members of the project team, identified by name and role, the team leader, and the hierarchical superior or sponsor responsible for the project.

In the Define phase, it is necessary to elaborate the **project timing** and any intermediate check points, sometimes defined as the **Gate Review Phase**. To this end, the project team develops a general **Project Rough Schedule** to then detail it for each phase, using project planning tools such as Gantt diagrams, or in rarer cases, PERT diagrams.

Figure 3.4 shows the Gantt diagram for the project, enriched by all of the useful information, including the potential budget anticipated to carry out the Define phase,

Figure 3.4 Example of Project Plan

Project Plan

PROJECT: Delivery Performance Improvement		Project leader: In. L. Maré	Unit: Plant Lyon (France)
Project Progress:	30/04/20XX	Status - Timeline - Milestones - Owner	
Next review:	30/05/20XX		

☐ = Original Plan • = Complete

Phase & Actions	Deliverable	Status	Planned Due Date	W14	W15	W16	W17	W18	W19	W20	W21	W22	W23	W24	Action Owner
DEFINE															
Problem definition		On time	05/04/20XX	•											OPS
VOC plan	*VOC Report*	On time	12/04/20XX		•										SALES
Suffering/Claims collection	*Customer Complains*	1 week Delay	22/04/20XX		•	•	•	•							SALES
CTQ definition	*CTQ Form*	1 week Delay	22/04/20XX				•	•							OPS
Process mapping	*Process map*	To be completed	22/04/20XX				•								PLANT
Flow diagram	*Flow Diagram*														PLANT
SIPOC definition	*SIPOC Report*														TEAM
Improvement goals definition	*Report by product line*														TEAM
Expected savings & benefits estimation	*Fin. & Contr. Estimate*														FIN & CONT
Project charter update	*Project Charter Update*														TEAM
DEFINE GATE REVIEW												X			TEAM

KPIs	AS-IS	TO-BE
OTIF	76%	90%
Avg Delivery Delay	6 days	max 2
Order lines complet.	95%	98%

People engaged in the project: Ing. Marzagalli, Dr.P. Stringa, Mr. XXXX.

Budget expenditure		€ ______	Authorized by:	Ing. S. Mattioli
Actual expenditure	% ______	€ ______		
Expected Benefits		€ ______	Signature:	
Project expected IRR or Pay Back: See Annexes			Date:	15/11/20XX-1

NOTES:

Figure 3.5 SIPOC framework

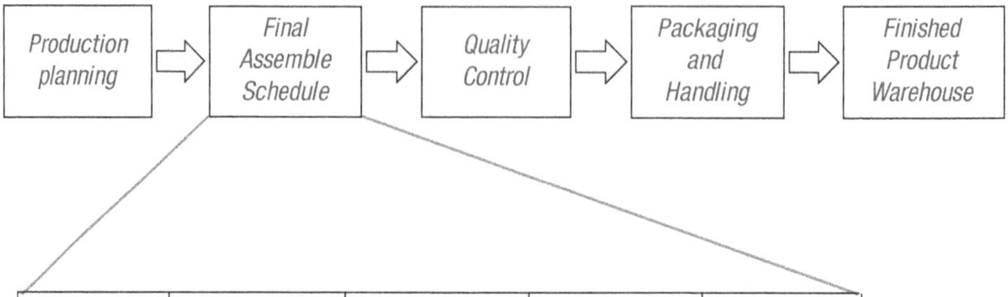

Suppliers	Inputs	Process	Outputs	Customer
Operations Planning Department	Demand forecast, Orders Book, On-hand stock	Final Assemble Schedule	Production Schedule reliability, Min. Orders backlog	Sales Department

the activity plan, the state of progress, and other information that should allow for defining: **why, what, when, who, where, and how**.

The Define phase is generally completed by the definition of the process observed, both in graphic terms, through **Activities Flow Diagrams**, and through the use of descriptive tools, constructed according to the **SIPOC** sequence – **Supplier, Input, Process, Output, Customer**. Figure 3.5 provides an example of the SIPOC framework, applied to the search for improvements in product deliveries by the production unit, with respect to what is scheduled.

As a function of the process observed, the Customer can be an **internal or external customer**, and it is crucial to fully understand its goals. The goals cannot be defined autonomously by the Project Team, but must clearly reflect those desired by the customer. To this end, all instruments are used that are useful to clearly express and quantify the customer goals. This refers to schemes, as indicated in Figure 3.6, that are able to:

- represent the **VOC – Voice of Customer**;
- define the characteristics of the product/service or the critical parameters that interpret the VOC, sometimes denominated **CTQ – Critical To Quality**;
- summarize operational aspects linked to monitoring of the evolution of CTQ, once again specifying **Who, What, When, and How**[1].

The second phase – **Measure** – is generally aimed at the description of the performance parameters used – As-Is KPIs – and the **Data Collection Plan** based on which they are quantified. This is to clearly state:

[1] To analyze these concepts, see Shankar R. (2009), Carrol T.C. (2013).

Figure 3.6 Voice of Customer and Critical to Quality

Voice of the Customer **(VOC)**	Need to improve logistics performances; improve delivery speed, delivery dependability and delivery completeness			
Critical to Quality **(CTQ)**	Improve reliability of delivery shedule; reduce the order backlog			
	Improve delivery completeness; reduce re-scheduling			
WHW of CTQ	- **WHAT:** On Time In Full Index			
	- **HOW:** Better Production Planning and Reliable Scheduling			
	- **WHEN:** Starting from Quarter III - Reach the target performance in Year N+1			
To-Be KPI		**Q.III**	**Q.IV**	**Q.II-Y+1**
	OTIF	80%	84%	90%
	Delivery Delay	6	4	max 2 days
	Order-lines Completeness	95%	97%	98%
As-Is KPI	OTIF: 76%			
	Delivery Delay: 6 days			
	Order-lines Completeness: 95%			

- **the domain**, i.e. the perimeter of collection of the data. For example, in the case in question, this is the specification of the products lines that are systematically late and the temporal interval in which they were collected or acquired;
- **the source of the data**, and potentially of the collection method used, stating whether the data comes from internal information systems, structured external sources, surveys, etc.;
- **the representativeness**, clearly stating if they are sample investigations, 100% analyses, the sample composition, etc.;
- **other information** considered useful to define the perimeter and the characteristics of the information used.

During measurement, further examination can be carried out through the stratification of the initial data aimed at representing its variability or studying some aspects considered to be key to better represent the VOIC and the CTQ measures. Those examinations will at times to be carried out in the subsequent phase, that of Analysis.

By way of example, **Figure 3.7** shows a stratification of delivery times, by deliveries.

The third phase – **Analysis** – sets the goal of interpreting the data collected and launching a process of diagnosis. Various tools are used in that phase, some specific and linked to particular improvement projects, such as VSM – Value Stream Map, used in Lean projects (see Chapter 17 on this point); others are more generic and thus more frequently adopted in projects of different natures, exemplified in **Figure 3.8**. We refer, for example, to:

Figure 3.7 Deliveries and Delivery Time

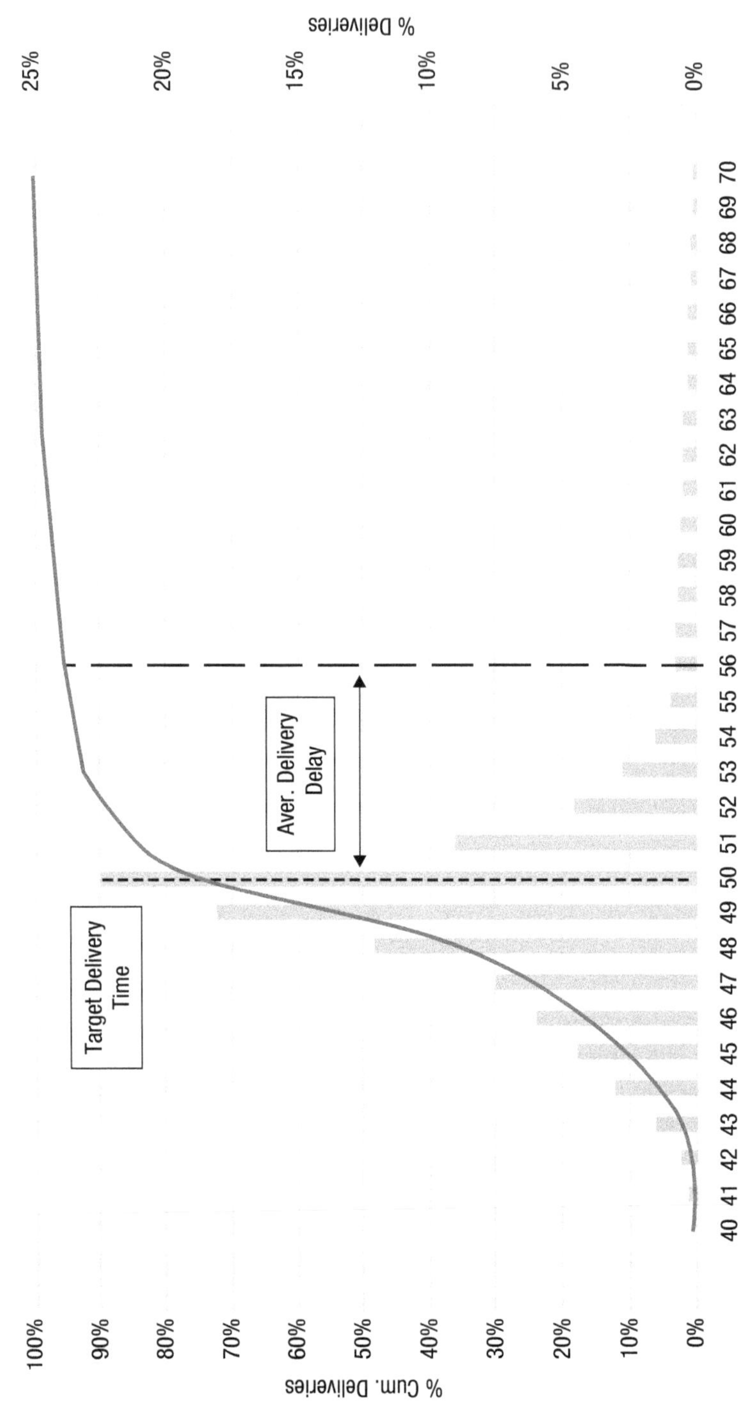

Data Collection Plan
- **Domain**: entire number of order lines delivered in the year by the Plant of Lyon
- **Source of data**: total list of orders recorded in the ERP (track record + auxiliary files/systems)
- **Representativeness of data**: 100% of received orders in 20XX-1

Figure 3.8 **Most frequently used tools in the analysis phase**

Ishikawa fishbone (cause-effect diagram)

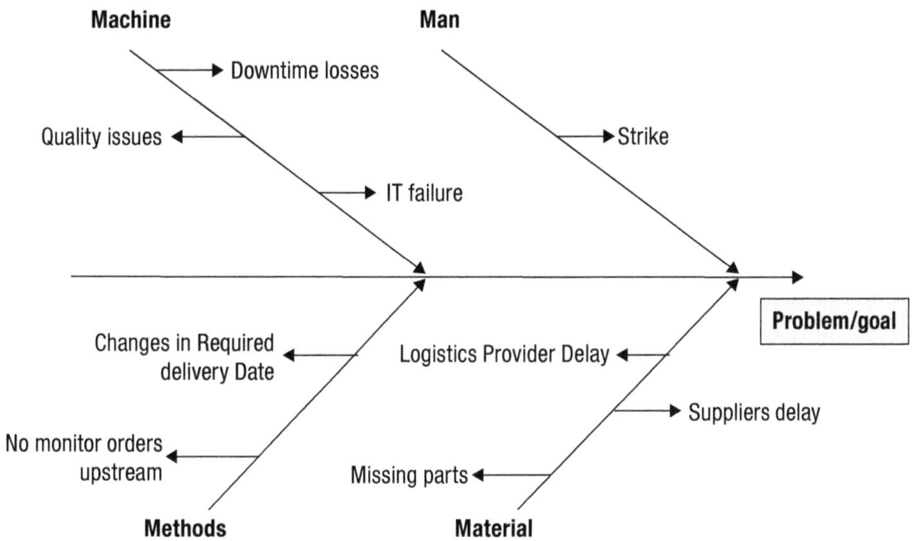

Failure mode

Failure mode	Code	Count	Categories
Changes of Required Delivery Date	a	12	Planning
Delays due to Machines downtime	c	11	Shopfloor perfor.
Production delay due to suppliers	g	9	SCM Delay
No report to monitor orders in upstream departments	h	5	Planning
Quality issues – Need to rework	i	5	Shopfloor perfor.
Logistics provider delay	b	3	SCM Delay
Information system failure	f	3	IT system
Need of re-schedule for missing parts	d	1	Planning
Other (strike)	e	1	

- **Ishikawa diagrams**, or Fishbone diagrams, also known as **4 M**. This is a tool used to understand the source of a problem, and thus move from the phase of analysis of a symptom to that of diagnosis of the causes that have generated it. By way of example, observe the case in which the problem of delays in delivery detected in the final accounting for weekly production can be traced back to one of the 4M (Materials, Methods, Manpower, Machines). Upon identifying the primary source(s), through a series of observations and investigations, we proceed with additional detailed analyses until the possible root causes are identified;

Figure 3.8 ***(continued)***

Pareto curve or ABC curve

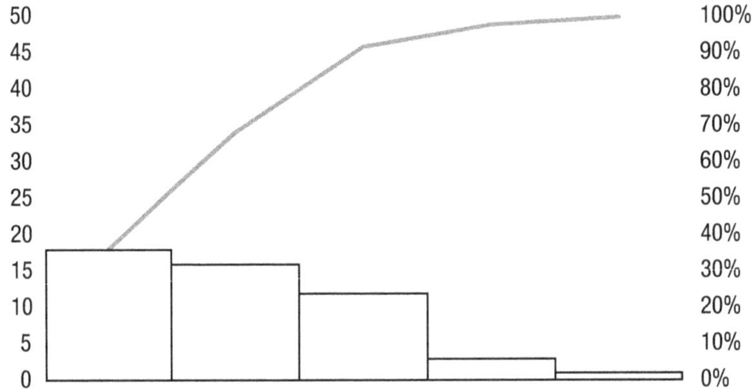

Category	Planning	SF Perfor.	SCM delay	IT system	Other
Count	18	16	12	3	1
%	36%	32%	24%	6%	2%
% Cumul.	36%	68%	92%	98%	100%

- **Failure Mode methods**, that have the goal of classifying the pathological phenomena observed as a function of the frequency of their occurrence, such as the frequency of delay analyzed with respect to the different product lines;
- **Pareto curves, o ABC curves**, also called 20-80, that are useful to represent the relevance of a distributed phenomenon, and as a consequence, to define some priorities for intervention;
- The approach defined as **Five Whys Gate Review**, that requires a retrograde analysis to identify the root cause of the problem observed, through different cause-effect links. On this point, see the example shown in Figure 3.9, which again refers to the same case of delays in delivery. Imagining that the causes of delay are principally attributable to problems of planning and shop floor performance, as shown by the diagnostic tools used previously, the use of the Five Whys Gate Review allows for tracing the probable root-causes in order to design improvement actions or solutions.

The methods described briefly here, and the many others that can be found in the literature and in business practice, have in common the ability to favor **Focused Management** and **Complexity Reduction** approaches, and in that way, to develop e coherent managerial mind set.

The fourth phase – **Improve** – is characterized by approaches and tools that cannot be generalized, because they are project-specific. This involves selecting managerial, technological, or organizational interventions, aimed at improving the process and services in question. In the rest of this book we will present multiple management techniques that are useful for improvement, such as the 5S technique

Figure 3.9 Example of the Five Whys Gate Review approach

PROJECT: Delivery Performance Improvement	Project leader: In. L. Maré	Unit: Plant Lyon (France)	Last Project Progress: …………
Main people engaged in the project:	Sub Project: Scheduling Delays Reduction		Next Review: …………

ROOT CAUSE ANALYSIS — CAUSES INVESTIGATION														ACTION PLAN — PLANED ACTIONS AND OUTCOMES		EXECUTION		
Why 1	CHECK POINT Y/N	Why 2	CHECK POINT Y/N	Why 3	CHECK POINT Y/N	Why 4	CHECK POINT Y/N	Why 5 (ROOT CAUSE)	CHECK POINT Y/N	MEN	MATERIAL	METHOD	MACHINE	Actions	Expected Impact	WHO Owner	Source of Inform. & Data	Status
Why we have sheduling delays	Y	Drop in the Efficiency Index	N															
		Drop in The Utilization Index	Y	Increasee of Set up Time	Y	Increase of mix produced	Y	Update Sert up procedures and technology	Y	X		X	X	Introduce SMED approach, Training workers, invest in dies		Engeneer. Director	See Report ED07/20XX	65%
				Increase of Failure downtime	Y	Increase of Breakdowns and minor stoppages	Y	Reactive Maintenance	Y	X		X	X	Introduce Planned or Predictive Maintenance		Mainten. head	See Report MD02/20XX	20%
				Increase of strikes	N													
		Drop in Quality Index	Y	Quality issues in Materials	N													
				Defective outputs	Y	Newly recruited workers	Y	Lack of Training	Y	X		X		Plan a Training program for newely recruited workers. Introduce mentoring program from Supervisors		HR Director	See Report HR15/20XX	50%

Notes:

in lean projects (Chapter 17), the Vendor Rating & Selection systems (Chapter 16) used in the evaluation of suppliers, the techniques for demand forecasting (Chapter 7) or managing stock (Chapter 10), the adoption of technological solutions oriented towards digital transformation (Chapter 20), or solutions for better planning the production (Chapter 6).

We will leave the examination of those techniques to the cited Chapters, and merely point out here the importance, during the actions, of using methods that facilitate the **clear definition of responsibilities and roles** during selection and implementation of the improvement actions. In that sense, many companies trust matrixes that codify the division of responsibilities within the project team, such as **RACI – Responsible, Accountable, Consulted, Informed**, matrices or its subsequent versions developed in business practice. So in a project, the distinction made by the RACI matrix – exemplified in **Figure 3.10**, allows for identifying:

- **Responsible**: the person(s) who have executive roles and assign specific activities;
- **Accountable**: the person who has full responsibility for the project and answers for the results;
- **Consulted**: the person(s) who assist the Responsible with support and assistance roles;
- **Informed**: the person(s) who must be informed regarding the execution of an activity.

Figure 3.10 Improvement projects and responsibility matrix

ACTION PLAN	Lyon Plant Manager	Planning & Scheduling	Head of Maintenance & Spare Parts	Process Quality Control	Information Systems & Technology	Assemble Department Head	Plant Human Resources Director
Introduce SMED approach	A	I	I	I		R	C
Train the newly recruiter workers	I			C		R	A
Develop Predictive Maintenance	I	I	A		R	C	
Define lot sizing parameters		A			I	R	
Monitor a missing part list		A	I	I	R	C	
Reduce the defective parts				A	I	R	
Monitor changes in the orders due dates	A	R			C	I	
....							

Key
R = Responsible
A = Accountable
I = Informed
C = Consulted

The roles identified by R, C, and I can be held by multiple people, while Accountability must be assigned to a single person.

The last phase – **Control** – has the goal of **monitoring the progress** of the project, through the updating of the Gantt diagrams and the control of the KPIs used at set control moments (check points). The control of a project can therefore be carried out through **performance measures**, whether financial or non-financial, generally linked to specific characteristics of the project, the analysis of the times of its development and the budget of resources used in its implementation. Below we will look at the principal elements of performance and the related performance indicators currently used in many industrial companies.

3.5 Operations Management and performance measures

As illustrated in **Figure 3.11**, the design solution originates, on the one hand, from the application of the method of **Problem Setting**, aimed at the identification of the spaces and alternatives pursuable, and on the other hand, by the application of the approach of **Strategic Alignment** and focus on the objectives of the logistics-pro-

Figure 3.11 The approach to the design of production systems

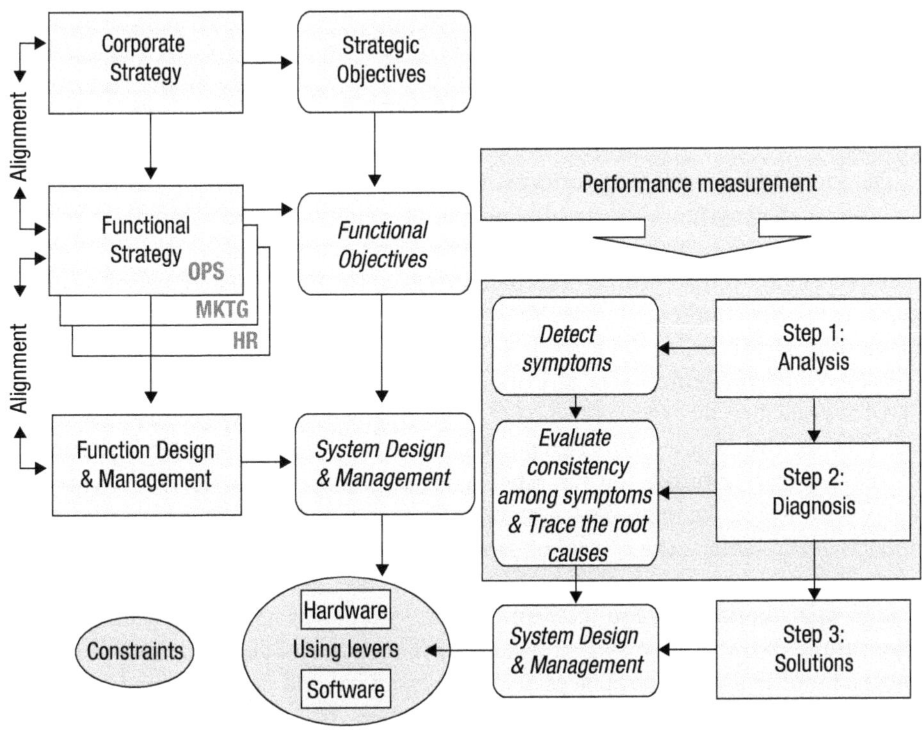

duction system, aimed at making the operations management choices coherent with the goals defined by the corporate strategy, in respect for the technical and economic constraints that are present within the company.

The issues discussed here principally regard the choices relating to the software levers, with the intention to limit the analysis to decisions linked to the management control of the production system.

We have stated that the detection of the symptoms constitutes the initial phase of the process of diagnosis of a production system; evaluating the state of functioning of the system, whether during full operation or in deteriorated conditions, thus translates into the identification and measurement of some significant performances, **As-Is KPIs**, in order to compare them with the expectations that arise from the process of definition of the goals assigned to the system itself, **To-Be KPIs**. For the most part, these are temporal and physical measures and other parameters that will be repeatedly referenced in the remaining Chapters. The need to have a clear and systematic picture of those aspects of performance has stimulated countless studies on the subject. First of all, it becomes necessary to make an initial distinction between:

- **External performance**, outside of the production system (or performance in a strict sense), i.e. linked to results measurable directly by the client or by the top management (at the business unit or corporate level);
- **Internal performance**, inside the production system (or measured within the sub-units responsible of the operations processes).

The first group refers to elements that explain customer satisfaction and reaching higher-level business goals, such as the ability to differentiate the product/service offered or to achieve results in terms of overall cost and the efficient use of production resources.

The second group refers to performance of a specific productive factor, the result of choices called system's functioning operating conditions (or configuration parameters), that is, performance and properties that characterize single productive factors. They contribute to determining internal performance (directly) and external performance (generally indirectly), since they represent the characteristics of the system's functioning, as illustrated in **Figure 3.12**.

External performances typically descend from the strategic business goals, and are aimed at evaluating the degree to which they are attained. For example, this may regard measures such as the level of service provided, the quality offered, product costs, etc.

Internal performance, on the other hand, expresses the effectiveness and efficiency with which the operations system is managed, and refers, for example, to the productivity of factors, the control of process lead-times, the flexibility of machines, the quality measured in acceptance checks, and so on.

Operating conditions are the result of the system's design choices and the sum of constraints to which the system is subject. They refer to elements linked to technological and management choices such as versatility and reconfigurability of machines, the manufacturing process throughput rate, plant calendars, the speed of

Figure 3.12 Internal and external performance and operating conditions of the production-logistics system

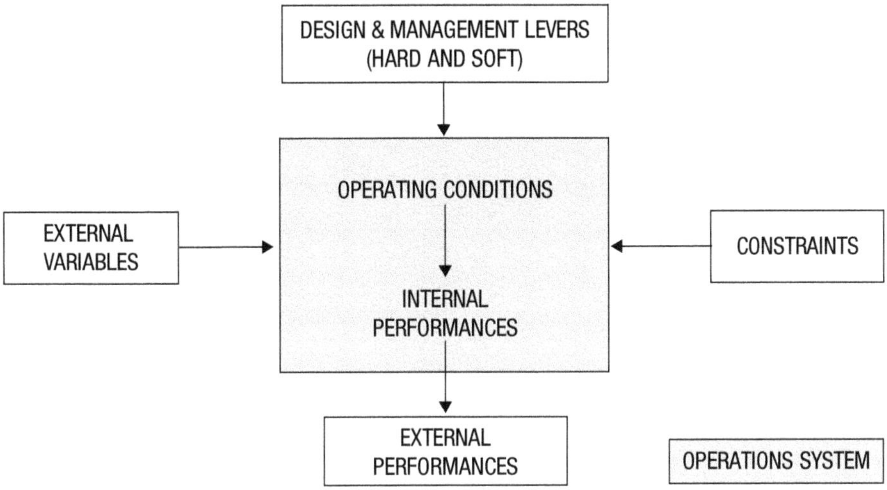

logistics flows, the utilizations and efficiency of the factors used, the inventory lot sizes or safety stock, setup times, etc. The Operations Management de facto has to constantly act on the operating conditions in order to improve internal performance, and thus external performance, the only aspect recognized by the market. In companies exposed to competition, since the customer is able to make a selection through the comparison between the performance offered by multiple manufacturers, the management and control of final performance becomes particularly critical: quality, service, flexibility, and cost, and an element added more recently, sustainability. In only a few cases, in which the supplier finds itself in the enviable position of being a monopolist, or in any event in a strong bargaining position, does the focus on internal performance prevail, with the search for greater efficiencies that are not necessary transferred to the market. This is the case of automobile companies, or suppliers of chassis to the manufacturers of special vehicles, who due to the bargaining power they hold, impose long delivery times (four or five months from the order), regardless of actual logistical needs, in order to optimize their own production schedule.

3.6 Performance measurement in Operations

Referring only to the types of performance illustrated in the previous section, those which directly interest the market or company management, it is possible to distinguish between **five groups or performance** families, respectively linked to concepts of **service, quality, costs** (factors productivity), **flexibility, and sustainability**. **Figure 3.13** shows the division of each category of performance into its principal components, deployed into multiple levels.

Figure 3.13 The breakdown of performance of a production-logistics system

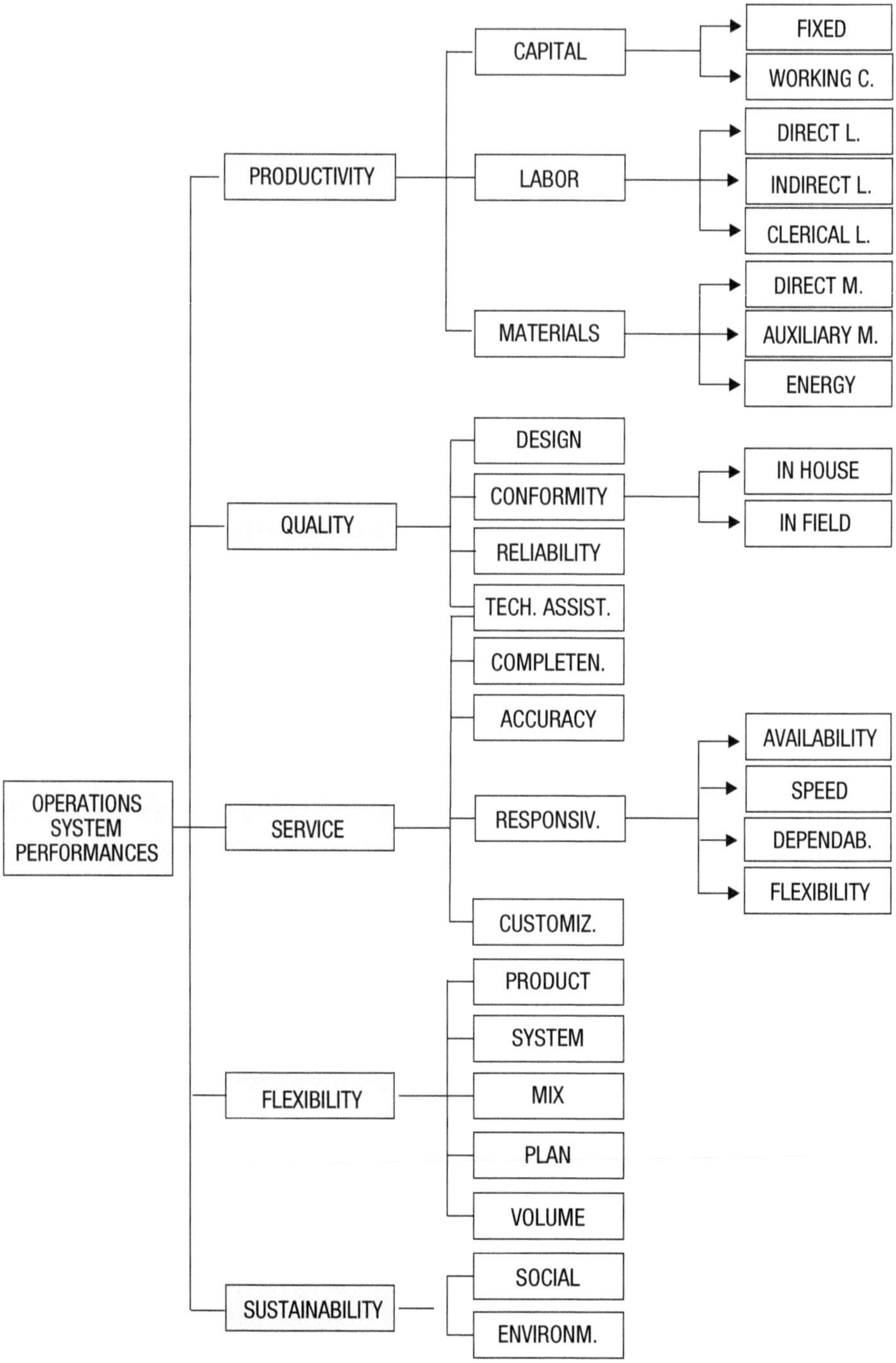

Service is expressed by the sum of technical assistance guaranteed during installation, at times entered into the quality parameters, the delivery responsiveness, completeness, and personalization with respect to customers' specific needs. The analysis will be conducted (Chapter 4) by dividing up the measure of responsiveness into its different elements of availability, speed, dependability and flexibility.

Quality is generally evaluated with reference to its components represented by the quality of the design, the quality of conformity, reliability and maintainability, and technical service. In the analysis of a production-logistics system, the measure of quality of conformity takes on particular importance, evaluated with reference to the area of its manifestation: internal or in-house, and external or in-field. Those measures will therefore be examined specifically in Chapter 4.

Total productivity, the expression of the ratio between output set to the standard industrial cost, and the sum of production factors, can be broken down into multiple partial measures of productivity, with reference to the single factors used: capital, labor, materials, and energy. In regard to capital, observed from a technical standpoint, we distinguish between fixed capital, invested in plant and equipment, and operating working capital, partially absorbed by stock of raw materials, semi-finished products, and finished products. We will observe the issue of plant productivity in Chapter 5, leaving the treatment of the measures of the inventory performances to Chapter 9, dedicated to that subject. As regards work and materials, it is possible to perform separations at multiple levels: the figure distinguishes, respectively, between direct, indirect, and clerical labor, for the former, and direct and auxiliary materials and energy, for the latter. We will analyze the productivity of labor and materials, limiting the discussion to the first level of aggregation, since the additional details add little from a methodological viewpoint.

Flexibility can be observed from various standpoints, all linked to the ability to deal with the greater or lesser variability expected in production flows and its impact on the production system in question. The generic concept of flexibility is thus broken down into multiple elementary aspects of flexibility, in terms of aggregate volume, mix, product, and system. Those dimensions will be observed in detail in Chapter 5, without failing to point out that some of the elements underlying product flexibility overlap with others, related to the personalization of the service, while system flexibility can be traced to some characteristics also attributed to readiness performance.

Lastly, **sustainability** is to be understood from the dual perspective of environmental and social sustainability, as economic sustainability is linked to the overall performance of the business. With reference to the issues of industrial production, environmental sustainability can be measured in terms of performance that refers to inputs and outputs. As for inputs, it is frequent to use measures that express energy consumption from non-renewable sources, or measures such as the consumption of water or other scarce resources, such as the percentage of virgin raw materials used, or materials with a large environmental footprint, such as polymers and plastics. As for outputs, the most commonly used measures refer to emissions of CO2, other greenhouse gases, waste water, or specific high-impact materials such as plastic packaging and other non-biodegradable or compostable components. The performance of social sustainability is measured in terms of impact of production activity on all of the company's internal and external stakeholders. Typical examples are safety indicators for factory operators, measured by the hours of work without accidents, classified based on severity, the number of hours dedicated to specific training, or all interventions that are frequently part of company welfare packages. As concerns outside subjects, we refer not only to suppliers and clients, with whom partnerships can be launched for improvement activities, but also local communities and the territories where the company's production units are located, that can benefit from dedicated investments such as schools, pre-schools, or other forms of support.

As we see from these brief remarks, it is not always possible to clearly assign a single type of performance to a specific family, since some measures can be included in one or more sets of performance. For example, think of technical assistance, sometimes included in quality performance and sometimes in service performance, or delivery responsiveness, attributable to service performance or flexibility. It should also be stressed that the indicated performance measures often have strict **links of interdependence**, and mutually influence each other. Take the example of a production system of mobile phones that, due to insufficient levels of reliability of the assembly process, generates defective products. If the non-conformities are intercepted and corrected before the entry of the products onto the market (in-house non-conformities), the need to repair operations or to substitute defective components will inevitably translate into an increase of costs. The client will obtain higher quality, but at the expense of a higher price or a lower margin for the manufacturer. Conversely, in the event that the quality problems are not intercepted in time and the products are sold on the market (in-field non-conformities), the possibility to limit the negative impact in financial or reputational terms is linked to the offer of suitable assistance services, repairs under warranty, or in extreme cases, substitution with new products.

In light of those considerations, it makes sense to deepen the analysis of those measures of performance, searching for suitable quantitative indicators for their measurement, although in the context of a management view. Therefore, we will discuss in Chapters 4 and 5 the principal measures used to monitor the state of functioning of a production-logistics system, highlighting the links existing between operating conditions, internal performance, and external performance. In that sense, it is possible to observe that:

- **the service** guaranteed to the customers or users for the most part derives from the ability to appropriately manage internal and external lead times and to provide the right product, in the right quantity, in the right way (times, places, and means of delivery);
- **the quality** recognized by the market is a function of the reliability of the process, the ability of labor, and the characteristics and intensity of control;
- **the factors productivity** depends on conditions such as throughput rate, availability, and the degree of utilization and efficiency of the same;

- **the flexibility** of response ensured by the system derives in significant part from elements such as the degree of versatility expressed by the machines and plants, the elasticity of production capacity, and the internal mobility and versatility of labor;
- **the sustainability** is linked to specific measures, often connected to the incidence of injuries, evaluated as a function of their degree of severity, and environmental impact, measured in terms of limitation of inputs (for example water or virgin raw materials) or outputs (such as CO2 and other pollutants).

For reasons of convenience, the analysis of those types of performance will be divided into two chapters: the measures of service and quality are analyzed in Chapter 4, while those of productivity and flexibility are addressed in Chapter 5. For sustainability performance, given the variety and specificity of measures adoptable, we refer to specialist contributions on the subject (Belvedere and Grando, 2017).

3.7 The concept of trade-offs and PwP (plant-within-a-plant) strategies

The identification of those types of performances allows corporate management to assign specific goals to the internal organizational structures, indicating the trajectory the efforts must follow to define the general strategic goals, and as a consequence, the sub-unit goals. These goals, however, can be numerous and often antithetical.

For example, consider a logistics system that attempts to jointly pursue limited investment in stock and high service levels, or a production process with the aim of maximum saturation and flexibility, or of high quality and low costs. As already mentioned, the traditional approach to Operations choices is thus based on the **concept of focusing**: in the presence of divergent goals, or **trade-offs**, one or a few need to be preferred and the others considered residually, as dependent variables. Competing on costs, for example, cannot translate into sacrifices in terms of quality and service. It appears evident that the attributes of high or low, used to describe the level of each type of performance, take on a relative value, with respect to the performance of the competition and market expectations. That relativity is also to be considered from a time standpoint: when faced with progress made by competitors or increasingly pressing demands from clients, a performance that is considered excellent today may soon turn out to be insufficient. Think of "same-day" delivery services such as Amazon Prime, currently considered a standard by many e-commerce companies, that just a few years ago would have been considered a useless excess, and possibly unattainable at an acceptable cost for customers.

It is evident that in the same company, or even in the same production or service unit, products and services can co-exist that are characterized by performance goals that are different, or even antithetical, because they are aimed at different market segments. To avert the **risk of defocusing**, i.e. of confusing performance goals, resources, and processes that should be kept separate, the **PwP** or **Plant-within-a-**

Plant approach was developed (Skinner, 1974), and then extended to OwO or Operations-within-an-Operation, also known as SwS or Shop-within-a-Shop (Slack et al., 2016). The intention of this phrase is to suggest isolating production processes dedicated to different product lines or customer segments within the same structure, whether a manufacturing plant, office, hospital, or large warehouse. This way, the personnel, resources, and technologies involved in a surgical unit must not be confused with those of a general medicine or maternity ward. Similarly, a factory that produces cold units or personalized refrigerated counters will not risk diluting the positioning of its products, mixing resources, processes and performance with the lines for assembly of standard refrigerators. Finally, even within a supermarket, the operators of the rotisserie, butcher, and cold meats counter, by selecting the products to "portion" on request from customers, have skills that are not to be confused with those who stack the shelves with standard prepackaged products. In many cases focusing pays off, despite limiting the versatility and sometimes the growth of personnel. The choice to design structures able to operate on multiple fronts, on the other hand, mediating existing trade-offs between products and services realized, translates into needs for flexibility that, despite offering the advantage of greater productive versatility, leads to more costly systems.

It is important to point out that, over time, the **different goals have gradually accumulated**, to the point that in the current situation they constitute a set of performance measures that all companies must jointly pursue in order to be competitive. Although one of these must prevail based on the positioning and historical phase in which a business is operating, the others cannot simply be considered irrelevant. To provide an example, an automotive company, despite aiming to produce reliable and high-quality cars, cannot offer its production at prices that are unaffordable for most people and with delivery times unacceptable for the market. Likewise, a low-cost airline must guarantee a minimum level of quality, reliability, and safety in the routes it covers.

The sum of the performance offered by an Operations System highlights the close interdependence and mutual conditioning between various areas: to attain specific levels of external performance, multiple internal elements of performance contribute, that are the result of considerations and choices made during the design of the Operations System.

3.8 Operations Management and Management Control

The need increasingly expressed by many companies to have effective and synthetic **Operations' performance control systems** has led to the design of "dashboards" that through the observation of a few aggregate indicators, are able to offer close monitoring of the key performance indicators of the operations system, its margins for improvement, and the direction it should follow.

It is known that decisions concerning the company's activities are traditionally located on **three levels**, distinguishing – due to the time horizon covered and the breadth of the problems treated – between choices aimed at:

- defining strategies;
- management control;
- operational control.

The subjects addressed again below will be discussed extending the analysis principally to decisions regarding the area of management control, consistent with the managerial approach of this book. The attempt to maintain a level of discussion that is equidistant between the strategic sphere and the strictly operational sphere is justified by the desire to interpret operations phenomena according to management logics, on the one hand avoiding the technicalities that characterize many operational choices and that therefore do not lend themselves well to generalization, and on the other hand, rejecting the simplistic assertions of those who believe they can identify strategic content in every business activity. The choices we will refer to below, thus come under the sphere of economic analysis for management decisions, and regard the acquisition and optimal allocation of available resources, for the pursuit of goals defined elsewhere, during strategic formulation. Once the process of analysis and diagnosis has been launched, in the choice between possible paths of conduct, it is necessary to focus on the reflections that such choices will have on the company's economic value. With reference to industrial companies, those changes are observed principally only in costs, although they are more frequently involving revenues as well, influenced by the performance of operations systems in terms of service, and investments, with particular reference to the dimension of fixed and working capital. Lastly, it seems useful to recall that managerial control rarely acts in deterministic contexts characterized by absolute certainty, but develops its decision-making processes within the unstable boundaries of risk and uncertainty, in which the expertise of the decision maker is crucial.

A final observation must be made with reference to the objective difficulty often found in correlating the operational performance attained by production-logistics systems, as an effect of improvement actions, and the success of the business overall, summarized by the increase of the company's market and economic-financial results. Many improvement actions and projects proposed in Operations find a weak response among the management due to the absence of indications that go beyond the increase of operational performance, and are aimed towards increased enterprise value. This is in large part justified, on the one hand, by the **time lag** generally found between the moment of the actions, i.e. of the production of immediately visible effects in terms of physical performance, and the more general economic-financial effects; on the other hand, there is the fact that economic-financial performance is based on contributions from other business functions – sales, administration, etc. – that make it difficult to pinpoint the specific contribution offered by operations. However, as shown in the final paragraph of Chapter 1, operations managers need to be aware that their actions have an impact on the overall economic value creation or disruption.

4 Measurement of Performance in Operations Management: Service and Quality

by *Alberto Grando*

4.1 The measures of service

As mentioned in the previous chapter, the **Service performance** can be observed from multiple standpoints, such as Responsiveness – divided into Availability, Speed, Dependability, and Flexibility –, Completeness, Accuracy and Customization seem most important.

With reference to **Responsiveness**, it is possible to make a distinction between **Availability**, typically linked to the characteristics of functioning of the **inventory system**, and the other dimensions – **Speed, Dependability and Flexibility –** that are strictly connected to the management of internal **lead times** (times for order acquisition and fulfillment, handling, manufacturing, packing and packaging, etc.) and **external shipping times** (transport times, stops at transit points or storage points, potential clearance through customs, etc.). Therefore, availability is strongly influenced by inventory performance, and thus can be measured by indexes that express the percentage of unfulfilled requests, and the incidence and persistence of stock-out. **Completeness**, in turn, is influenced by the ability to produce, manage in stock, and ship client's order lines, while **Accuracy** and **Customization** Performance refer, respectively, to the ability to perform processes – both administrative and delivery processes underlying the entire order cycle – precisely and without errors, and the ability to satisfy specific client needs – such as the type of packaging, transport, etc., and the ability to adapt the characteristics of the service to requests from clients.

4.2 The principal precautions in the construction of performance measures

To build reliable performance measurement systems, certain precautions should be taken, linked to various factors.

- **The Time Horizon of the observations**: the lead times taken into consideration must be measured and mediated with respect to the number of Orders, or better, of **Order Lines**, considered significant and thus observed over a time horizon consistent with the characteristics of the business in question. This means that for a company that produces machines on design, the observation horizon will last many months, as it must include an adequate number of deliveries. An e-commerce company that executes hundreds of daily deliveries in the world of fashion, consumer electronics, or food and beverage, may find appropriate measures weekly or monthly.

- **The selection of the reference data-set**: orders are to be extracted from the company's more general database that can be compared based on characteristics and expected delivery. So it does not appear advisable to "pollute" the measures by summing custom orders and standard orders, orders for deliveries in a limited geographic range – locally, for example – with deliveries of the same articles ordered by clients located abroad. The definition of the data-set is crucial to avoid mediating times that are very different because they refer to different orders in terms of product-service characteristics or location of the client.
- **The cleaning of the data used**: the persons responsible for operations must collaborate with those working on the information systems to periodically verify the reliability and integrity of the databases used. In the case of data-sets used for logistical measures, it is necessary to correct or remove data polluted by different phenomena and behavior, such as assigning dates that have not been updated, fictitious orders entered by the sales functions to commit shares of availability not covered by actual client orders, notification of changes of dates imposed by the client, etc.
- **The setting of the time unit**: in general, measurements are made in days. However, for companies that operate with long lead times, such as shipbuilding, large plants, or producers of some parts for mechanical industries, etc., the measure is often denominated in weeks, while for companies that commit to providing deliveries in set time slots, such as in the world of e-commerce, a measure in hours can be adopted. In general, in B2C situations days are used, while in B2B situations the use of weeks is frequent.
- **The selection of the other units of measure**: in general, in the measures of availability and stock-out the measures used are expressed in quantities, order value, and order lines. Measures expressed by quantities have the merit of not considering the variation of values over time, such that they lend themselves more to intertemporal comparisons. Measures expressed in value allow for determining the differences between goods and products with different cost-price or revenue-price, and thus are more useful in cases of broad product ranges with significantly different values. Measures based on orders have the benefit of simplicity and are better in cases of orders with one or only a few order lines or for those for which the key dates (Required Delivery Date, Planned Delivery Date, etc.) are the same among the order lines, i.e. directly referring to the order. Measures based on order lines have the benefit of greater analytical nature, with respect to the prior ones, and are preferable in the case of orders consisting of multiple lines, especially if there are different key dates for those lines.

Having carried out those interventions, the measure of service performance can be developed each time through temporal measures, physical measures linked to quantities ordered, or indicators that consider both of the perspectives together.

It is evident that the identification of the priority elements of performance that the operations system must focus on, is related to the method of response to demand (Chapter 2) that characterizes the operations management model.

- Systems that operate with a **Make To Stock** (**MTS**) logic are more focused on Availability, stock-out, and persistence of stock-out.
- Systems that operate with an **Assemble, Make, Purchase or Engineering To Order** (**ATO, MTO, PTO, ETO**) logic give more attention to Speed and Punctuality performance.
- Lastly, all production systems, regardless of the logic of response to demand, must equally monitor service performance of **Completeness, Accuracy and Customization.**

4.3 The measure of performance prevalent in MTS systems

In MTS systems, in addition to the measures used for controlling inventory, such as the Average Inventory Coverage Period and Average Inventory Turnover (Chapter 9), measures are used to control performance of Availability and its inverse, i.e. stock-out and the persistence of stock-out.

Inventory Availability (IA), expresses the ability of the operations system to guarantee adequate inventory coverage to face internal requests, coming from the departments that use the materials, components, and semi-finished products, or external requests, coming from the market, of finished products and spare parts. For example it can be measured through the indicators shown in Table 4.1.

4.4 The measure of performance prevalent in ATO/MTO systems

In systems that work To Order, the previous measures are useful for the inventory control of stock located in decoupling points or in upstream warehouses. Towards the client, greater importance is taken on by time-measures, those linked to the ability to respond to client orders. In particular, we note the following types of performance:

- **Delivery Speed**, aimed at measuring the promptness of deliveries;
- **Delivery Dependability**, that monitors average delays in delivery times;
- **Delivery Flexibility**, used to evaluate the margins of variation allowed with respect to the terms or characteristics of the deliveries agreed upon.

Table 4.1 Examples of Stock Availability KPIs

Unavailability

- $\% \text{ of orders backlog} = \frac{\sum \text{back orders}}{\sum \text{orders}} \%$
- $\% \text{ of order lines backlog} = \frac{\sum \text{back order lines}}{\sum \text{order lines}} \%$

Stock-out

- $\% \text{ of stock-out} = \frac{\sum \text{stock-out items}}{\sum \text{ordered items}} \%$
- $\% \text{ of stock-out quantities} = \frac{\sum \text{stock-out quantities}}{\sum \text{ordered quantities}} \%$
- $\% \text{ of stock-out periods} = \frac{\sum \text{stock-out periods}}{\sum \text{periods}} \%$

Stock-out persistence

- $\% \text{ of stock-out persistence} = \frac{\sum \text{days of stock-out} \times \text{stock-out items}}{\sum \text{days} \times \text{items}} \%$

All of these are strictly linked to the management of lead-times. The measure of the indicators linked to the time-dimension of the service is carried out by identified some **key dates** along the order time cycle and measuring the time intervals between then. On this point, see the example shown in **Figure 4.1**.

The company information system maps the order cycle through the parameterization of some key moments in time linked to specific events, defined below as **Key dates**:

- **OD – Order Date**, the date the order is received from the client;
- **OCD – Order Confirmation Date**, the date the supplier confirms the possibility to fulfill the order received from the client;
- **PSD – Production Start Date**, the date the production of a specific order is started;
- **PED – Production End Date**, the date the production of a specific order is completed;
- **RDD – Required Delivery Date**, the delivery date requested by the client at the time the order was issued;
- **PDD – Planned (or Promised) Delivery Date**, the delivery date that, following negotiation and based on the bargaining power of the parties, is agreed on between the client and the supplier;
- **ADD – Actual Delivery Date**, the date of actual delivery by the supplier. The delivery can be divided into more than one moment, considering potential partial deliveries. By way of example, imagine an initial delivery made on the ADD_{FD}, Actual Delivery Date – First Delivery and on the ADD_{BD}, Actual Delivery Date – Balance Delivery.

The key dates identifiable in the company information system can also be much more detailed. For example, if the company intends to measure a series of internal lead times linked to planning, production, and control cycles, multiple dates, or moments in time, can be identified, such as the date of entry of the product code required in planning, the production start date, the warehouse entry date, the date set for quality control, that of preparation for shipment, etc.

Having defined the key dates, the construction of the indicators originates from the measurement of the **time intervals** defined each time:

- **Lead times**, if referring to times consistent with planning;
- **Delays**, if referring to delays with respect to planned dates.

Figure 4.1 provides examples of:

- **Order Confirmation Time**, equal to lead time between OCD and OD;
- **Production Lead Time**, equal to lead time between PED and PSD;
- **Planned (or Promised) Delivery Time**, equal to lead time between PDD and OD, or **Planned Delivery Speed**;
- **Planned Delay**, equal to delay between PDD and RDD;
- **Actual Delivery Time** – measure of **Actual Delivery Speed**, equal to lead time between ADD and OD (or OCD), in turn divisible into First Actual Delivery Speed and Balance Actual Delivery Speed;

Figure 4.1 Order cycle and measure of deliveries speed and dependability

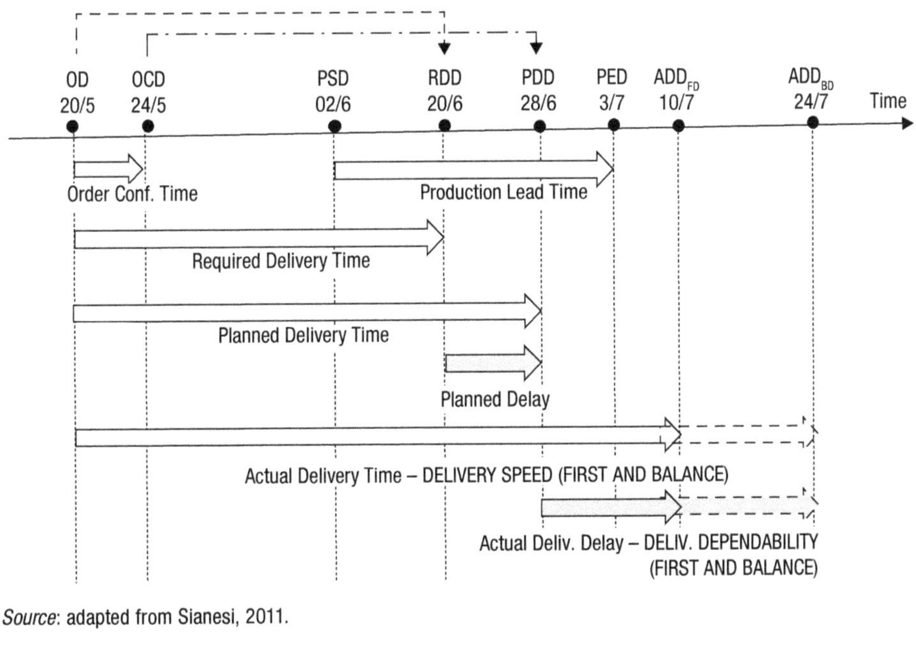

Source: adapted from Sianesi, 2011.

- **Actual Delivery Delay** – measure of delay in delivery or of **Delivery Dependability** – equal to the lead time between ADD and PDD, distinguishable between First and Balance, similar to Delivery Speed.

In systems that operate To Order, it is therefore possible to come to a series of measures, expressed in time units or in percentage values, those for which examples are provided below. From a time standpoint, responsiveness performance can be divided into three elements.

Delivery Speed measures the speed of the operations system's response to a client order. It is quantified by the extension of the time interval (Delivery Time) between the date of issue of the client order or Order Date (OD) and that in which the client obtains availability of the goods, or Actual Delivery Date (ADD). It is measurable through the indicators shown in **Table 4.2**.

Delivery Dependability measures the ability of the operations system to limit delivery delays and respect the Planned or Promised Delivery Date (PDD); it is measured in terms of time deviation between the latter and the Actual Delivery Date (ADD). It can be measured through the indicators illustrated in **Table 4.3**.

Table 4.2 **Examples of Delivery Speed KPIs**

- $\text{Orders Delivery Speed} = \dfrac{\sum ADD - OD}{\sum \text{orders}}$ (average delivery days)
- $\text{Order Lines Delivery Speed} = \dfrac{\sum ADD - OD}{\sum \text{order lines}}$ (average delivery days)
- $\text{Order Lines Delivery Speed}_{FD} = \dfrac{\sum ADD_{FD} - OD}{\sum \text{order lines}}$ (average first delivery days)
- $\text{Order Lines Delivery Speed}_{BD} = \dfrac{\sum ADD_{BD} - OD}{\sum \text{order lines}_{BD}}$ (average balance delivery days)
- $\text{Order (or Order Lines) Delivery Speed} = \dfrac{\sum \text{orders delivered in standard days}}{\sum \text{orders}}\%$

Table 4.3 **Examples of Delivery dependability KPIs**

- $\text{Orders Delivery Dependability} = \dfrac{\sum ADD - PDD}{\sum \text{orders}}$ (average delivery delay days)
- $\text{Order Lines Delivery Dependability} = \dfrac{\sum ADD - PDD}{\sum \text{order lines}}$ (average delivery delay days)
- $\text{Order Lines Delivery Dependability}_{FD} = \dfrac{\sum ADD_{FD} - PDD}{\sum \text{order lines}}$ (average first delivery delay days)
- $\text{Order Lines Delivery Dependability}_{BD} = \dfrac{\sum ADD_{BD} - PDD}{\sum \text{order lines}_{BD}}$ (average balance delivery delay days)
- $\text{Order (or Order Lines) Delivery Dependability} = \dfrac{\sum \text{orders delivered on time}}{\sum \text{orders}}\%$

Delivery Flexibility measures the ability of the operations system to accept potential variations in the delivery date requested by the client, and can be measured through the indicators shown in **Table 4.4**.

The indicators illustrated above refer to the time perspective of the service. The example shown in **Table 4.5**, however, makes it evident that deliveries made in two tranches, with identical date but different quantities (or value) delivered, cannot be considered equivalent in terms of service.

Table 4.4 **Examples of Delivery Flexibility KPIs**

- *Delivery Flexibility = Min N. of days before the PDD to accept delivery changes* (days)
  where delivery changes can be referred to volumes, dates, articles, etc.
- *Delivery Flexibility = Min N. of days to accept urgent orders* (days)

4.5 The measure of common performance among different production systems

The measures of availability and time factors, that are prevalent in MTS and ATO/MTO systems, respectively, must be accompanied by other measures that, taking into account the volumes delivered and the overall quality of processes underlying the order cycle, are common to any production-logistical system. In that case, we speak of **Delivery Completeness**, **Accuracy**, and **Customization** performance. Delivery Completeness can in turn be measured in quantity, value, orders, or order lines, and is generally expressed by a percentage of the ratio between what is delivered and what is ordered over a suitably defined time horizon.

Delivery Completeness measures the ability of the operations system to deliver what is contained in the client order, in terms of order lines and volumes or values, and can be measured through the indicators shown in **Table 4.6**.

Delivery Accuracy refers to the ability to guarantee the absence of errors in the various phases of the shipping process. In particular, in order to obtain diagnostic indications on the possible causes of errors, it is advisable to codify the different reasons for errors and stratify the indicator by causes (administrative, invoicing, picking errors, etc.). Among the many indicators, the most commonly used are reported in **Table 4.7**.

Lastly, **Delivery Customization** refers to the ability to respond to requests for customization from the client. Examples can be delivering in formats not foreseen, palletized or loose, performing kitting or rackings, urgent deliveries, etc. In that case, specific measures are developed, that are often expressed in percentages, as exemplified in **Table 4.8**.

Table 4.5 Example of deliveries with equal time performance, but different volumes and values

ORDER	OD (week)	ORDER LINE	PDD (week)	ADD_{FD} (week)	ADD_{BD} (week)	OQ (units)	V (€/unit)	AQ_{FD} (units)	AQ_{BD} (units)	OV^*_{FD} €	OV^*_{BD} €
A	10	1	30	42		1000	4	1000		4000	
		2	30	30	31	2000	20	1200	800	24000	16000
		3	39	39		800	10	800		8000	
						3800		3000		36000	
B	22	4	40	40		500	10	500		5000	
		5	40	52		4000	20	4000		80000	
		6	51	51	52	2000	4	300	1700	1200	6800
						6500		4800		86200	

	ORDERS	
	A	B
Order line fill rate	100%	100%
Order line delivered – First Delivery	66.7%	66.7%
Delivery $speed_{FD}$	27 weeks	27 weeks
Average $delay_{FD}$	4 weeks	4 weeks
$Completeness_{FD}$ (Quantity)	78.9%	73.8%
Quantity delivered in late/Total Quantity	21.1%	26.2%
Value delivered in late/Total value	30.8%	7.3%

** OV = Order Value.*

Table 4.6 **Examples of Delivery Completeness KPIs**

- $Order\ (or\ Order\ Lines)\ Delivery\ Completeness = \frac{\sum orders\ delivered}{\sum orders}\%$
- $Order\ Lines\ First\ Delivery\ Completeness = \frac{\sum order\ lines\ fulfilled\ on\ FD}{\sum order\ lines}\%$
- $Order\ Quantity\ (or\ value)\ Delivery\ Completeness = \frac{\sum quantity\ delivered\ on\ time}{\sum ordered\ quantity}\%$
- $Average\ delivery\ number\ per\ order\ (or\ order\ line) = \frac{\sum deliveries}{\sum orders}(N)$

Table 4.7 **Examples of Delivery Accuracy KPIs**

- $Order\ lines\ Delivery\ Accuracy = \frac{\sum delivery\ process\ errors}{\sum order\ lines}\%$
- $Order\ lines\ Delivery\ Accuracy = 1 - \frac{\sum contested\ order\ lines}{\sum order\ lines}\%$
- $Order\ lines\ Delivery\ Accuracy = 1 - \frac{\sum order\ lines\ with\ a\ X\ error}{\sum order\ lines}\%$

Where X, Y and Z are specific causes (or categories) of possible mistakes.

- $Order\ lines\ Return\ Ratio = \frac{\sum order\ lines\ returned}{\sum order\ lines}\%$

Where the returns are related to Accuracy issues.

Table 4.8 **Examples of Delivery Customization KPIs**

- $Order\ lines\ Delivery\ Customization = \frac{\sum urgent\ delivered\ order\ lines}{\sum delivered\ order\ lines}\%$
- $Orders\ \ Delivery\ Accuracy = 1 - \frac{\sum out\ of\ standard\ delivered\ quantity}{\sum delivered\ quantity}\%$

It is also frequent to use compounded indicators that make it possible to **aggregate multiple indicators** considered relevant in a single, easier-to-read measure. In the case of logistical systems, one of the best-known is represented by **OTIF – On-Time In-Full**. This measure combines in a single measure the indicators of Delivery Dependability (time measure) and Delivery Completeness (volume/value/order lines measures) of deliveries.

That indicator can also be built by adapting it to the various control needs, introducing **different degrees of "flexibility"** in the measure of performance to be considered acceptable. On this point, see the example shown in Table 4.9. Imagine a set of deliveries in which information is gathered relating to the Delivery dates (promised and actual) and the related volumes (ordered and actual).

Table 4.9 **Examples of calculation of OTIF**

ORDER LINE	OQ	PDD	AQ	ADD	ADD ≤ PDD	AQ/OQ	OTIF I	OTIF II	OTIF III
1	1000	Apr 15	1000	Apr 20	N	1.00	0.00	0.00	0.00
2	500	Apr 15	500	Apr 15	Y	1.00	1.00	1.00	1.00
3	1200	Apr 20	800	Apr 20	Y	0.67	0.00	0.67	0.67
4	2000	Apr 22	900	Apr 24	N	0.45	0.00	0.00	0.45
5	400	Apr 26	400	Apr 26	Y	1.00	1.00	1.00	1.00
6	800	Apr 26	700	Apr 26	Y	0.88	0.00	0.88	0.88
7	1400	May 4	1000	May 4	Y	0.71	0.00	0.71	0.71
8	2000	May 6	2000	May 8	N	1.00	0.00	0.00	1.00
9	300	May 10	300	May 10	Y	1.00	1.00	1.00	1.00
10	2800	May 12	2400	May 12	Y	0.86	0.00	0.86	0.86
11	700	May 12	700	May 18	N	1.00	0.00	0.00	0.00
12	1500	May 18	1500	May 18	Y	1.00	1.00	1.00	1.00
						TOTAL	4.00	7.12	8.57
						OTIF	**33.3%**	**59.3%**	**71.4%**

The data on the left side of the table allows for calculating various indicators, aimed at measuring progressive degrees of flexibility admissible in delivery performance. For example:

- **OTIF I**, in which, in order to contribute to the calculation of OTIF, the requirement underlying Delivery Dependability and Delivery Completeness performances must be fully satisfied for both, as shown by the logical operator "and." Thus, the calculation takes place in two steps:

$$\text{OTIF I} = \mathit{if}\,(ADD \le PDD \text{ and } AQ = OQ) = 1, \rightarrow 0$$

$$\text{OTIF I} = \frac{\sum 1}{\text{N. order lines}}\%$$

- **OTIF II**, in which, in order to contribute to the calculation of OTIF, the requirement underlying Delivery Dependability and Delivery Completeness performance must be fully satisfied for only the first of the two elements, while for the second a margin of tolerance is allowed, as shown by the logical operator "then." Thus:

$$\text{OTIF II} = \mathit{if}\,(ADD \le PDD) = AQ/OQ, \rightarrow 0$$

$$\text{OTIF II} = \frac{\sum AQ/OQ}{\text{N. order lines}}\%$$

- **OTIF III**, in which, in order to calculate OTIF, the requirement underlying Delivery Dependability and Delivery Completeness performance must be satisfied considering a margin of tolerance for both elements, as shown by the logical operator "if ... ± *X*", where X represents a fraction of the PDD (e.g. 2 days in **Table 4.7**). Thus:

$$\text{OTIF III} = \mathit{if}\,(ADD \le PDD \pm X) = AQ/OQ, \rightarrow 0$$

$$\text{OTIF III} = \frac{\sum \text{AQ/OQ}}{\text{N. order lines}}\%$$

The equations shown make it evident that the ideal OTIF is 100%. That indicator can be further adapted to take into account performance considered relevant in the specific distribution system. In some companies, for example, an indicator is calculated with the name OTIF-IC, or On-Time In-Full Invoice Completed, to take into account the payment made by the client and the consequent closing of the financial cycle.

4.6 Mapping of flows and lead times

Delivery delays can derive from excessive lead times with respect to the needs imposed by competition. In that case, it is advisable, always with diagnostic goals, to analyze its components to identify the priorities that can be compressed. The study of lead-times is essentially an effective method to keep under control, and where possible to improve, the operations system's response capacity. The observation of the times able to influence the service level offered to the client is generally conducted through the dissection of the order cycle into its principal phases, with the goal of reaching an overall reduction, obtainable through the limitation of the times of each phase or interventions for overlapping of the same.

The phases of the cycle can be divided into:

- **order transmission**, measured by the time interval between the date the order is issued by the client and the moment it is received; today this phase is limited through the adoption of data transmission systems, such as the sales-force automation devices or customer-suppliers connections, that are extremely rapid thanks to the progress of telecommunications networks;
- **order processing**, quantified as the time used to process the order, consisting of accounting audits concerning the solvency of the client, the formal correctness of documents, interrogations of inventory archives, aimed at assessing the state of stock availability, and the preparation of the documentation authorizing withdrawal; the current information systems integrate forms called **ATP – Available To Promise or CTP – Capable To Promise** (Chapter 11), able to automatically produce the possible delivery date starting with the analysis of inventory availability and orders acquired; if the material is not available, the availability of the

components necessary to produce the finished products and the production capacity to perform the related phases are calculated going backwards; then the relative production lead times accumulated are calculated in order to reach a reliable estimate of a delivery date;

- **production**, depending on the technological characteristics of the specific processes and management logics adopted. In the presence of ATO systems, it is sufficient to calculate assembly times, while in MTO, PTO and ETO cases, the lead times necessary to complete each phase of the Pull section (pulled by the client order) of the production process will be gradually accumulated;
- **delivery preparation**, consisting of the phases of withdrawal, packaging, labeling, and packing. Those phases may be accelerated by the adoption of bar codes and optical readers, the use of RFId (Radio Frequency Identification) technologies and innovative IoT (Internet of Things) applications, as well as the use of warehouse management systems, and automated manipulation and handling systems;
- **shipping and transport**, measured by the times between the moment of exit from the warehouse or plant gate and that of delivery to the client. In reality, each of the phases illustrated can, in turn, be analyzed in greater detail from a time standpoint, reaching a definition and measure of the standard lead times for each micro-phase of the cycle; through the observation of the single times it is often possible to identify the process bottlenecks or the principal areas of inefficiency in the management of the physical flows, and this way, to implement appropriate organizational, management, and technological interventions, aimed at their removal.

By way of example, think of the process illustrated in Figure 4.2, regarding a company that produces pigments for industrial coloring agents used in the textile-clothing industry. The analysis of the single lead times (As-Is mapping) showed a bottleneck in the chemical laboratory phase, that with the launch of each batch, had the task of checking some product samples with respect to the **color card** used by sales and

Figure 4.2 **As-Is and To-Be mapping of pigment production process**

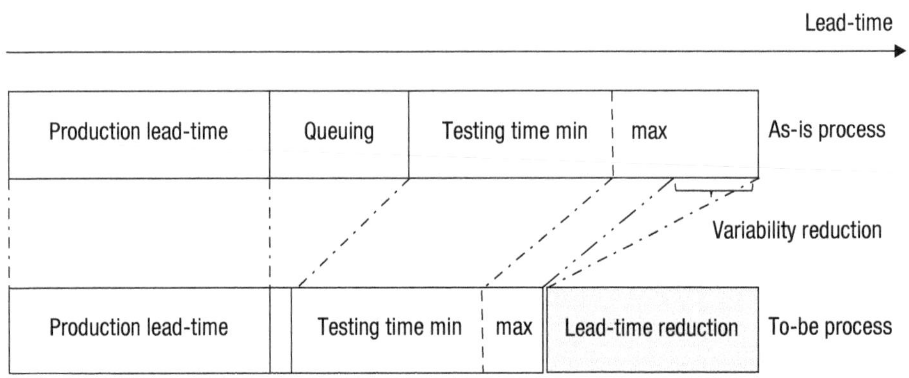

some parameters linked to the chemical recipe. It was found that while production operated in 3 shifts for 7 days a week, the laboratory was organized with a single shift and operated from Monday to Friday. The batches produced at the end of Friday afternoon waited until Monday morning when they were analyzed by the laboratory. In some cases, the instability of the recipe, linked to the use of organic raw materials, made adjustments or dilutions of the product batch necessary, further lengthening the overall times and reducing reliability. The adoption of some technological interventions (sensors able to detect the configuration parameters of the production processes in real time) and management interventions (increase of laboratory shifts and selection of higher quality organic raw materials suppliers) allowed to limit overall lead time (To-Be mapping) and to reduce variability.

4.7 Measures of conformity and quality

The **quality** of a product or a service is expressed by its ability to satisfy the needs of the client, i.e. to be fully suitable for the use for which it was designed and produced. The achievement of that performance is the result of various actors located in a cycle that, starting with an express or latent need recognized by marketing functions, is concluded with the offer of a product or service able to fully satisfy it. As illustrated in Figure 4.3, there are four moments in the cycle in which the information collected on the market (the client's voice) is transmitted and repeatedly re-codified into the languages of marketing, design, production, and sales.

Only by virtue of a high level of cohesion and syntony between the subjects involved in the cycle is it possible to realize a quality product; vice versa, in the event such harmony between the parties interested in satisfying the need does not exist, discontinuities can arise that can be represented in terms of:

- **perception gap**, due to the inability or untimeliness with which the marketing or sales functions are able to recognize and interpret the need, or to condense it into design indications, frequently defined as briefs;

Figure 4.3 **The conformity cycle**

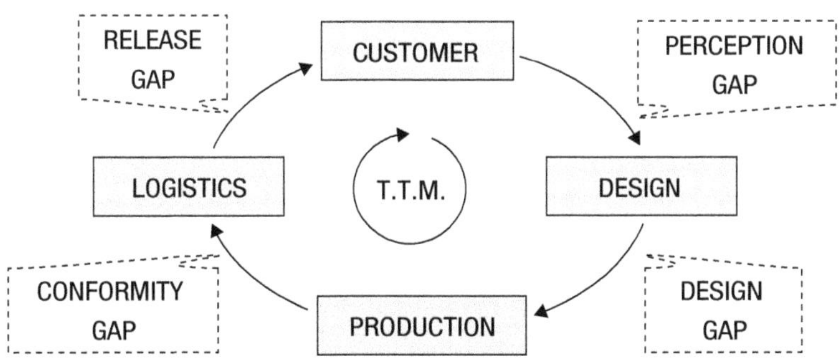

- **design gap**, that manifests itself when the entities in charge of design (research & development, engineering, technical offices, style offices) are not able to transfer the client's wishes, condensed in the brief produced by the sales functions, into project technical specifications, whether due to inability on their part, or poor integration and communication with the marketing functions; in that location, frequently and especially for complex products and services, we see the phenomenon of the "Triple O – Overdesign-Overspecification-Overequirements," the offer of too many futures or performances, that by making the product or service too complex and redundant, end up deluding the client with excess costs, delivery delays, etc. (Coman and Ronen, 2009; Belvedere et al. 2013);
- **conformity gap**, that is visible in cases where production does not appear able to produce a product conformant to the technical requirements and specifications contained in the design;
- **release gap**, in which the product, despite being consistent in terms of functions and features, is offered to the client with methods, prices, and times incompatible with market requests.

The presence of one or more gaps inevitably leads to the production of a defective or unsatisfactory product. It seems just as evident that the Conformity Cycle must take place in increasingly limited time intervals, linked to the growing need to reduce **Time To Market**, on the pain of a competitiveness loss in terms of the ability to industrialize and sell new products aligned with market requests. The gradual reduction of product life cycles in almost all sectors, from mobile telephony to automobiles, from laptops to industrial machines, requires companies to compress their time to market, and as a consequence exposes them to the risk of underestimating the gaps linked to an insufficient period to adequately test the different phases of the conformity cycle.

4.8 The elements underlying quality measures

Not being able to address the very vast issue of quality management here, but wanting to limit the field of investigation to the measurement of performance, it is possible to break down the generic **performance of quality** into multiple factors (**design quality**, **conformity quality**, **reliability** or maintainability, and **technical assistance**), and for each element to an identification of the suitable indicators.

Design quality is generally measured with specific indicators depending on the characteristics of the product, such as the hardness and resilience of a metal, the elasticity of a fiber, the power of an electric motor, or the viscosity of a lubricant.

The **quality of conformity** is defined by how the product responds to the design specifications and is measured differently depending on whether it is measured within the factory (**in-house** conformity quality), or after the sale of the product (**in-field** conformity quality). In the first case, indicators are used that express the incidence of discards and defective pieces out of the total volume produced or the recovery and reworking times out of the total work times, while in the second case *ad hoc* indica-

Table 4.10 The measures of reliability, maintainability, and technical assistance

Performance	Indicator	Measure
Reliability/Maintainability	MTBF MTTR	Mean Time Between Failures Mean Time To Repair
Technical Assistance	Intervention Time	Mean time between a technical assistance call and the intervention Mean time between the order and the shipment of a spare-part

tors are adopted such as the number of hours of assistance under warranty, in relation to the total number of units sold, or the frequency of complaints and requests for product substitution under warranty.

Concerning **reliability** and **maintainability**, we refer to the tendency of a product to maintain its functional specifications unaltered over time, or to easily restore them. Here, technical assistance is considered in the traditional sense of the term, as activity of installation and repair in the case of breakdowns, such that the performance indicators are constructed evaluating the promptness of the interventions.

From the viewpoint of the operations system, the subject of this study, the evaluation of **in-house conformity quality** appears the most important, as it is hard to generalize the design quality measures and those relating to the area of customer satisfaction such as in-field quality. As for the characteristics of reliability, maintainability, or reparability, and technical assistance, considering the ample literature on the subject, we will limit ourselves to briefly citing the most commonly used indicators, described in Table 4.10.

Despite limiting ourselves to presenting the principal measures of in-house conformity quality, below we will embrace the entire problem through the elements of its economic profile, based on the analysis of quality costs.

4.9 In-house conformity quality: Notions and measures

Non-conformity consists of an anomaly, detected through an appropriately structured observation (by sampling or 100% inspections), that expresses a discrepancy (or non-correspondence), between a measure or characteristic detected along the production process and the analogous measure defined in the design phase. Two considerations worthy of mention emerge from that simple definition.

- First of all, we speak of **anomalies detected along the process**, thus referring to the indications expressed by the principles of Total Quality Control, for which the measure is not carried out only on the product (system output), but must be carried out, starting with the acceptance check, evaluating the capability of the entire production process.

- Secondly, we need to distinguish between **control by variables and control by attributes**, which differ based on the models of detection of the characteristics to be measured; the control by variables is applied in cases in which the quality characteristics are measurable along graduated scales and the outcome of the control is expressed as a value of the scale used. This is the case of size measures, dimensions, tolerances, tensions, etc. Controls by attributes, on the other hand, consist of a selection made by observing the binary quality of a characteristic, with a "good or bad" judgment, or "passes or fails," etc. The type of control, by variables or attributes, influences the technique of samples and the instruments adopted, but can be carried out indifferently at any point of the production process. Among the many instruments used in the conformity control, below we describe three indicators, selected because they express three critical moments of the process: the entry phase, characterized by an acceptance check; the central phase, on production proper, in which multiple inspections can be performed aimed at detecting and eliminating any undesired defects; and the exit phase, prior to delivery to the client. With reference to the flow diagram shown in **Figure 4.4**, we stress the placement of the three controls: **AQC – Acceptance Quality Control, CDI – Corrected Defects Index, and QI – Quality Index**.

Figure 4.4 **Flow diagram and examples of quality control indexes**

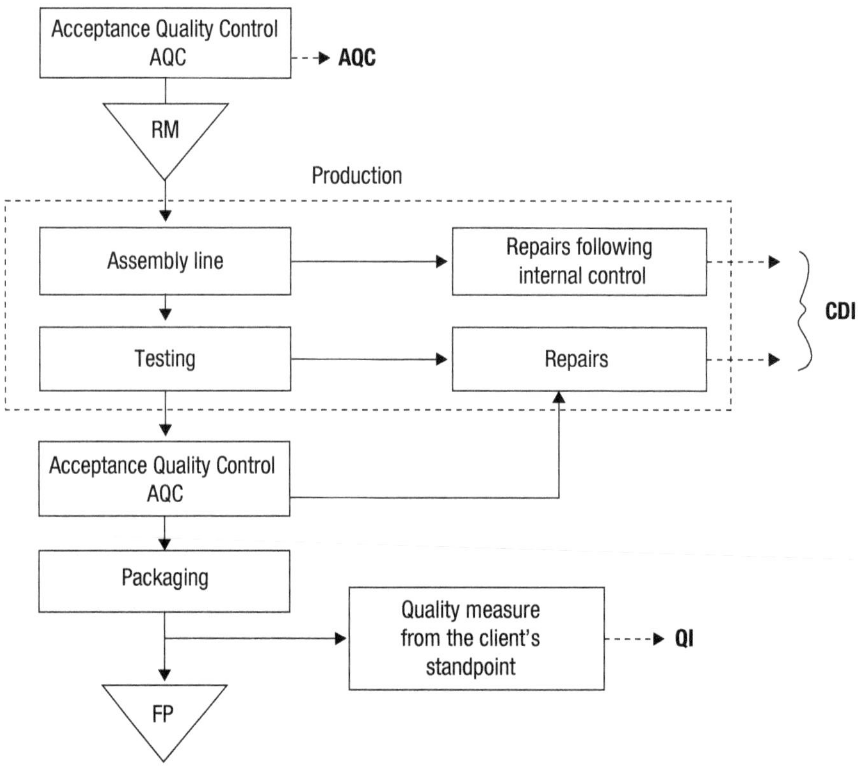

The **Acceptance Quality Control (AQC)** is generally performed on materials and components coming from a supplier, entering the raw materials warehouse, but can also be conducted upstream of a semi-finished products warehouse, for work processes performed by contractors, or downstream of a work phase. The AQC consists of a measurement performed on a sample n, extracted from a batch N, and determines the fate of the batch: if the outcome is positive, i.e. the measurement of defects detected does not exceed a preset limit, called the **Acceptable Quality Level (AQL)**, the batch is considered good and thus accepted, although in that way the statistical risk is assumed relating to the significance of the sample and the presence of false positive. If, on the other hand, the quality level detected in acceptance is lower than the acceptable limit, the batch is rejected and a selection is conducted of 100% of the pieces (to avoid the risk of false negative).

The **Corrected Defects Index (CDI)** expresses the total number of defects detected and corrected in a specific period of time, along the whole process, compared to the number of units produced in the same time interval.

$$CDI = \frac{\sum_t d_t}{N_t} \% \qquad [4.1]$$

where:

d = total defects detected in period t;
N = units produced in period t.

This is a synthetic index, regarding a process, that despite not allowing for a comparison between different processes and products, lends itself to an aggregate measure of the natural defect rate of a process and depends on the efficacy of the control systems implemented. The corrected defects can be identified in the inspection, or along the different phases of the production cycle, with specific on-line and off-line detection methods, designed as a function of the characteristics of the product and process itself. The presence of corrected defect indexes with low values can indicate a natural process reliability or an insufficient or ineffective intensity of controls; in that case, incontrovertible diagnostic indications can derive from a comparison with measures expressing the level of in-field conformity.

Lastly, the **Quality Index (QI)** expresses the quality detected through sampling the product, from the client's standpoint, or with respect to its expectations. Extracting a sample n from a population N, the product is subject to a functional inspection, performing all of the operations and maneuvers the client can perform, such as checking the integrity of the packaging, the presence of any instructions, installation, start-up, and use for defined time intervals. This way non-conformities are detected based on specific tests, dividing them into three categories.

- **Critical non-conformity** (level A), that makes the product not safe for the user or lacking one or more essential feature or performance.

- **Serious non-conformity** (level B), that translates into the absence of an important aspect of performance, whether functional or aesthetic.
- **Minor non-conformity** (level C), that is associated with a non-functional inconvenience (for example aesthetic) or a functional imperfection.

The ratio between the number of non-conformities, divided by classes of gravity, and the number of products making up the sample, is defined as the QI.

$$QI_x = \frac{\sum d_x}{n} (\%) \qquad [4.2]$$

where:

d_x = level A, B, or C non-conformity.

Table 4.11 shows an example of calculation of the indicators illustrated.

Table 4.11 Analysis of non-conformities of an assembly line

DATA COLLECTION:			
Daily production		2000	(units/day)
Selected sample (measure QI)		30	(units/day)
Working days		22	(days)
Non Conformities:			
	Level A	1	(#)
	Level B	5	(#)
	Level C	12	(#)
Number of defects identified and corrected:			
	Components weldering	300	(#)
	Product assembling	280	(#)
	Product final testing	140	(#)
Screens supplied		50000	(units/month)
Selected sample (measure AQC)		100	(units/batch)
Detected non-conformities in acceptance		5	(#)
	AQL	7	%
	Qia	1/660	= 0.15%
	Qib	5/660	= 0.76%
	Qic	12/660	= 1.82%
	CDI	720/44.000	= 1.63%
	AQC	5%	< AQL

4.10 The costs of quality

Leaving aside physical measures, a different way to evaluate the quality of a production system lies in the analysis of the costs associated with it. Those elements are aggregate economic measures, useful for the analysis of deviations between estimated and actual values. The analytic observation of the costs of quality sets the goal of increasing the awareness of the problems at the company, placing in the right light the importance of the issue and its extent in every spending category. To that end, it is by now consolidated practice to divide the costs of quality into four distinct categories:

- **prevention costs**, consisting of all the costs sustained to prevent the manifestation of a defect. This category includes the costs relating to the study, planning, and implementation of the system that ensures the conformity of the product, and the training of personnel, the activity for increasing awareness, communication, and spreading of the principles of quality, the design and preparation of the equipment used in controls and inspections, assistance programs, evaluation and selection of suppliers, etc.;
- **evaluation costs** (or inspection or control costs), sustained to evaluate the degree of product conformity with the project specifications. In that case, the reference is to tests, controls, inspections, or audits aimed at preventing the release of defective pieces to the market, blocking them before they leave the system;
- **costs of internal defects** (in-house non-conformities), detected on materials, parts and products, found to be non-compliant before their transfer to the client. These can consist of the cost associated with the production of the defective pieces, whether discarded or downgraded, and necessarily substituted, or by the cost of supplementary work processes and resumptions, necessary for their restoration;
- **costs of external defects** (in-field non-conformities), detected on materials, parts, and products, found to be non-conforming after their transfer to the client. In that case, there can be repairs or substitutions under warranty, additional assistance costs, etc.

Concerning the costs of defects, whether internal or external, it is appropriate to stress the dual nature, both direct and indirect, that characterizes them. In addition to the costs cited, that are directly measurable and not hard to quantify, it is in fact necessary to consider additional costs, that are uncertain and sometimes can only be determined approximately, such as the increase induced by discards and reworking on the inventory and on the necessary production capacity, the need to reschedule production plans, drops in productivity and lower yields (due to internal defects), and also intangible elements linked to the impact of poor quality on the client (due to external defects), such as complaints, delayed payments, loss of clients, and decline of company image and reputation that are hard to quantify. The constant comparison of those costs, carried out over time and by unit of analysis, as illustrated in Table 4.12, offers valid suggestions for identifying and overcoming potential alterations of the quality system.

The analysis of the dynamics of the different categories of costs over time can offer useful suggestions for interventions, aimed at minimizing the costs of non-quality.

Table 4.12 The costs of quality

	January	%	February	%	March	%	April
1. PREVENTION COSTS 1.1. Definition of Product quality specifications 1.2. Quality Function expenditures 1.3. Pilot production 1.4. Suppliers qualification 1.5. Quality controls planning 1.6. Workers training							
Total							
2. EVALUATION (INSPECTION) COST 2.1. Acceptance quality control 2.2. Lab tests on purchased materials 2.3. Control tools and devices 2.4. Testing and inspections 2.5. Process control 2.6. materials for destructive testing							
Total							
3. INTERNAL DEFECTS (IN-HOUSE) 3.1. Semi finisced products scraps 3.2. Finisced products scraps 3.3. Reworking and repairing 3.4. Product downgrading 3.5. Product wastes							
Total							
4. EXTERNAL DEFECTS (IN-FIELD) 4.1. Quality returns 4.2. Under warranty returns 4.3. Complaints 4.4. Penalties payments							
Total							
TOTAL COST		100		100		100	

5 Measurement of Performance in Operations Management: Productivity and Flexibility

by *Alberto Grando*

5.1 The measures of productivity

The measures of **productivity** are generally expressed in terms of value of outputs compared to the value of inputs of the system observed. In this case, we speak of **overall productivity**, to distinguish it from the **single factor productivity**: labor, machines, materials and energy. It should also be stressed that, like the other measures analyzed here, productivity measures can be detected at the level of the single line or department, the factory, or production unit, and can be expressed in physical measures or value (Grando and Turco, 2005). Before going into the details of the measures illustrated below, we need to provide some introductory considerations, of a general nature.

- First of all, in the analysis that follow, we will refer to **defined types of production systems**, such as single machines or sets of machines, since it is not possible to separate the characteristics of the process from the methods of detection and measurement of its performance.
- Secondly, being aggregate measures, useful for the sizing of a plant's capacity, the workforce of a department, or evaluations of efficiency that are not episodic, **the observations are carried out for medium to long-term time horizons**, assigning little or no significance to short-term variations.
- Thirdly, the measures **are generally expressed in physical units**, potentially expressed in equivalent units in multi-product situations. In case of an analysis of heterogeneous aggregates of productive factors, the measurements **can be conducted only in value**.
- Lastly, it is necessary to stress that the significance of some of those measures depends on the degree of detail and the reliability with which the measurement system (on board the machine, in the department, etc.) is able to operate. It is in fact important to remember that the validity of the measure obtainable from a production process depends on at least two elements: **the level of detail** and **the reliability of the measurements**. For example, the use of manual systems (production declared on cards or on devices such as laptops or handheld PCs) or automatic systems (piece counters on the machine or department terminals integrated in the lines, or more sophisticated systems, such as MES - Manufacturing Execution Systems) can lead to detailed and precise measurements as a function of:

- on the one hand, the **degree of complexity of the coding plan** of the causes used to justify the commitments of production or non-production time (actual working time with respect to time lost for tests, inspections, maintenance, etc.);
- and on the other hand, of the **reliability of those declarations**, generally subject to control by the department head and to periodic balancing.

Productivity indicators, especially if expressed in value as in the measurements conducted by management control, that are certainly useful for control purposes, at times appear too aggregate to offer useful suggestions from a diagnostic standpoint. In order to understand the causes underlying productivity values different than expected, it seems useful to divide the productivity indicator into its principal elements. In order to perform that disaggregation, it is necessary to have structured data relating to times and volumes deducible from actual measurements and from the definition of standard values. Therefore, below we will introduce the concepts of **Production Rhythm** (or **Throughput Rate**) and **Plant States**, to then reach the determination of the two principal components of the productivity indicator, i.e. the indicators of **Resource Utilization** and **Working Efficiency** (or Yield).[1]

5.2 The Key Performance Indicators linked to productivity

Defining **T Plant Calendar Time** (also called **Plant Opening Time**, **Total Operations Time or Gross Available Time**, equal to the paid hours of theoretical presence, for labor), as **ST, Solar Time** or Calendar Year (365 days, 24 hours a day in measurements based on a year of operations) net of **Unscheduled Time** for causes such as holidays, vacation, or reduced hours or planned stoppages, less than three shifts, etc., it is possible to set out a series of times, fractions of time T, corresponding to definite states of a specific resource, machine or labor (Grando, 1995):

- $PTcq_i$ = Production Time used to produce CQ, quantity of good or conforming products;
- $PTdq_i$ = Production Time used to produce DQ, quantity of reject products or defective quantity;
- TST = Total Setup Time, time spent for indirect production tasks such as set-ups, tooling, cleaning, etc. and the overall minor stoppages time resulting from resets, adjustments, etc.
- FDL = Time Losses due to machine Failure Downtime;
- MDL = Time Losses due to machine Maintenance Downtime;
- LDT = Time Losses due to Lack of Demand (or lack of orders in the production plan);

[1] The indicators described refer to a resource used in single product production. In Chapter 8 we will analyze the mix (Mix Throughput Rate) of the determination of the available capacity of the plant in multi-product contexts.

- *LMT* = Time Losses due to Lack of Materials in stock or Energy;
- *LWT* = Time Losses due to Lack of Workers (Manpower due to strikes, absenteeism, trade union meetings, etc.);
- *OTL* = Time Losses due to Organizational issues such as micro-absenteeism, shift changes, man-machine interactions (work organization losses), etc.
- *TTT* = Time Losses due to Technical Tests, training of workers and manufacturing of non-marketable products (prototypes, samples, etc.).

The list provided above has only the purpose of offering a non-exhaustive overview of the **possible causes associated with time losses or non-productive times**. It goes without saying that within a company, there is generally a concentration on a few significant causes, obtained through systematic measurements of departments or through appropriate sampling. Those losses can be grouped by logical categories. The classification commonly distinguishes between:

- **Planned Shutdown**, i.e. stops and losses of production for planned activities such as ordinary maintenance, technical tests, personnel training, union meetings, planned relayouts, etc.;
- **Unplanned Downtime** (or **Downtime Losses** in a strict sense), linked to other stoppage causes indicated above such as unplanned or extraordinary maintenance, failure downtime, lack of demands, materials or energy, etc.

According to this aggregation, it is possible define a cascade of times:

- **Net Available Time**, equal to Gross Available Time (or Total Operations Time) net of Planned Shutdown; and
- **Actual Operating Time**, equal to Net Available Time reduced by Unplanned Downtime.

Actual Operating Time, equal to the sum of ($PTcq_i$) Production Time used to produce CQ, quantity of good (or conforming) products, ($PTdq_i$) Production Time used to produce DQ, quantity of reject products or defective quantity and (*TST*), Total Set-up Time present in multi-product contexts, is thus equal to Plant Calendar Time, net of the various times absorbed by the different classified plant states. On this point, see Figure 5.1.

A frequently used measure in ex post analyses aimed at understanding the physiological or pathological causes of the use of a resource is given by the **Utilization Coefficient**. That indicator is generally expressed by a ratio between two times, calculated net and gross of times expressing particular plant states, as a function of the degree of detail of the measurement system adopted and the characteristics of the situation observed, as already stressed. Recalling the notations relating to the causes presented above, the **Utilization Coefficient U** of a generic resource (machine or labor) can thus be expressed as follows:

Figure 5.1 Possible plant states

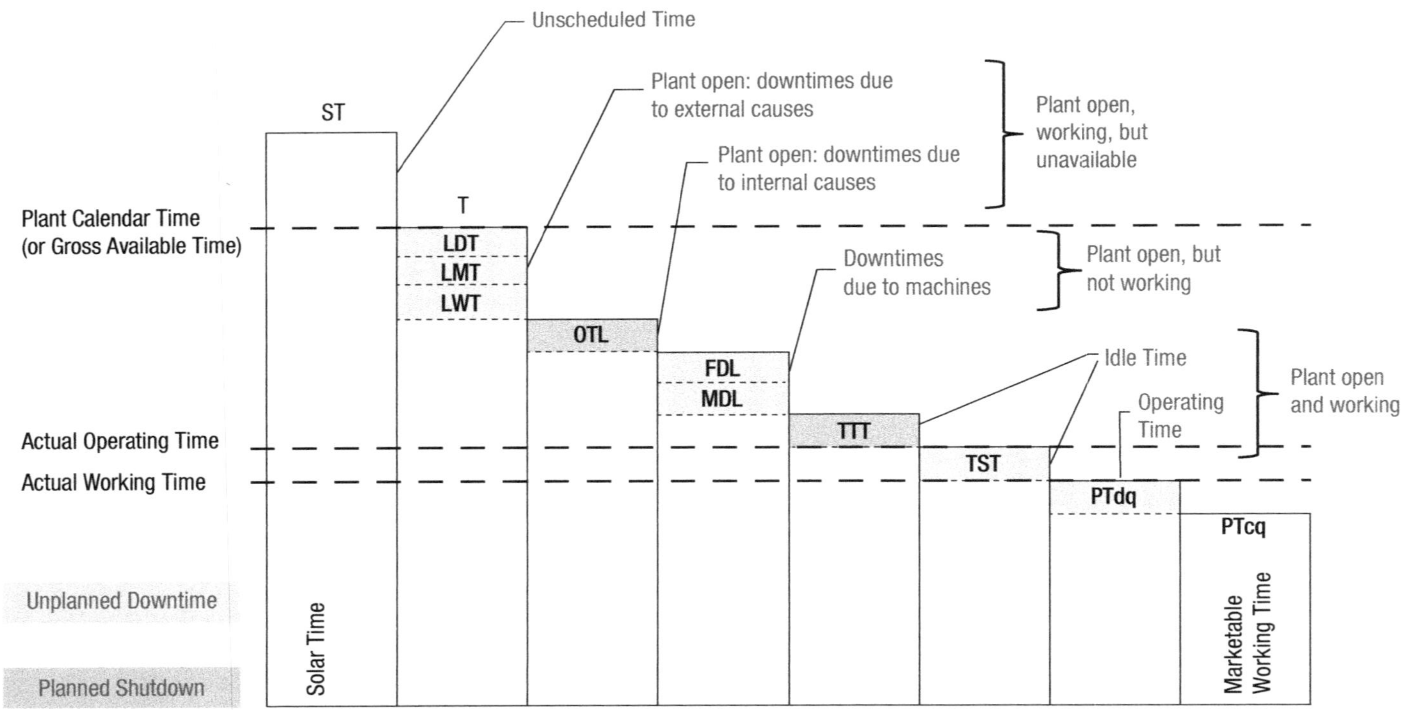

Source: adapted from Grando and Turco, 2005.

$$U = \frac{Actual\ Operating\ Time}{Plant\ Calendar\ Time}$$

$$= \frac{T - LMT - LDT - LWT - OTL - FDL - MDL - TTT}{T} \qquad [5.1]$$

or:

$$U = \frac{\sum_i [(PTcq_i + PTdq_i) + TST]}{T} \qquad [5.2]$$

In that case, for example, subtracting from the numerator the time losses required for tests and inspections (TTT) expresses the desire to calculate a net index, representing plant use for only potentially sellable production (CQ), even if it turns out to be partially discarded subsequently (DQ). The **Utilization coefficient can also be calculated in units**. With reference to a generic interval of time *i* and a resource that operates in a mono-product context on a generic product x, it is possible to obtain, through periodic measurements performed by the Engineering Department, the **Standard Throughput Rate $\overline{TR}$x** that expressed in units/time, indicates how many units relating to the product x can be obtained in the unit of time in normal operating conditions, that is, with operators with average abilities and no special incentives, in normal conditions of machine functioning and availability and quality of materials used. The Standard Throughput Rate – also called Rhythm, Pace or Flow Rate – is a flow measure and is obviously **the inverse of the Standard Unit Time** or **Standard Throughput Time ($\overline{TT}$x)**, i.e. the standard time to produce a unit of product such that:

$$\overline{TT}x = \frac{1}{\overline{TR}x}\ \text{(hours/unit)} \qquad [5.3]$$

Multiplying the Utilization Coefficient thus calculated by the Standard Throughput Rate ($\overline{TR}$x) – or dividing by the Standard Throughput Time ($\overline{TT}$x) – we obtain:

$$U = \frac{\sum_i [(PTcq_i + PTdq_i) + TST] \times \overline{TR}x}{T \times \overline{TR}x} \qquad [5.4]$$

that is, the standard production (volumes) obtainable based on the hours actually worked, related to the theoretical production (volumes) obtainable in time T.

Or also, in the case of actual data, comparing the production capacity used to produce the actual volumes to the installed capacity, defined by using the actual Throughput Rate; this is equal to the actual good and reject volumes produced compared to the time actually used to produce them $\frac{\sum_i (CQ_i + DQ_i)}{[\sum_i (PTcq_i + PTdq_i) + TST]}$, such that:

$$U = \frac{\sum_i (CQ_i + DQ_i)}{T^* \left(\frac{\sum_i (CQ_i + DQ_i)}{\left[\sum_i (PTcq_i + Pdq_i) + TST\right]} \right)} \qquad [5.5]$$

In this regard, see the example presented in **Table 5.1**, referring to a generic machine for a defined time horizon, such as a month, in which it worked 21 days for 2 daily shifts of 7.5 hours each. The plant produced 204,800 units, of which 4,800 were subsequently found to be non-conforming. Note that, as the context is mono-product, the set-up times are equal to zero.[2]

Table 5.1 Example of calculation of utilization coefficient

Detected variables	Values
T - plant opening (calendar) time (hours)	315
$PTcq_i$ – Good production time (hours)	250
$PTdq_i$ – Reject production time (hours)	6
TTT – Time dedicated to tests and inspections (hours)	10
MDL – Time dedicated to maintenance (hours)	14
LMT – Downtime due to lack of materials	19
OTL – Downtime for organizational causes (hours)	16
TST – Actual set-up time (hours)	0
$\overline{TR}_{xi}$ – Standard Throughput Rate (units/hour)	820
CQ – Quantity of good production (units)	200,000
DQ – Quantity of reject production (units)	4,800

In that case, the time used for good and reject production, as well as lost time, are **actual times**, because they are the result of measurements ex post. Therefore:

$U = (315 - 59) / 315 = 0.8127$, applying the equation [5.1]

or

$$U = \frac{(250 + 6)}{315} = 0.8127, \text{ applying the equation [5.2]}$$

[2] The same determinations can be carried out in multi-product situations. In that case, the Standard Throughput Rate ($\overline{TR}_x$) is substituted by the mix Throughput ($\overline{TR}_{mix}$). On this point, see the considerations presented in Chapter 8.

and obviously, if expressed in volumes, using the Standard Throughput Rate

$$U = \frac{(250+6)\times 820}{315\times 820} = 0.8127, \text{ applying the equation [5.4]}$$

or in the ex post measurement of volumes, using the Actual Throughput Rate deducible from the available data

$$U = \frac{204{,}800}{315\times 800} = 0.8127, \text{ applying the equation [5.5]}$$

The concept of Utilization must not be confused with that of **Working Efficiency** (or **Yield**). The latter refers to the ratio between volumes actually produced with respect to the standard production theoretically producible in the hours actually used.

It is therefore possible to calculate the **Yield** as:

$$\eta = \frac{\sum_i (CQ_i + DQ_i)}{\left[\sum_i (PTcq_i + PTdq_i) + TST\right] \times \overline{TRx}} = \frac{\textit{actual production}}{\textit{std prod. based on actual hours}} \quad [5.6]$$

or:

$$\eta = \frac{\sum_i \left[(CQ_i + DQ_i)/TRx_i\right] + \overline{TST}}{\sum_i (PTcq_i + PTdq_i) + TST} = \frac{\textit{actual prod. expressed in std hours}}{\textit{actual operating time}} \quad [5.7]$$

that is equal to:

$$\eta = \frac{\sum_i (\overline{PT}cq_i + \overline{PT}dq_i) + \overline{TST}}{\sum_i (PTcq_i + PTdq_i) + TST} = \frac{\textit{standard hours produced}}{\textit{actual operating time}} \quad [5.8]$$

where the notations placed in the numerator (marked) assume standard value. Using the numerical example shown in Table 5.2 and noting that the set-up times are equal to zero as it is a mono-product process, we obtain:

$$\eta = \frac{200{,}000 + 4{,}800}{(250+6)\times 820} = 0.9756, \text{ where we apply [5.6]}$$

or:

$$\eta = \frac{[(200{,}000 + 4{,}800)/820]}{(250+6)} = 0.9756, \text{ where we apply [5.7] and [5.8].}$$

The **product of the coefficients of Utilization and Efficiency** offers a first measure of the **Productivity** of a resource, such that:

$$P = \frac{\textit{actual production in standard hours}}{\textit{plant calendar time}}$$

or, recalling the equations of U [5.2] and of η [5.7]

$$P = \frac{\sum_i [(PTcq_i + PTdq_i) + TST]}{T} \times \frac{\sum_i [(CQ_i + DQ_i)/\overline{TRx_i}] + \overline{TST}}{\sum_i (PTcq_i + PTdq_i) + TST} = \qquad [5.9]$$
$$= \frac{\sum_i [(CQ_i + DQ_i)/\overline{TRx_i}] + \overline{TST}}{T}$$

and using the data reported in **Table 5.2**,

$$P = U\, x\, \eta = \frac{(250+6)}{315} \times \frac{[(200{,}000 + 4{,}800)/820]}{(250+6)} = 0.7928,$$

equivalent to the two indicators calculated, that is: 0.8127 × 0.9756 = 0.7928.

5.3 The productivity of factors: notions and measures

Abandoning the hypothesis of a generic resource, it is possible to observe the breakdown of the **Productivity indicator** for the three principal factors – **labor, machines, and materials** – in the respective Utilization and Efficiency indexes. That division has diagnostic value since it allows for examining the causes underlying productivity performance that deviate from the expected values. In fact, the indicators generally used to assess factor productivity are too aggregated and not very useful for the purpose of interpretation.

Productivity indicators such as labor expressed by the ratio between volume produced and number of Full Time Equivalent (FTE) workers, or of a machine calculated as volume produced related to available production capacity, or also that of materials measured for example by the volume of shoes produced related to the consumption of leather used, are certainly correct, representing a ratio between an output and an input, but not very useful in the perspective of their decoding.

It is therefore possible to express the productivity of factors, disaggregating it into its principal components, through adimensional indexes, as seen in the example in **Table 5.2**.

In the case of **labor**, productivity is expressed by the ratio between volumes produced and labor used, or by the paid hours of theoretical presence to produce those volumes.[3] In this case, it is possible to break down the aggregate indicator into **utilization of labor**, given by the ratio between hours actually worked (actual operating time), declared on the timecard, and theoretical paid hours of presence, and **yield** of

[3] In the presence of significant incidence of strikes, the preference is to use the number of hours of expected presence since in that case, the paid hours would be less than time T and this could make the comparisons less readable.

Table 5.2 **The breakdown of productivity of factors**

Factor	Utilization	Yield	Productivity
Labor	Actual working hours / Paid hours of theorical presence	Volumes produced expressed in standard hours / Actual working hours	Volumes produced expressed in standard hours / Paid hours of theoretical presence
Machines	Actual production hours / Plant calendar hours	Volumes produced expressed in standard hours / Actual production hours	Volumes produced expressed in standard hours / Plant calendar hours
Materials	Theoretical consumption / Actual consumption	Actual volumes produced / Theoretical consumption	Actual volumes produced / Actual consumption

the same, expressed by the ratio between volumes produced in standard labor-hours, and hours actually taken to produce those volumes.

In the case of labor, in order to increase the diagnostic capability of the observations, the Utilization coefficient can be distinguished into two components: **Gross Utilization**, equal to the ratio between hours of presence (or attending hours) and paid hours of theoretical presence, and **Net Utilization**, equal to the ratio between hours actually worked and hours of presence. The product of Gross Utilization and Net Utilization leads to the utilization index already mentioned, that is sometimes called **Overall (or Global) Utilization**. The increased value that emerges from the breakdown of the aggregate index appears evident: faced with a gradual drop in productivity of labor of a factory, for example, the observation of the Utilization index (gross and net) and Efficiency index can improve the reading and interpretation of the phenomenon, increasing the effectiveness of corrective measures. In the case of Efficiency aligned over time, and decreasing Utilization, it is advisable to orient the investigation towards the interpretation of causes such as strikes, absenteeism, lack of orders and materials, and other elements of unavailability of the factor observed. To the contrary, in the presence of Utilization coefficients stable in time and decreasing Efficiency, the causes of the drop in productivity may reside in the methods of incentivizing the workers or the manifest inadequacy of the standards applied, at times made rapidly obsolete by the evolution of the products and technologies and the methods of the process.

Similar considerations can be offered for the **Productivity of machines**, equal to the volumes produced related to installed production capacity, in which the **Utilization index** compares the machine hours (production capacity actually used) with the plant opening hours (installed production capacity), while the **Efficiency (Yield) index** is the ratio between volumes produced expressed in standard machine-hours and the machine hours actually required to produce them.

The case of **Productivity of materials** is slightly different, where the aggregate indicator, more correctly defined as **Yield of the materials**, is equal to the ratio between volumes obtained and materials actually consumed. In this case, it is possible to break the indicator down into **Theoretical or Gross Yield**, given by the ratio between theoretical consumption of materials (standard) and actual consumption, and material **Efficiency (Yield)**, obtained by the comparison between volumes obtained and theoretical consumption, all expressed in homogenous physical measures. To provide an example, this is the case of a footwear factory that faced with a standard consumption of assigned materials (leather), has higher actual consumption, due to the incidence of higher scrap than expected (over-scrap), and obtains a lower number of uppers than anticipated, due to the presence of waste during the quality control performed upstream of assembly. In that case, we stress that the phenomenon of higher scrap than the standard, being related to the natural characteristics of the material, evokes the concept of Gross Yield (in the example, lower than one), while the phenomenon of waste, linked to the more general production process, can be associated with the concept of Efficiency.

As already illustrated, the Utilization and Efficiency coefficients, and thus the Productivity indexes, can be constructed with different degrees of detail, separating the different times expressing different states of the plant, as a function of the degree of detail of the measurements and the goals of the control system. The Efficiency indicators, for example, can be considered gross or net of rejects, as a function of the methods of quality control implemented (on the machine, directly by the operators, or subsequently in a quality control station). When the production system observed allows for separating the conformant production from the reject one, attributable to the production unit observed, with the goal of producing a measurement net of defective production, the Productivity indicator can be disaggregated highlighting the additional component of **Quality (Q)**, given by the ratio between good production, i.e. total production, net of defective volumes and total production. For example, think of the productivity of a department in which the scarce resource, that is measured, is given by labor. In that case, the measure of Productivity can be divided into the indicators illustrated in **Table 5.3**.

Lastly, in the case the observation is carried out in aggregate terms, for a production unit or a business, the analysis of factor productivity can be traced to the usual precepts of corporate analysis, abandoning physical measures to use the values inferable from the accounting and non-accounting documents, as shown in **Table 5.4**.

5.4 The measures of Versatility and Flexibility

The issue of **Flexibility** and its appropriate measures, despite being amply discussed in the literature, still appears to be partially unresolved, with the widespread conviction that it is not possible to assign a potential value to that concept, as it is very difficult to assess *a priori* in economic-financial terms. Despite this, the issue has taken on increasing importance, from both the standpoint of the problem as a whole, and ultimately from the decisive standpoint of the phases of preparation and adjustment of an operations system, of technology and plant choices. The generic concept of Flexibility is commonly associated with different subjects: businesses, production

Table 5.3 Labor productivity and its components

$$\textit{Gross utilization} = \frac{\textit{attending hours}}{\textit{paid hours of theor. presence}}$$

×

$$\textit{Net utilization} = \frac{\textit{actual working hours}}{\textit{attending hours}}$$

×

$$\textit{Yield} = \frac{\textit{volumes produced in std hours}}{\textit{actual working hours}}$$

×

$$\textit{Quality} = \frac{\textit{conforming product volumes}}{\textit{total product volumes}}$$

$$\textit{Productivity} = \frac{\textit{conf. vol. expressed in std. hours}}{\textit{paid hours of theor. presence}}$$

Table 5.4 Analysis of company productivity

$$\text{Global Productivity} = \frac{\text{Sales} + \Delta\ \text{Inventory}_{PF+WIP}}{\text{Labor} + \text{Materials} + \text{Capital} + \text{Other inputs}}$$

$$\text{Labor Productivity} = \frac{\text{Sales} + \Delta\ \text{Inventory}_{PF+WIP}}{\text{Labor}}$$

$$\text{Materials Productivity} = \frac{\text{Sales} + \Delta\ \text{Inventory}_{PF+WIP}}{\text{Materials}}$$

$$\text{Capital Productivity} = \frac{\text{Sales} + \Delta\ \text{Inventory}_{PF+WIP}}{\text{Capital}}$$

$$\text{Other inputs Productivity} = \frac{\text{Sales} + \Delta\ \text{Inventory}_{PF+WIP}}{\text{Other inputs}}$$

systems, plants and machinery. This is where we find one of the elements that make it difficult to identify appropriate measurement methods. So it is necessary to discuss the specification of the subject to which we refer when we speak of Flexibility. We intend to stress that **Flexibility appears as a set of responses** that a company system as a whole generates when faced with explicit or latent requests from the market, and that said attribute is only partially based on the properties of the production system, as the system must be supplemented by appropriate organizational structures, information systems, sales and financial policies, and know-how, etc.

As a first approximation, referring to only the production-logistical system, Flexibility can be defined as the ability to respond to external and internal changes. An example of the first is the specific requests from clients, changes in the volumes and mix absorbed by the market, changes in the delivery plan, in the specifications of the materials used or the abilities of the available labor. The latter can include the launch of new products, breakdowns of plant and machinery, impediments in the transport system, variability in work times, or unexpected peaks of absenteeism.

Multiple **Flexibility** strategies can be pursued not only acting on the characteristics of the machines, but also through the use of management systems, such as Just in time techniques, or through effective modular planning approaches for configurable products, or also, acting on appropriate supply relationships. With reference to machinery, the attribute of **Versatility** seems more appropriate, a condition that is generally necessary but not sufficient to guarantee high levels of Flexibility. Plant choices that modify the degree of Versatility of machines, do not always translate into increased (or reduced) levels of Flexibility of the production system. In fact, Flexibility derives from the adaptation of the different components of a production system, in which the plant, however significant, is only one element. For example, if an investment in a flexible manufacturing system does not correspond to a coherent arrangement of other levers (layout, logistical and information flows, management methods and techniques) the Versatility obtained, that is high in relative terms with respect to potential alternative technical solutions, does not succeed in increasing the Flexibility of the overall Production System.

Consider the following examples, aimed at demonstrating that the Flexibility of a Production System is not linked to the degree of versatility of its machineries. The first is the case of a company operating in the food sector of roasting and packaging of coffee, characterized by a production process divided into three lines of production and packaging, that are extremely rigid and dedicated to three distinct products: 250 gram packages of ground coffee, 3 kilogram packages of coffee beans, and coffee in pods. The differences in terms of volume and the granulometry of the products translate into technological and packaging systems that are very dissimilar, absolutely not versatile, or fungible. Nevertheless, due to the high mobility of a multi-skilled labor, that can be assigned to different lines at each shift change, and due to the significant overcapacity of each line, the availability of packaging material, and a significant Flexibility in the Master Production Schedule, the system as a whole appears extremely flexible, guaranteeing the fulfillment of orders of any nature and size within two days. The second example is that of an engine production plant for an automotive company. Due to considerable investments in flexible automation, the process is able to be retooled in almost zero time, and to work through a varied mix of versions of the same engine, guaranteeing high performance in terms of reconfigurability of the line. Despite this, the extreme integration of the process, its volume rigidity (also known as inelasticity) and the need to amortize an investment characterized by a long payback, strongly constrain the degree of convertibility of the system, that proves to be inadequate for the introduction of new models, inevitably lengthening the company's Time to Market. In that case, the Versatility of the plant, understood in the design phase only in terms of productive mix, does not correspond to a full response to the needs of Flexibility demanded by the new vectors of competitiveness in the sector, linked to the need to introduce new models, reducing Time to Market and expanding engine options.

Table 5.5 **Flexibility needs and responses**

Flexibility Requirements	Machinery (Versatility)	Labor	System
Mix	Rinconfigurability	Multiskill/Mobility	Availability (safety stock)
Volume (Elasticity)	Overcapacity	Working time flexibility	Reactivity (factory BEP)
Product	Convertibility	Learning curves	Promptness (Time to Market)
Plan	Rinconfigurability/Overcapacity	Multiskill/Mobility/Flexibility	Responsiveness (Frozen Period)

These brief considerations lead us to state that, to obtain satisfactory measures from an operational standpoint, it is necessary to further explode the concept of Flexibility, seeking more significant dimensions or relevance. By attributing the concept of Flexibility the meaning of a characteristic required of the production system as a whole, it is possible, as illustrated in Table 5.5, to distinguish between dimensions of the request for Flexibility, dividing it into specific needs, and dimensions of the Flexibility response, generated by the factors that contribute to it being addressed effectively. By breaking the phenomenon down in this way, it is less complex to identify appropriate measures for each aspect identified.

5.5 Flexibility: notions and measures

It is thus possible to identify the different dimensions of Flexibility required of the operations.

- **Mix (or range) Flexibility**, defined as the ability to reassort a broad range of products, or in other words, to produce a mix of products at acceptable costs, within a defined range; in essence, the market expresses a "request for variety of known production."
- **Product Flexibility**, represented by the ability to industrialize or engineer a new product, i.e. to enter a new article into the range of products offered at acceptable times and costs; this is a "request for variety of new production."
- **Volume Flexibility**, more properly defined as elasticity, understood as the ability to absorb the variations requested by the market from a quantitative standpoint while maintaining adequate levels of efficiency, induced by phenomena of seasonality, cyclicity, or instability of demand; this is a "request for variability of known production."
- **Plan Flexibility**, determined by the ability to modify the sequences of production contained in the master production schedule, even considerably, to deal with unforeseen and urgent changes; these are "joint requests for variety and variability of known production."

In all of the definitions of Flexibility proposed, we see the characteristics of relativity typically assignable to the concept, that therefore assume the value of **attribute of potential qualification** of a production system with respect to a need, ultimately coming from the market. With this we wish to stress that, as for other aspects of performance, such as quality, service, and so on, the attribute of Flexibility can in no case take on an absolute value, as it reflects relative performance, to be evaluated in a differential manner, with the surrounding conditions being equal. As a consequence, the measures proposed also assume a value that cannot be generalized, but needs specific contextual references. The space for operation of those definitions can therefore not be defined by the variety and the variability of the rules of competition, such that the requests for Flexibility described can find effective and satisfactory responses only in solutions that link their feasibility to requirements of **"reasonable" times and costs**, where the latter are the result of efficiency and quality performance. It is clear, in fact, that in the absence of budget constraints and over a long horizon, any production system, however rigid or inelastic, if detached from the competitive context in which it operates, can be flexible. Moreover, a machine or an operator able to produce a broad range of products at the expense of low efficiency and high rejects, cannot be considered flexible.

In response to those requests for Flexibility coming from the market, the operations system seeks appropriate responses, shaping the factors as a whole. The Flexibility solutions generated by the system are thus to be observed, in order to be able to measure them, distinguishing the contribution offered by the following components: machine, labor, and the more general Operations System, for which it is possible to define specific measurements. In any event, the measure of Flexibility can be expressed by an indication relating to a range of articles producible by a machine (and thus by its versatility) or by an operator (and thus by the breadth of its skills). At the Operations System level, additional variables come into play, such as the possibility to produce an article on alternative machines belonging to the same system (rerouting) or the possibility to allocate personnel to different production lines, thanks to their degree of internal mobility and skills.

Machineries, for which, as mentioned, it seems more correct to speak of **versatility**, respond to the needs of Flexibility expressed by the market by leveraging specific properties (Belvedere, 2007; Beamon, 1999).

- **Mix Flexibility** can be addressed with different levels of efficacy, a function of the degree of reconfigurability, that is measurable through the range indicator (F_{m1}), and through the estimate of the total set-up time (or the cost) requested to produce a given mix constituted by N different articles (F_{m2}):

$$F_{m1} = N(t)$$
$$F_{m2} = \sum_{i=1}^{N} \overline{UST}_i \qquad [5.10]$$

where $\overline{UST}$ is the standard unit set-up time.

The indicator F_{m1} de facto constitutes the outcome of an evaluation regarding the maximum number of sequences of production (and related articles) that can be produced in a time interval t, given a minimum goal of efficiency and saturation of production lines. As a consequence, the true criticality in the measure of this performance consists of considering the sum of the circumstances that can affect the amount of the set-up time. In this regard, it is necessary to consider:

- which products within the range of the market offering make up the **mix considered**, since each of them can require different set-ups;
- the **sequence** with which these articles are produced, that can influence the duration of the single set-up operations of the machine.

To take these circumstances into account, it seems useful to use the Mean Sensitivity to Change (MSTC) indicator, that is inversely correlated with the Mix Flexibility, that better responds to the problems just highlighted. That indicator is obtained through a weighted average of the unitary set-up times (UST) relating to a total N of different products, obtained as followed:

$$MSTC = \sum_{i=1}^{N} P_i \times (1 - P_i) \times UST_i \qquad [5.11]$$

where P_i represents the probability that the i_{-th} product is placed into production and the relevant set-up is carried out. However, that circumstance actually occurs only if the article produced immediately afterwards is different than the previous one, which happens with a probability equal to $P_i \times (1 - P_i)$. Obviously, the sequences of production can be defined as a function of a need for efficiency and minimization of machine stoppages for the set-ups. However, in production situations managed with a pull logic, in which production plans are a function of the way in which demand is manifested, as in mixed-model lines, the variable P_i is to be understood as the probability that the specific product will be requested at that moment by a client. From this viewpoint, the MSTC indicator expresses a measure of average set-up time not distorted by the intervention of management and able to evaluate the Mix Flexibility as a function of the goal of alignment of production with demand.

- **Product Flexibility** can be addressed based on the degree of **convertibility**, measurable by the percentage of recovery of existing machines during redesign or the cost of adaptation of an existing machine for new production. Product Flexibility measures the ability of the company to launch production of a new article in respect for time and cost needs. From a formal standpoint, this measure can be expressed as follows:

$$F_{p1} = T \qquad F_{p2} = C \qquad [5.12]$$

where T expresses the average time necessary launch a new product and C the relative cost to be sustained.

However, it appears entirely evident that the measures just shown are the result of both logics of the new product development process (for example, the application of principles of product modularity), and of the operations design choices and related operational conditions (such as the degree of automation, the versatility of the machines, and the Flexibility of labor). A discussion on the measurement of product Flexibility can thus not avoid an examination of the factors that determine it.

- **Volume Flexibility or elasticity** can be addressed due to the degree of production **overcapacity**, and is measurable with Utilization indexes, analyzing the breadth of the range of volume within which the company can operate in economical conditions, whose extremes consisted of the break-even point and the production capacity installed in a given interval of time. The measure of the range of volumes variability can thus be formalized as follows:

$$Range_V = V_{max} - V_{min} \quad [5.13]$$

It appears entirely evident that in the case of the same range, the true Flexibility performance of a production system, precisely because it is an attribute with a potential nature, depends above all on the variability of market demand. In sectors in which sales are stable and predictable, modest changes can be dealt with even by a rigid production system, appropriately scaled as a function of the expected trend of demand. To the contrary, in contexts characteristics by high uncertainty of sales, a broad range does not always correspond to a high probability of meeting demand. So it is necessary to integrate the informative content of the measure of *range* with the data regarding the variability of sales, to reach a quantification of volume Flexibility able not only to express the interval in which the company produces in economical conditions, but also the probability of meeting market demand. To this end, we can suppose that demand is distributed according to a normal curve, with average $\overline{D}$ and variance S_D^2, calculated as follows (Belvedere, 2019; Beamon 1999):

$$\overline{D} = \frac{\sum_{t=1}^{T} d_t}{T}$$

$$S_D^2 = \frac{\sum_{t=1}^{T} (d_t - \overline{d})^2}{T-1} \quad [5.14]$$

where:

dt = demand in period t
T = total number of periods considered

Volume Flexibility (Fv) can thus be expressed as the probability that demand D is included in the range illustrated above, varying in an interval from 0 to 1. From a formal standpoint, this implies the following:

$$F_v = P\left(\frac{V_{min} - \overline{D}}{S_D} \leq D \leq \frac{V_{max} - \overline{D}}{S_D}\right)$$

or also:

$$F_v = \Phi\left(\frac{V_{max} - \overline{D}}{S_D}\right) - \Phi\left(\frac{V_{min} - \overline{D}}{S_D}\right) \qquad [5.15]$$

- **Plan Flexibility**, accumulating requests for variety and variability, can be dealt with as a function of the degree of **reconfigurability** and **overcapacity** of the machineries and is measurable by the extension of the so-called **frozen plan**, i.e. the portion of the time horizon of planning below which the plan is frozen and cannot be varied, without incurring high costs and inefficiencies.

Analogously, for **labor**, it is possible to observe that:

- **Mix Flexibility** finds an adequate response in the characteristics of skills and mobility, that for example are measurable with tools aimed at assessing the range of workers' abilities, in general detected through **skills matrixes**;
- **Product Flexibility** is satisfied by the workers' ability to learn, and can be quantified through the study of **learning curves** or training times;
- **Volume Flexibility** or elasticity is dealt with by modulating the degree of elasticity of work hours, measurable in terms of incidence of overtime, possibility of doubling shifts, frequency of part-time and shifted hours, etc.;
- **Plan Flexibility** is based on the sum of the elements just described.

Finally, the effectiveness of the response of the operations system to the different Flexibility needs expressed by the market is not linked exclusively to the characteristics of the factors illustrated, but must also be based on the sum of elements that, together with machine and labor, guarantee the operations of a production unit: information system, supply system, availability of materials, technical support functions, etc. In that case, being a group of elements, it does not seem possible, and in any event convenient here, to list detailed specific measures; so while putting off the description of those specific indexes until the considerations presented in the following chapters, it does however seem useful to mention aggregate measures, that express the ability to respond to needs of Flexibility, implemented by the factory as a whole, defined as "system" for the sake of brevity in Table 5.5:

- **Mix and Volume Flexibility** generate considerable stress which the operations system must counter with **availability**, in terms of safety stock, and reaction capacity, in taking rapid corrective actions such as the reduction of lead times, management of urgent procedures, rescheduling of departments and machines, etc. Synthetic indicators of the ability to face Mix and Volume Flexibility can be found, respectively, in the measures of the degree of disassortment of the finished products inventory, for the former, in the utilization index of machines and in factory break-even, for the latter, and in the worsening of certain service level measures for both;

- **Product Flexibility** requires a prompt response, that can be found in the industrialization and engineering competencies of the production unit, and can be synthetically measured from **time to market**, i.e. from the time that passes between the definition of the product specifications, its industrialization, the production ramp-up, and the start of series production;
- **Plan Flexibility** translates into responsiveness of the system, that are difficult to evaluate in aggregate terms, except through approximate measurements, such as the periodicity of the scheduling process, and as mentioned, the duration of the **frozen period** of the plan.

6 Production Planning and Control System

by *Alberto Grando*

6.1 Introduction

The central aim of planning activities is to harmonize customer requests (expressed by a demand forecast, or consolidated in an orders book) with the available capacity of the production system. This entails observing the constraints expressed by the scope of the mix required, the pace of demand and the terms of delivery, on the market side, and by the needs for saturation of machines' capacity, limiting of investment in stock, and specific supply relationships, on the supply side. The **Production Planning and Control system** is thus correlated with the other systems that characterize company management: the commercial, financial, and organizational systems, for example, and interacts with them in the constant search for overall equilibrium.

To address this issue in a structured manner, we must consider the specific type of production process to be analyzed, and its operational characteristics. The observation of the planning method also lends itself to an analysis divided into **phases** that can be identified as **long-term, medium-term, short-term, and immediate.** So, after having illustrated the influence of the different types of processes on the characteristics of planning, below we will deal with the question of formulating a Production Plan (or Aggregate Plan), leaving the illustration of the MPS – Master Production Schedule and the Production Scheduling and Control, to Chapters 8, 11 and 14.

6.2 Production planning and types of production processes

In observing the possible classifications used for the analysis of production processes (Chapter 2), it seems evident that the characteristics of the product creation process (by parts or by process), the type of operations (manufacturing or assembly), the volumes handled (single pieces, batches, and continuous flows) and the method of response to the market (to order or to stock), present radically different problems in terms of production planning and control.

With regard to the management of a production planning system, the different types of production can be grouped as follows.

Continuous flow processes (chemical, oil, paper, steel, etc.), that are generally characterized by a limited number of simple (or not very complex) products, few processing phases, and essentially uninterrupted production flows, that are relatively easy to schedule. The products in question are defined as simple because of the low number (2 or 3) of levels of the bill of materials (also known as the formula or recipe). Elements of complexity may arise in the event of joint productions, in which the technologically-restricted and unmodifiable production mix differs from the mix requested by the market. This is particularly frequent in petrochemical production, in which due to *cracking* processes, each unit of product derived corresponds to defined units of other joint or co-produced products, the same proportion of which are not necessarily requested by the market. In the manufacturing area, yarn is very similar, in which each spinning mill can produce items with defined elements of complementarity that must be appropriately accounted for in the business plan.

Batch flow production (heavy machinery, mechanical production, shoes, furnitures, etc.), on the other hand, is a more complex environment from a planning standpoint. Especially in the case of functional layout, this is due to the numerous work centers, the interaction of long and complicated work cycles, the different duration of the single operations, and the presence of complex products, characterized by countless levels and part numbers in the bill of materials.

Line flow operations (for assembly or manufacture) are in an intermediate position with respect to the two previous cases, having a higher variety of products than the process industry, but longer series with respect to production in batches. The greater constraints in those production operations may arise, however, in the event of a rapid growth in needs, due to the possibility of situations where certain materials or components are lacking or unavailable when requested. If those processes operate on parts and components purchased externally, as is the case with assembly in particular, they are in fact subject to the risk inherent in supply markets, and in the absence of adequate stock, they can significantly condition the need for synchronization of the flows of components to be assembled.

Companies that work **To Order** (single or repetitive orders) face different problems and solutions. In that situation, the criticality of the planning process focuses on the control phase of the orders being processed, and the state of progress of each order. On this point, it seems appropriate to distinguish between **projects in fixed locations** (large projects, work sites, buildings, etc.) in which the control of resources, times and costs is critical, and **production to order**, that in terms of complexity can be compared to production by batches. In the former, especially for complex products or large works, scheduling is carried out through the use of Project Management techniques (PERT, Gantt, etc.), with the principal goals of monitoring respect for times, costs and the use of resources as planned, while Aggregate Planning is used to estimate the amount of capacity and size of work site teams.

An initial examination of the different complexity profile of the Planning process thus comes from the context in which that activity is carried out, represented by the intersection of the following variables:

- **characteristics of the product offer**, based on the number of articles (finished product part numbers) and the breadth and depth of their bills of materials;
- **degree of predictability**, distinguishing between Make to Stock and Make to Order;
- **type of processes and layout**, distinguishing between parts productions (in job shops, by batches, and in line) and continuous processes.

Below, despite not omitting references to other types, we will refer to the most common situations in the manufacturing sector, characterized by a high level of product complexity, production to order or to stock, with processes that operate by parts, whether they are manufacturing or assembly. It also seems evident that, if from a theoretical standpoint the Production Planning and Control Process shows different degrees of complexity, derived from the context in which the process takes place; the multiple constraints expressed by the market, the risks related to supplier conduct, and the unpredictability of certain phenomena linked to transformation processes, can considerably modify those assumptions.

6.3 Constraints and economic-financial profiles of planning choices

Before starting a detailed analysis of the single phases of the Production Planning and Control Process, it is necessary to further clarify its complexity and temporal development.

Different production planning choices cannot be independent of the technological and management characteristics of the processes governed and produce inevitable repercussions that go beyond the merely technical/operations environment, to impact **economic and financial** needs. In short, the planning process appears as an activity aimed at optimizing a detailed and complex system of variables:

- **physical variables**, in terms of product volumes and mix;
- **temporal variables**, concerning the delivery date and advance production policies, linked to the management of lead times;
- **economic and financial variables**, from the standpoint of relevant and differential costs, and of the degree of structurality of the financial needs generated.

We will discuss optimization in physical terms (volumes and mixes) and temporal terms (lead times) later, as it represents the essence of Production Planning and Control Process. We can, however, mention the economic and financial implications of the choices made, with the case of planning alternatives in the presence of cyclical or seasonal demand. In fact, based on demand characterized by variable rates of increase, there are different alternatives for managing production capacity, as shown in the graph on the left in **Figure 6.1**.

Production structures characterized by high levels of elasticity can find it convenient to develop plans (A) following the course of demand (D), as best they can, based on their volume flexibility. In such cases, planning gives preference to the adoption

Figure 6.1 Sizing of production activity and cyclical demand

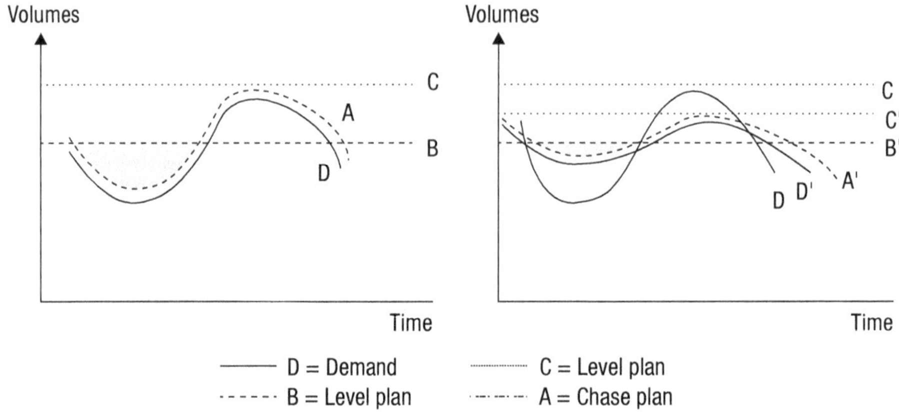

of "**chase plans**." Rigid structures, on the other hand, must follow strategies for leveling production plans, obtaining average values (B), or peak values (C), with plans commonly defined as "**level plans**". The choice between the three options must be made through an analysis of the **relevant differential costs** (Vollman et al., 1984; Chase et al., 2006). In the first case, significant costs for adapting production capacity will emerge, as a consequence of set-up and adjustment of production pace or the activation of second or third shifts, overtime, subcontracting relationships, and so on. In the second case, by choosing to produce in advance and following saturation policies, inventory costs will emerge (organizational stock due to inelasticity). Lastly, the third case, driven by obvious service needs, will entail significant unsaturation costs, due to fixed costs and investments that on average are in excess. This last case occurs with the provision of services for which, not being able to make use of the opportunities offered by inventory, the solution is oriented towards the appropriate sizing of production capacity, as a function of the market and service goals identified.

The three cases indicated all regard interventions aimed at adapting **Available Production Capacity** to the variability of the Demand. In the not infrequent cases in which that is impracticable or inconvenient, an attempt can be made to intervene on the demand side, forcing market requests in order to bring them back within the limits of the elasticity dictated by the production system. This is the case, represented in the graph on the right, where marketing levers are used (countercyclical discounts or price discrimination, promotional campaigns, etc.) with the goal of leveling, or at least smoothing, the cyclical nature of demand (D'), thus limiting the interventions for adjustment of production capacity (A'), inventory costs (case B') or lack of saturation (case C').

Therefore, if the essence of planning lies in having to deal with the variety and variability of requests, its optimization cannot but be traced to the observation only of the relevant costs and the minimization of the overall cost associated with the formulation of the plan. As already stated, in addition to time, there are three constraints that are always present, and closely related, in operations management decisions: **technical feasibility**, linked to resource availability and intrinsic carachteristics, **time feasibility**, and **economic/financial convenience**. In the case at hand, the issue is posed both in terms of evaluation of a plan, and in the case of selection between two alternative plans. During Production Planning, the costs considered, only if they are relevant and differential, are essentially the following:

- production costs, whether they are variable, like raw materials, labor and energy, or fixed, only if influenced by the choices made, such as special equipment (e.g. molds) or maintenance or other indirect costs;
- labor overtime costs;
- subcontracting costs;
- stock holding costs;
- stock-out costs;
- set-up costs.

Therefore, the production **variable costs** must include the costs relating to raw materials, direct consumables, direct maintenance, direct labor (if flexible and mobile), and energy. **Fixed costs** are supervision, scheduled maintenance, direct labor if not flexible and mobile, amortization rates and general production costs.

The **costs of overtime labor** arise if it becomes necessary to work outside of regular hours and are generally accompanied by greater costs and charges for possible drops in productivity.

Subcontracting costs are linked to the extra-costs generated in the event volumes are contracted out to third parties. In that case, the transfer price negotiated must be increased by the costs relating to more careful incoming quality control, the cost of evaluation and selection of the supplier, any shipping costs, costs for equipment offered for use, samples and pre-series, and greater safety stock due to higher risk involved in deliveries (see Chapter 9).

The **stock holding costs** and the cost of potential stock-out are traceable to availability or unavailability, respectively, based on a request from the market. The cost of holding stock is represented by the sum of the unit holding costs associated with a period, for all of the periods in the inventory, multiplied by the number of units in stock. In this regard, we recall that only the costs actually fronted are relevant, i.e. differential costs, those that are typically variable (see Chapter 9).

Stock-out costs are generally proportionate to the lost contribution margin correlated with the lack of sale. This occurs in the absence of a backlog, i.e. when it is impossible to recover the stock-out with late delivery. In addition to those charges, there are additional costs that are very difficult to assess, such as the loss of image and reputation, loss of client loyalty, and so on. In cases of production where the temporal factor is important, especially for the execution of large jobs, contractual clauses establish additional costs in the event of delays, in the form of penalties.

Lastly, there is **set-up cost**, the cost linked to a change in production pace or of the article in production. Set-up cost is generally divided into two distinct components: the "out-of-pocket" set-up cost and the cost due to "lack of production." The former is an actual expenditure and includes the costs for the workers that retool the production line, any consumables used in the operations, and wastes and rejects linked to restarting the lines, if they cannot be recovered. The second represents an opportunity cost sustained in the time interval between the production of the last unit of the processed batch and the first conforming unit of the subsequent batch to be processed. In that sense, the cost is linked principally to the period of time required by the operation, that as already stressed, contributes to the reduction of Available Production Capacity. Thus, it is clear that this factor is determined by the level of use of capacity, or Saturation level. Due to its nature as an opportunity cost, in this case it is necessary to refer to the situation in which the cost arises. Although in operational practice set-up costs are often included in product cost for each batch produced regardless of its method of construction, we can distinguish between two types of situations, depending on the saturation level of the plant:

- in a **non-saturated capacity**, the out-of-pocket cost is relevant, since the cost of lack of production is zero if the labor in charge of set-up is dedicated to the plant, or at the most, it is equal to the cost of labor directly used in the set-up, if it could be used otherwise;
- in a **saturated capacity**, if the quantities not produced during the set-up period translate into lack of sales, in addition to the out-of-pocket costs, the set-up cost must take into consideration the contribution margin lost during the stoppage of the plant.[1]

Based on the unpredictability of the demand in terms of volumes and mix, in theory, there are essentially **two approaches** an enterprise can pursue:

a. maintain elastic or excess production capacity (overcapacity) and/or an extremely flexible organization, so as to be able to promptly react to any changes in requests, thus operating with a **Make to Order** logic, generally with **chase planning policies**;
b. anticipate the variability of the market by producing for stock, with a number and assortment that guarantees satisfaction of anticipated demand, thus operating with a **Make to Stock** logic, generally seeking efficiency with **level planning policies**.

There are two possibilities: react promptly or speculate and act in advance. It seems superfluous to stress that these two possibilities cannot always be alternatives, due to the specific aspects of production systems and the context in which they operate. In addition, they lead to organizational, technical and managerial solutions that are entirely different, with different repercussions from an economic standpoint (degree of cost variability) and a financial standpoint (entity and structural level of financial requirement). An example of the two Planning Policies is reported in the **Box 6.1**.

---

[1] Moreover, from the standpoint of the theory of differential costs, in a saturated plant, if the quantities not produced during the set-up time can be recovered with the use of overtime work or subcontractors, the set-up cost would be equal to the relative cost; in that case, the out-of-pocket cost is expressed with overtime, while for subcontracting, it will be included in the related negotiated price.

Box 6.1 Example of Planning Policies

To face a high seasonal demand, planners have to evaluate the impact of the two possible plans, represented in the following table.

	Demand (units/period)		Plan (units/period)			
Periods	Per period Demand	Cumulat. Demand	Per period		Cumulated	
			Plan A	Plan B	Plan A	Plan B
Jan w1	200	200	200	400	200	400
Jan w2	200	400	200	400	400	800
Jan w3	300	700	350	400	750	1200
Jan w4	400	1100	350	400	1100	1600
Feb w1	500	1600	550	400	1650	2000
Feb w2	600	2200	550	400	2200	2400
Feb w3	400	2600	400	200	2600	2600
Feb w4	200	2800	200	200	2800	2800
Mar w1	200	3000	200	200	3000	3000
Mar w2	200	3200	200	200	3200	3200
Mar w3	200	3400	200	200	3400	3400
Mar w4	200	3600	200	200	3600	3600

In evaluating the two options, the following information have been taken into consideration:

- Selling price: 25 € / unit;
- Nominal Production Capacity 1st shift: 200 units / period;
- Nominal Production Capacity 2nd shift: 200 units / period;

This capacity can only be exploited in full (on / off solution); in case of further capacity requirements, it is possible to utilize:

- Overtime: up to a maximum of 60 units / period;
- Outsourcing without capacity limitations.

Production Capacity of 1st shift, 2nd shift and overtime can generate a maximum of 95% sellable product, given an average of defective volumes of about 5%.
In planning the weekly total volumes, two constraints must be taken into account:

- The product is sold in package units of 50 pieces each, therefore minimum lot of 50 pieces can be planned (only multiple of 50 pieces are allowed) as global volumes;
- Plan B: no more than 4 set-ups of production speed changes (volumes/week) during the investigated quarter.

Assuming the relevant costs as follows:

- production overheads: 3,850 € / year;
- annual amortization rate: 1,800 € / year
- stock holding cost: 1 € / unit-period;
- stock out cost (equal to the contribution margin lost): 12 € / unit (no backlog);
- Direct production cost (labor, maintenance, etc.): 2 € / unit;
- Overtime cost: 2.6 € / unit;
- Outsourcing: 2.92 € / unit, up to a maximum of 100 units / period; for outsourcing higher quantities, the whole lot is burdened by 0.5 € / unit for the occurrence of increased costs associated with contact and selecting new suppliers, samples control, 100% incoming quality control, etc.
- Production in the 2nd shift and overtime are penalized 0.3 € / unit for the connected efficiency reduction (more waste);

Every change of production speed (volumes/period) raises a cost of 515 € / change over, due to line balancing. This cost is also assumed in the plant down-hypothesis, because of the necessary extraordinary cleaning and maintenance operations.
In the following tables, the stock volumes and costs (holding and stock out cost), as well as the overall internal external costs related to the alternative plans, are represented.

STOCKS AND COST OF STOCKS										
	Stock (units/period)		**Available Stock (cumulated units)**		**Stock Out (units/period)**		**Stock Cost (€/period)**		**Stock Out Cost (€/period)**	
Periods	Stock A	Stock B	Stock A	Stock B	A	B	A	B	A	B
Jan w1	0	200	0	200	0	0	0.00	200.00	0.00	0.00
Jan w2	0	200	0	400	0	0	0.00	400.00	0.00	0.00
Jan w3	50	100	50	500	0	0	50.00	500.00	0.00	0.00
Jan w4	-50	0	0	500	0	0	0.00	500.00	0.00	0.00
Feb w1	50	-100	50	400	0	0	50.00	400.00	0.00	0.00
Feb w2	-50	-200	0	200	0	0	0.00	200.00	0.00	0.00
Feb w3	0	-200	0	0	0	0	0.00	0.00	0.00	0.00
Feb w4	0	0	0	0	0	0	0.00	0.00	0.00	0.00
Mar w1	0	0	0	0	0	0	0.00	0.00	0.00	0.00
Mar w2	0	0	0	0	0	0	0.00	0.00	0.00	0.00
Mar w3	0	0	0	0	0	0	0.00	0.00	0.00	0.00
Mar w4	0	0	0	0	0	0	0.00	0.00	0.00	0.00
							100.00	**2200.00**	**0.00**	**0.00**

COSTS OF INTERNAL AND EXTERNAL PRODUCTION (€/period)																
	First Shift		Second Shift		Overtime		Efficiency Fall		Changes		Internal Working Cost		External Working Cost		Total Working Cost	
Periods	A	B	A	B	A	B	A	B	A	B	A	B	A	B	A	B
Jan w1	380.00	380.00	0.00	380.00	26.00	52.00	3.00	63.00	515.00	515.00	924.00	1390.00	0.00	0.00	924.00	1390.00
Jan w2	380.00	380.00	0.00	380.00	26.00	52.00	3.00	63.00	0.00	0.00	409.00	875.00	0.00	0.00	409.00	875.00
Jan w3	380.00	380.00	0.00	380.00	148.20	52.00	17.10	63.00	515.00	0.00	1060.30	875.00	352.26	0.00	1412.56	875.00
Jan w4	380.00	380.00	0.00	380.00	148.20	52.00	17.10	63.00	0.00	0.00	545.30	875.00	352.26	0.00	897.56	875.00
Feb w1	380.00	380.00	380.00	380.00	148.20	52.00	74.10	63.00	515.00	0.00	1497.30	875.00	386.46	0.00	1883.76	875.00
Feb w2	380.00	380.00	380.00	380.00	148.20	52.00	74.10	63.00	0.00	0.00	982.30	875.00	386.46	0.00	1368.76	875.00
Feb w3	380.00	380.00	380.00	0.00	52.00	26.00	63.00	3.00	515.00	515.00	1390.00	924.00	0.00	0.00	1390.00	924.00
Feb w4	380.00	380.00	0.00	0.00	26.00	26.00	3.00	3.00	515.00	0.00	924.00	409.00	0.00	0.00	924.00	409.00
Mar w1	380.00	380.00	0.00	0.00	26.00	26.00	3.00	3.00	0.00	0.00	409.00	409.00	0.00	0.00	409.00	409.00
Mar w2	380.00	380.00	0.00	0.00	26.00	26.00	3.00	3.00	0.00	0.00	409.00	409.00	0.00	0.00	409.00	409.00
Mar w3	380.00	380.00	0.00	0.00	26.00	26.00	3.00	3.00	0.00	0.00	409.00	409.00	0.00	0.00	409.00	409.00
Mar w4	380.00	380.00	0.00	0.00	26.00	26.00	3.00	3.00	0.00	0.00	409.00	409.00	0.00	0.00	409.00	409.00
	4560.00	4560.00	1140.00	2280.00	826.80	468.00	266.40	396.00	2575.00	1030.00	9368.20	8734.00	1477.44	0.00	10845.64	8734.00

TOTAL COSTS			
	Cost of production plans		0%
Cost Items	**A**	**B**	**A-B**
Stock	100.00	2200.00	-2100.00
Stock Out	0.00	0.00	0.00
First Shift	4560.00	4560.00	0.00
Sec. Shift	1140.00	2280.00	-1140.00
Overtime	826.80	468.00	358.80
Efc. Red.	266.40	396.00	-129.60
Changes	2575.00	1030.00	1545.00
Outsourc.	1477.44	0.00	1477.44
Total	10945.64	10934.00	11.64
Unit Prod.	3600.00	3600.00	
Unit Cost	3.040	3.037	

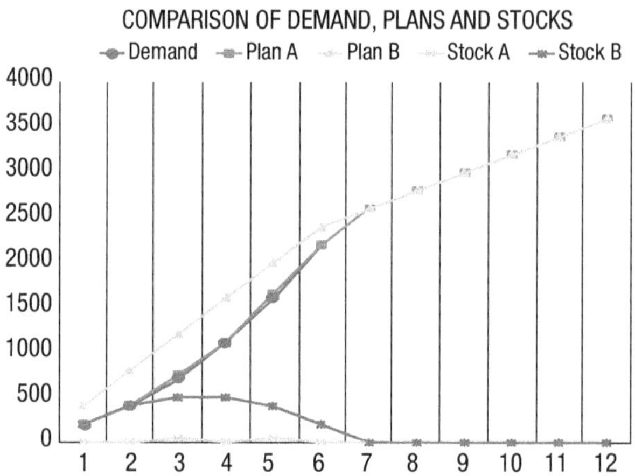

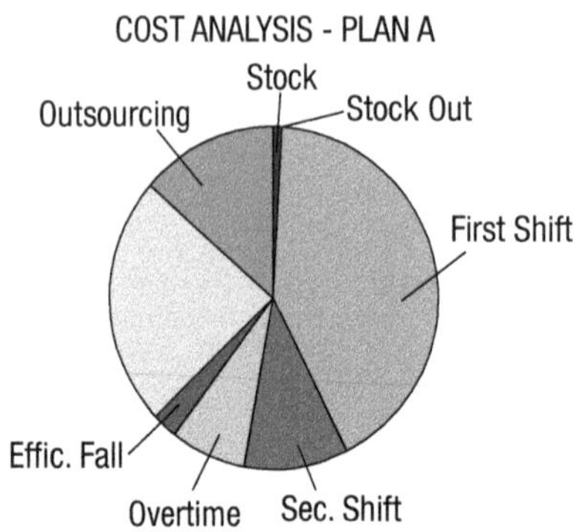

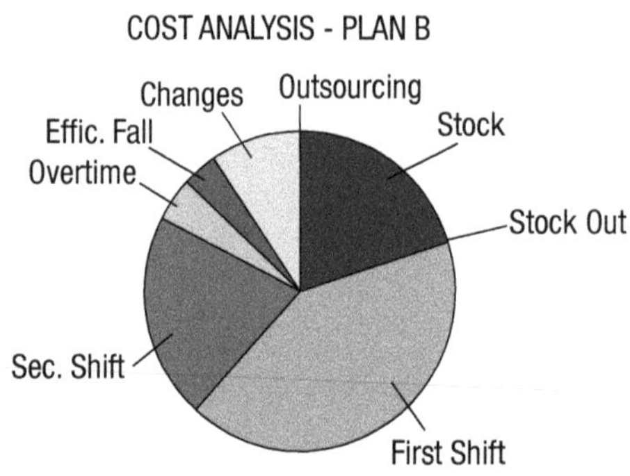

The above reported figures represent the cumulative volumes related to the Demand, Plan A and B. Even if the two plans produce the same total volumes at almost the same unit cost, its composition is different and leads to different operations issues and policies.

6.4 The Production Planning and Control Process over different time horizons

In order to describe the Production Planning and Control Process, although synthetically,[2] it first seems appropriate to analyze the issue by dissecting the process, on the one hand longitudinally, represented by the cited long-term, medium-term and short-term time horizons. Secondly, by grouping together the activities carried out in the three logical sub-sets of **planning, scheduling, execution and control**, as described in Figure 6.2. Conceptually, for each activity considered in the **long-, medium-** and **short-term**, respectively, the phases of planning, scheduling, execution and control must be equally carried out, with gradually increasing degrees of detail, following a sequence in which the feasibility control and the release of an authorized plan on a higher level constitutes an input for the plan, at lower level, from which it derives. Furthermore, considering the continuity of company processes, it goes without saying that said activity takes place without interruption, on a **rolling basis** with methods, visibility horizons, and paces specific to each plan: **PP – Production Plan**, **MPS – Master Production Schedule**, and **Scheduling.**[3]

As illustrated in Figure 6.2, for each time horizon, the Production Planning and Control Process tends to **reconcile production requirements**, triggered by sales forecasts and by consolidated demand in the orders book, **with the Available Production Capacity** of internal resources (labor and machines) and external resources (suppliers and contractors) (Grando, 1995).

In the **long-term**, that reconciliation effort is conducted from an initial planning phase that in terms of time horizon embraced, degree of detail of the processes, and resources involved, draws informational input from the company's budgeting and sales planning activities. The **Demand Plan**, developed by the sales functions, draws from the sales forecasts and the existing order book. In this phase, it is necessary to consider the effect produced by promotional and advertising policies, as well as the cyclical and seasonal nature of sales, the impact on the market of new products launched by competitors, etc. Those elements are dealt with in the **Production Plan – PP**, which has the aim of sizing and planning the use of internal resources (production capacity), at the same time evaluating their limits in terms of elasticity, and critical external resources, or the supplies characterized by restrictions on capacity or long lead times.

Precisely because this phase represents a moment of conjunction between the sales perspective and the production perspective, it is now commonly called **Sales and Operations Planning – SOP**. By comparing the **Required Production Capacity – RPC** thus defined, since it is necessary to satisfy estimated demand and/or the existing orders book, taken from the Demand plan net of on hand inventory, with **Available Production Capacity – APC**, calculated based on specific indexes that

[2] For further discussion, see the exhaustive work by De Toni and Panizzolo (2018).
[3] On this point, see Da Villa (2000); Vollmann et al. (1997).

Figure 6.2 **Conceptual diagram of scheduling and control process**

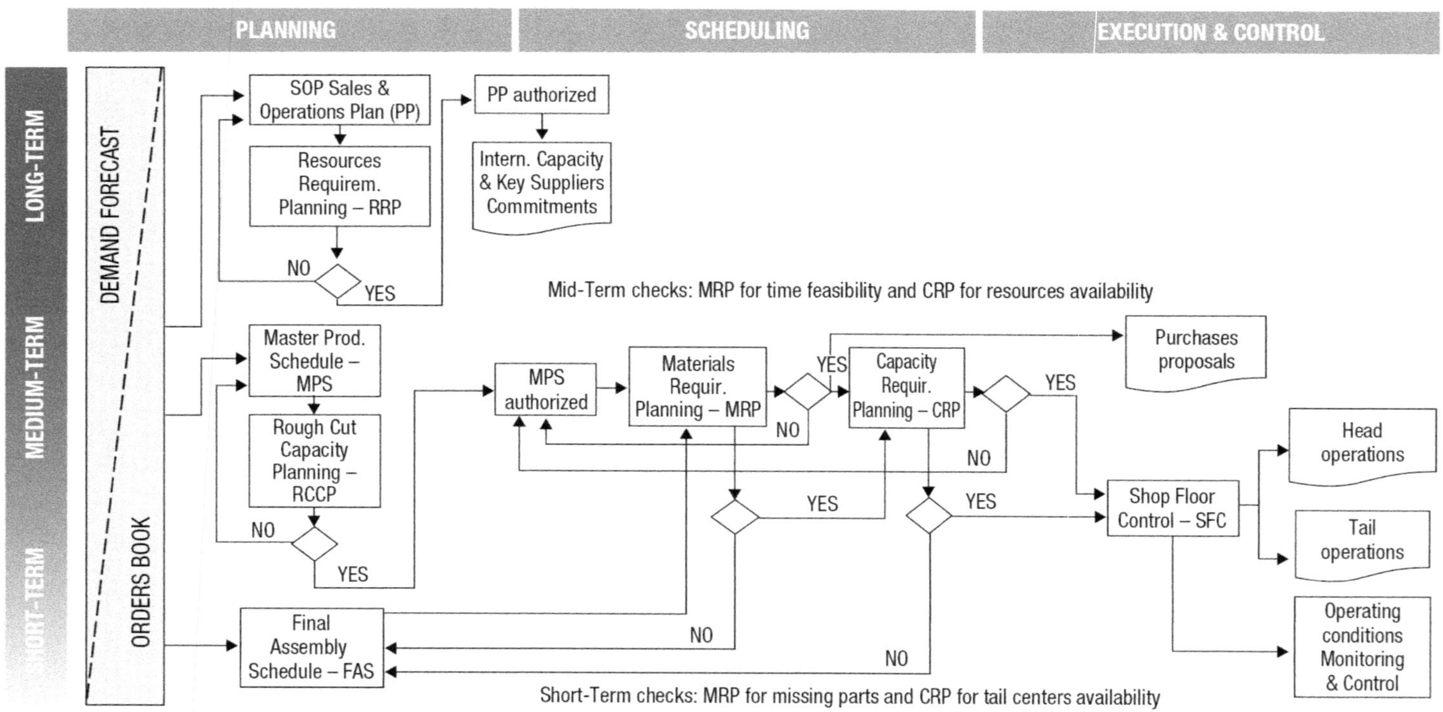

Source: adapted from De Toni and Panizzolo, 2018; Da Villa, 2000.

we will discuss in Chapter 8, an initial determination of feasibility or load is generated. In some applications, that phase is supported by the use of **Resource Requirements Planning – RRP** systems. If the result of that analysis is negative, it is necessary to conduct re-planning, searching for ways to expand production capacity or diminish sales goals, through a review of sales plans. If the result is positive, the **Production Plan is authorized**, making the commitments in terms of internal capacity and issuing orders to suppliers, in the form of defined orders or open orders, that are similar to commitments to use external capacity.

In the **medium-term**, the subsequent planning phase is developed, launched with the **Master Production Schedule – MPS** (Chapter 11).

In that phase, the sequencing of production is defined, along with the size of the production batches (lot sizing), with attention to saturation of production capacity and leveling of work loads. Such operations clearly entail a simultaneous analysis of available capacity and critical components, with the use of systems generally defined as **Rough Cut Capacity Planning – RCCP**, conducted through indexes that express the availability of critical resources. Once those "rough" checks of internal capacity have been carried out, the planning phase can be started, through the explosion of the MPS based on the information contained in the Bill of Materials – BOM and in other files which will be mentioned later, and on the needs for materials and components, using **Material Requirements Planning – MRP** systems and modules (Chapter 13). The main goal is to assess the feasibility of the MPS from a temporal standpoint, based on the consistency between the planned order receipt dates and production and purchase lead times. At the same time, it is necessary to evaluate the production capacity expressed by the Work Cycles, using **Capacity Requirement Planning – CRP** systems. As mentioned for long-term planning, if the results of those feasibility checks – in terms of times and capacity – are positive, the production plan is authorized and thus the orders for production and purchasing are issued. The latter are received by **Shop Floor Control** and the **Purchasing Function**, that while evaluating the possibilities of grouping, modification, deferment, etc., dispatch them to the internal resources and external suppliers, constantly monitoring the progress of production and arrivals from suppliers.

In the third **short-term** phase, Production Planning involves the scheduling activities, characterized by the **Final Assembly Schedule – FAS**, for the final or "tail" activities, and by Scheduling for all other operations, including "head" operations (Da Villa, 2000). MRP/CRP elaboration process is needed at this level as well, for the constant verification of any missing materials and components and for control of short-term availability of internal resources. Once the availability of all of the components and production capacity has been verified, the **execution phase** is started, with the gradual release of production and assembly orders, on the one hand, and purchase orders, on the other. In this phase, the assignment of priorities between the various orders and lots is important, which is conducted following precise short-term goals (deadlines, saturation, unavailability) (see Chapter 14). The last link of this cycle is represented by **Production Control**, intended to monitor the proper work in progress, the occurrence of bottlenecks, the accumulation of backorders, the production of defective products, or other anomalies that can affect the possibility to achieve the planned goals of efficiency and service. The control activity develops informative inputs on the status of the machines and processes that allow for measuring the performance of the operations system (see Chapters 4 and 5), and the reformulation of subsequent planning cycles.

Lastly, we must mention the temporal development of the overall planning process. On this point, regardless of the temporal horizon covered, we refer to **rolling plans**, as mentioned above, characterized by multiple levels of constant visibility over time, which are correlated with increasing degrees of detail, as presented in **Figure 6.3**.

6.5 Planning Horizons and rolling plans

A further consideration must be presented concerning the time horizon covered by the planning process. On this point, the Planning horizon is divided into different **Time Fences**: Tentative (or planned), Confirmed (or firm) and Frozen Planning Horizons (Da Villa, 1991; De Toni and Panizzolo, 2018).

Figure 6.3 Rolling plans and time horizons

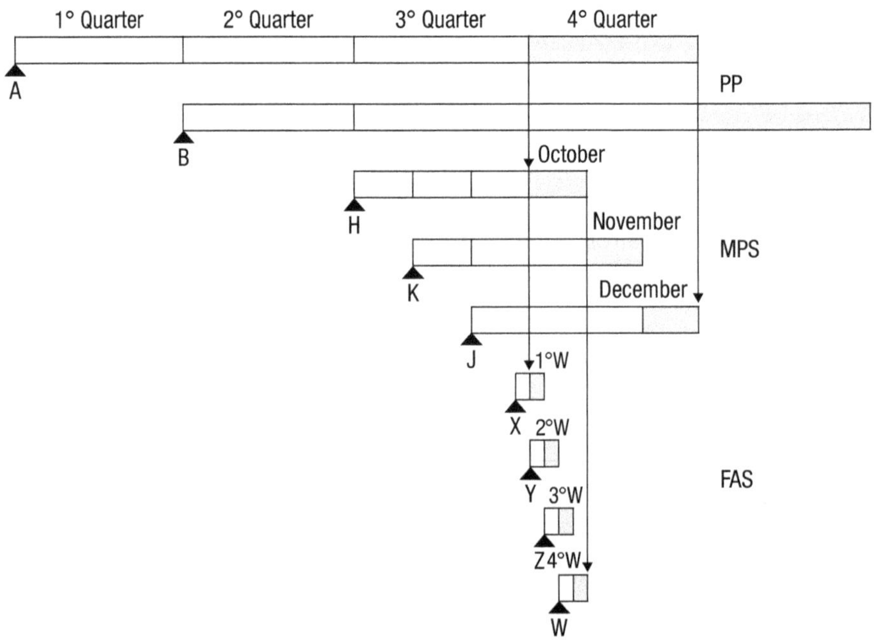

Source: Da Villa, 1991, p. 191.

The **Planning Horizon**, common to all of the part numbers subject to planning, consists of the sum of the Tentative (or planned or provisional), Firm (Confirmed) plans, the latter specific for each product and the Frozen plan. The Firm Plan is Frozen for each product for the portion covered by its lead time, and upon verifying the availability of materials (MRP) and production capacity (CRP), detailed at the level of the single article and phase. On if this check returns a positive result is the plan authorized, with the consequent gradual issue of the purchase and manufacture requests (or proposals), according to timed sequences and short time horizons.

- In the **Tentative (or planned) Horizon**, the planner may conduct different simulations of the plan and enter changes to the resources: machines, personnel, and supplier orders. At the **PP** level, it is necessary to simulate options for sizing production capacity, considering the feasibility and times of adaptation, and of the agreements with suppliers. For the **MPS**, taking into consideration the constraints on production capacity set by the PP and operating on forecasts and planned orders, alternative plans and lot-sizing options are simulated, until they enter into the firm horizon. At the **FAS** level, the constraints considered regard the volumes set by the MPS, lot-sizing and sequencing alternatives are simulated as well taking into account the due dates and the delivery times requested by the clients.

- In the **Firm Horizon**, the planner must take into consideration the capacity and time constraints set by the different higher-level plans. At the **PP** level, the aggregate volumes at the family level are confirmed. At the **MPS** level, since the planning cycle is within the cumulative lead time of the product, each change or variation must find positive confirmation in terms of the expected availability of materials and components, on the one hand, and of production capacity, on the other. The purchase orders and commitments of production capacity have in fact already been issued, such that any change to the plan will entail additional costs, as a function of the degree of flexibility of the production system in question. At the **FAS** level, the work instructions for the departments are confirmed, relating to the single finished products.
  The firm horizon must extend for a number of periods equal to at least the cumulative lead time of the articles subject to the plan, on penalty of temporal unfeasibility, verified through the MRP explosion. That unfeasibility becomes manifest in cases in which, following the drafting of a tentative plan, it is found that as a result of the anticipation – or precession – process linked to the lead times, launches of purchase or production orders are suggested on dates prior to the moment of planning t_0 (today). These are orders that, in other words, to respect the final delivery dates, should have been issued in the past or will somehow have to be accelerated. By way of example, consider the case of production of faucets for residential use, produced in ATO (Assemble to Order) mode. The final phases of assembly, packaging, and inspection are carried out on order, while planning regards the codes of semi-finished products to be assembled. If we suppose the simple timed work cycle shown in **Figure 6.4**, it is evident that a planning horizon

Figure 6.4 **Simplified work cycle**

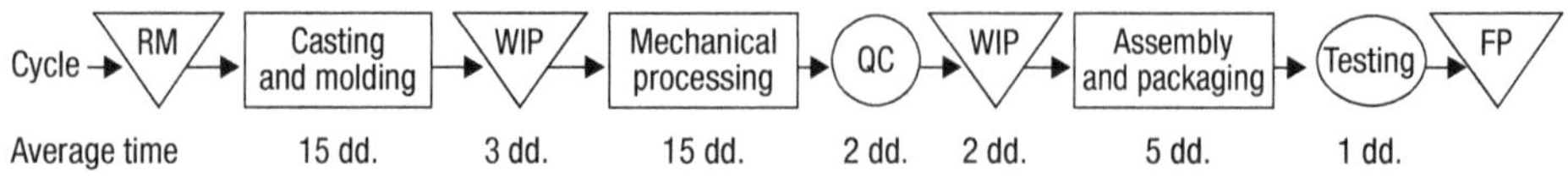

of 20 days would prove to be inadequate and a source of unfeasibility, allowing for insufficient advance time to plan purchases and work operations characterized by greater lead times.
The removal of the time constraint on critical resources is a necessary but not sufficient condition to proceed with authorization of the plan, that is subject to further controls, not the least the availability of all of the component materials and the necessary production capacity resources. Unlike what happens in the Aggregate Production Plan, in the preparation of the Master Production Plan, the verification of feasibility is no longer conducted in aggregate form according to Resources Requirements Planning – RRP procedures, but initially through "rough" methods, with Rough Cut Capacity Planning – RCCP systems, on the basis of representative technical data or aggregate standards.

- The **Frozen Horizon**, equal to the product lead time, does not allow for any changes to the plan – except in the case of particular exceptions – because since the orders of the parent code have already been issued, each subsequent change entails high costs for the blockage or rescheduling of orders already launched or urgent purchases of unavailable components, the use of overtime work shifts, over- or underutilization of machines with respect to plans, etc., such that in general an attempt is made to limit exceptions as much as possible (De Toni and Panizzolo, 2018).

The preparation and issuance of the Master Production Plan, as well as the Aggregate Plan, from which it derives, and of the short term Schedule that it drives (FAS or head operations scheduling), takes place through the gradual **change of the plan status** (tentative, firm, or frozen plan) as it gradually becomes part of the horizons described above, divided by Time Fences, according to the **rolling logic** illustrated in the example shown in Box 6.2.

Box 6.2 **Example of Planning Policies**

Imagine the case of a company that produces three families of vehicles for travel and vacations – Campers, Motorhomes, and Motorcaravans – that must develop its production plans, divided into PP, MPS, and FAS, respectively. In the simplified example, that assumes the same lead times for each product, imagine being at the end of the first quarter of year 1 with an overall plan to produce 3,700 units over the next 12 months.
As illustrated in the following diagram, the planning develops with a rolling logic, such that each level of planning has its own horizon and a refreshment and release frequency: the **PP** plans the three

ROLLING PLANS — PLANNING HORIZON

PRODUCTION PLANNING (PP)

PRODUCT FAMILIES	YEAR I: I QUARTER	II QUARTER	III QUARTER	IV QUARTER
Camper	300	300	350	400
Motorhome	200	200	300	300
Motorcaravan	200	300	300	200

	YEAR I: II QUARTER	III QUARTER	IV QUARTER	YEAR II: I QUARTER
Camper	300	300	300	250
Motorhome	200	300	300	300
Motorcaravan	300	350	400	400

MASTER PRODUCTION SCHEDULE (MPS)

MODELS	APRIL	MAY	JUNE
MC - 2000 - G	25	30	35
MC - 3000 - G	30	30	40
MCS -2500 - D	30	40	40

	MAY	JUNE	JULY
MC - 2000 - G	30	35	35
MC - 3000 - G	30	40	50
MCS -2500 - D	40	40	40

	JUNE	JULY	AUGUST
MC - 2000 - G	35	35	30
MC - 3000 - G	40	50	40
MCS -2500 - D	40	40	40

FINAL ASSEMBLE SCHEDULE (FAS)

VARIANTS	W I	W II	W III	W IV
MCS - 2500 - D - I	10	5	5	5
MCS - 2500 - D - F	-	5	5	5

	W II	W III	W IV	W I
MCS - 2500 - D - I	5	5	7	5
MCS - 2500 - D - F	5	5	3	5

Legend: FROZEN · FIRM · TENTATIVE

product families on a horizon equal to four quarters and a refreshment frequency per quarter, the **MPS** plans the three **models** of the Motorcaravan family (characterized by different engine sizes and gasoline (G) or diesel (D) fuel) over a quarterly horizon and a monthly frequency, while the **FAS** plans the assembly of two **variants** of the 2,500 diesel Motorcaravan, for the Italian (I) and French (F) markets, respectively, over a horizon of 4 weeks with a weekly frequency.
The progress of the planning system over time takes place through the change of the status of the plan. For each level of planning, in the longer horizon a tentative plan is simulated, in the intermediate horizon a firm plan is drawn up, and in the immediate one, a frozen plan is prepared. The example shows that at the PP level, the first quarter of year 1 is final, while the second quarter is frozen, the third firm, and the fourth tentative. The visibility horizon of the business plan goes until the first quarter of year 2 that will enter the planning horizon once the second quarter of year 1 is also final.
Following the example reported in the table, the frozen PP plans to produce 800 units, split among 300 Campers, 200 Motorhomes and 300 Motorcaravans, in quarter II; the firm PP accounts for a total of 950 units, split among 300 Campers, 300 Motorhomes and 350 Motorcampers, in quarter III; the tentative PP simulates to produce 1,000 units, in turn split among product families, in quarter IV of year 1.
Likewise, at the MPS level, April is frozen, May is firm, and June is tentative. In terms of volumes, on April the frozen MPS plans to produce 85 units of the Motorcaravan product family, in turn split among models: 25 units of the 2000 Gasoline model, 30 units of the 3000 Gasoline model and 30 units of the 2500 Diesel one. A total of 300 units of the Motorcaravan product family is planned in April, May and June.
As time moves forward, the plan advances in a rolling fashion: the figures for the frozen periods are final, those firm are frozen, and in turn, the tentative ones are firm. The same takes place at the FAS level, where the horizon is monthly and the frequency is weekly. Volumes can change at tentative plan level, as shown at FAS level in week IV, in terms of mix between Italian and French variants.
Thus, each article is subject to planning multiple times, in increasing detail both as regards the profile of the planned object (product family, model, variants) and for the disaggregation of the volumes in increasingly detailed horizons (quarter, month, week).

6.6 The Demand Plan

The **Demand Plan** represents the fundamental input in the **Sales and Operations Planning** phase, and is aimed at estimating production needs and determining **Required Production Capacity**, i.e. the production volumes requested in a specific period of time, expressed in units of production capacity, adjusted by the delta stock of the observed item: Demand in period n + target inventory at the end of period n – on hand inventory at the beginning of period n. The methods of defining the demand plan differ in terms of time horizon covered and level of detail, as a function of the type of market, the sector, the product life cycle, and other details. Regardless of those specific aspects (see Chapter 7), each company requires forecasts in order to orient their production planning. Long-term forecasts (3-5 years or more), and medium or short-term forecasts (1-3 years), are critical for investment policies, setting up production capacity, the choice of systems and technologies, the approximate definition of workers' requirements, the selection of suppliers and contractors, and policies for new product development. Limiting ourselves to Operations Management here, medium and short-term forecasts (1 year) are important: such forecasts are the basis for the Production Planning and Control Process. As concerns the issue of Demand Planning of finished products, components and materials, two aspects should be stressed:

a. not all requests must be forecast, some can be calculated. This refers expressly to the distinction between **independent demand goods** and **dependent demand goods**, which will be further discussed regarding stock management (Chapters 10 and 13). The former include articles for which demand cannot be deduced directly from that for higher level items (parents) in the BOM, but at the most estimated (finished products, spare parts, consumables). On the other hand, dependent demand goods are those for which the demand, or to be more precise the requirement, is derived arithmetically from the demand for other products in a hierarchically superior phase, whether the latter is calculated or estimated;
b. as a function of the time horizon, as the aims of the forecast change, **the level of aggregation subject to estimation and the related risk vary**. Long-term forecasts are in fact carried out on broad sets generally represented by turnover, by market segment, or product line; medium-term forecasts are carried out on subsets, that are typically product lines or families; and lastly, short-term forecasts can reach high levels of detail (on single products, variants, or SKUs), pushed to the limits of significance and reliability. Over time, multiple demand forecasting techniques have been developed, that are generally divided into **qualitative and quantitative techniques**. The former refer to subjective evaluations by the sales force, expert opinions, at times organized according to the Delphi Method, market research, etc. The latter are based on econometric models or regression approaches, the techniques based on historical statistics, such as the moving average or exponential smoothing, that may be corrected based on seasonal and/or trend coefficients (Chapter 7).

7 Demand Forecasting

by *Valeria Belvedere*

7.1 Introduction

This chapter illustrates the methods for developing sales forecasts, that are necessary for drawing up the Demand Plan. In particular, we will present qualitative techniques, based on subjective evaluations by individuals, and quantitative techniques, divided into time series methods and causal methods. We will present the indicators aimed at accurately evaluating the forecasts, and lastly, we will address the issue of the choice of the most adequate forecasting technique with regard to the specific characteristics of a product and its demand.

7.2 The Demand Plan

The Demand Plan represents the starting point for the production planning process, because by providing qualitative and quantitative indications on future sales, it allows for planning the production capacity necessary to satisfy demand and orient the choice of the production mix as a function of the requests anticipated from clients (Vollmann et al., 1984).

In this chapter, we will look at **forecasting demand for independent demand goods**, such as the finished products requested directly by the market. On the contrary, we will not discuss the determination of the quantities of dependent demand goods (such as components, subsets, or raw materials) necessary to supply the production process, whose entity is de facto derived from the estimates of demand for finished products. For further examination of these aspects the reader can refer to Chapters 9, 10 and 13.

Over time, several demand forecasting techniques have been developed, that are generally distinguished between qualitative and quantitative techniques (Sanders, 2017). On this point, see Figure 7.1. **Qualitative techniques** refer principally to executive opinions, market research, and expert evaluations obtained through the Delphi method. **Quantitative techniques** include causal methods and those based on the analysis of time series. In this chapter, we will illustrate the main qualitative and quantitative forecasting techniques. Subsequently, we will present the forecasting accuracy indicators most used in the business world, and we will outline the principal criteria for selecting the most appropriate techniques based on the specific characteristics of the product and its demand.

Figure 7.1 **A classification of the demand forecasting methods**

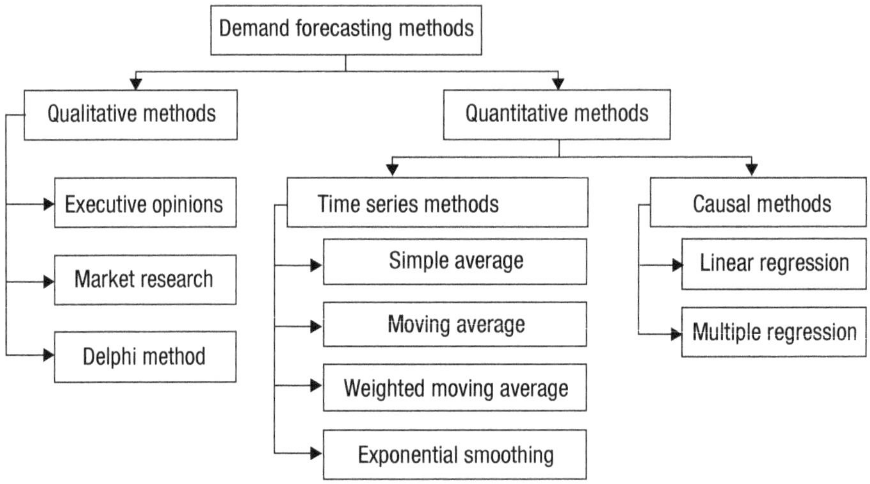

7.3 Qualitative techniques

Qualitative forecasting techniques are generally used when the determinants of demand for a good are not known and/or no time series exist to analyze through appropriate statistical models, as occurs following the launch of new products for a company (Goodwin and Wright, 2014; Sanders, 2017). They are also useful in the case of significant changes in the context, that radically transform the profile of demand or generate a need for very different articles than in the past, as happens with the development of new technologies. In these cases, the analysis of the time series (if available, for example in relation to previous versions of the product) is not effective because the market has been radically transformed. To deal with these difficulties, companies make use of qualitative sales forecasting methods, based on the idea of exploiting knowledge of the market and the competitive context possessed by specific persons, able to identify, collect, and interpret qualitative ("soft") information that is relevant for the quantification of demand. These techniques, also known as "**judgemental forecasting**," thus have the merit of leveraging experience and skills often present in the organization, that can be difficult to incorporate in a statistical model (Goodwin and Wright, 2014). However, they are not devoid of limits that can reduce their effectiveness, linked to both the risk of intentional manipulation and the presence of unconscious mechanisms that distort people's decision-making. The first case occurs especially when it is possible to obtain economic benefits through the subjective calculation of a forecast data point with respect to which the decision-maker will subsequently be evaluated. For example, this can happen when the sales force receives a performance bonus based on a positive difference between actual sales and sales forecasts.

The second case is related to numerous mechanisms that distort decisions ("**cognitive biases**"), and unconsciously influence the individual (Goodwin and Wright,

2014), for example inducing him/her to overestimate the sales of a product that the person has contributed to developing ("wishful thinking"), to attribute excessive importance to more recent sales data ("anchor and adjustment"), or overvalue the view of opinion leaders ("sunflower effect").

The most frequently used qualitative techniques are the following:

- executive opinions;
- market research;
- the Delphi method.

The first technique consists of organizing a **team of managers**, often from different departments of the company, who are asked to formulate sales forecasts based on their knowledge of the market and the most recent information regarding the circumstances that could affect demand in a short time period, as happens with special promotions by the company or its competitors. Sometimes, this solution can also be adopted to modify the forecasts generated by statistical models, in order to take into due account some important information, whose effect on sales is difficult to estimate through adjustments to the forecasting model (for example, the launch of a new product by a competitor or a new law that can affect the market).

As illustrated in **Table 7.1**, this method has the benefit of being able to exploit market knowledge and skills present in the organization. However, it is particularly exposed to the risk of cognitive biases in the evaluation process, and in particular to the "sunflower effect," if the group includes charismatic individuals, who are able to influence the judgment of others.

Table 7.1 **Qualitative demand forecasting methods**

Method	Characteristics	Advantages	Disadvantages
Executive opinions	Development of forecasts by a group of managers	Usable in the case of the launch of a new product. Possibility to exploit the potential of qualitative information on market and sector trends	Subject to risk of cognitive biases, especially the "sunflower effect"
Market research	Extensive surveys or focus groups with customers	Effective for understanding the degree of approval of a product and the drivers of customer satisfaction	Difficulty to gather an adequate number of observations through the survey. Time-consuming. Difficulty in elaborating effective questionnaires or research protocols for gathering reliable opinions of customers
Delphi	Search for consensus among experts in the sector or on the technology	Appropriate in the case of technological innovations with long-term impact on demand	Time consuming. Risk of not reaching a consensus among experts

Source: adapted from Sanders, 2017.

A second alternative is represented by **market research**. This tool can be particularly useful in the case of the launch of new products, since it allows for identifying the factors that most affect customer satisfaction, determining the overall level of potential approval, and measuring purchase intentions. Operationally, market research can be conducted through extensive investigations in the form of **surveys**, or at times through focus groups. The first solution allows for acquiring large quantities of data relating to samples of consumers representing the target population, that can be the subject for statistical analysis aimed at extracting useful information to orient choices regarding the sizing of production volumes. However, surveys are often time-consuming, since obtaining reliable results for data analysis depends on the size of the sample, and collecting a high number of questionnaires implies consumption of considerable human resources and time. Moreover, the effectiveness of this method depends heavily on the quality of the questionnaire. If, for example, the questions are not formulated clearly or provide a group of preset responses that are incomplete, the results of the survey will not provide an accurate picture of the consumer's attitude toward the product.

To overcome these limitations, sometimes it is preferable to use **focus groups**, which are structured moments of discussion with a limited group of target consumers, to whom the product whose sales potential is to be evaluated is presented. This solution appears particularly effective to identify possible product criticalities, that could jeopardize its commercial success. This type of outcome can lead the company to decide to suspend the launch of the new product, or sometimes, to change it in order to remove the causes for a lack of customer satisfaction. While on the one hand focus groups appear to be a useful and effective tool, on the other, it is difficult for them to provide support for the demand forecasting process, since the data relating to purchasing propensity collected this way is often too limited to be used for statistical procedures with reliable results.

The **Delphi method** is used to formulate long-term forecasts relating to major changes in the scenario, induced by significant technological innovations or other phenomena with a widespread impact, such as certain scientific discoveries. In these cases, a team is created of experts in the sector, whose identity remains mutually concealed in order to ensure the evaluation is not conditioned by knowledge of the other participants. Those individuals are given a questionnaire with questions aimed at understanding their position in regard to the issue under discussion. Once the first round of feedback is gathered, the impressions are collected in a single document, that is submitted to the experts, who based on this new input can reconsider their original position or confirm it. This process is repeated until a consensus of opinion is reached. The main risk linked to this technique is the possibility that a unitary, shared position may never be reached among everyone. In addition, even when that happens, it may take place after a high number of repetitions of the process, that make it long and burdensome to complete. Lastly, it should be stressed that although this technique has demonstrated a high degree of reliability of results when a consensus is reached, there is always a margin of error that is difficult to evaluate ex ante.

In general, qualitative methods are characterized by performance that on average is worse than that of quantitative techniques, since they are unable to correctly identify the characteristic patterns of demand curves and thus end up overestimating the entity of unexplained variance. They also cannot be used in situations characterized by very broad product range, due to the excessive time necessary to formulate a sales forecast for each product. Despite this, qualitative techniques are broadly used in businesses due to their ease of implementation and the high importance assigned in some organizations to subjective knowledge and skills in the sector.

7.4 Quantitative techniques

In the area of quantitative techniques, usable for forecasting demand, a distinction is made between methods based on time series analysis and causal methods (Sanders, 2017). The former are based on the assumption that the analysis of sales (or consumption) data collected in the past can constitute a useful reference for forecasting levels of demand in the future. Causal methods, on the other hand, apply when the value to be forecast is a function of specific environmental variables.

7.4.1 Time series methods

These methods allow for identifying significant demand patterns, through the analysis of actual sales data from the past, that can be classified as follows (Siemsen, 2015; Chase, 2016):

- increasing or decreasing trends (T);
- seasonality (S);
- random variations (I).

Therefore, according to an "additive" approach, the demand for a good in a period t can be considered the sum of these elements, formalized as follows:

$$d_t = T_t + S_t + I_t \qquad [7.1]$$

Figure 7.2 shows an example of a demand profile in which it is possible to identify the components of trends, seasonality, and random variations.

Time series methods have increasing levels of complexity, as a function of the presence or absence of forms of trends and seasonality. The simplest situation is that relating to perfectly stationary demand, that is, where the probability of assuming value x in the period t_1 is equal to the probability of assuming the same value in the period t_2:

$$p_{t_1}(d = x) = p_{t_2}(d = x) \qquad [7.2]$$

Figure 7.2 Demand profile

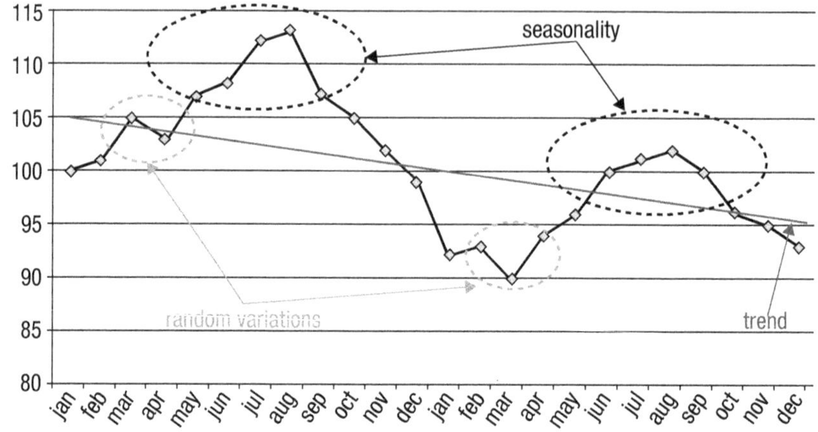

In this case, and thus with demand lacking any form of variability, it is possible to calculate the **simple average**, obtained as follows:

$$f_{t+1} = \frac{\sum_{i=1}^{n} d_i}{n} \quad [7.3]$$

where:
f_{t+1} is the forecast sought for the period t+1, set as equal to the simple average of the actual demand recorded in the n available periods;
d_i is the actual demand recorded in period i;
n is the number of periods for which an actual demand (or consumption) figure exists.

This technique seems appropriate for articles with a long sales cycle, stable demand, and for which the acquisition of more data over time allows for gradually improving the quality of the forecast.

However, in the calculation of the simple average, the most recent values have an impact equal to that of the farthest values in time, which represents a limit especially when the more short-term dynamics begin to deviate significantly from those farther away in time. In these situations, demand begins to take on forms of remarkable variability that can be measured through the standard deviation (σ), whose entity increases with the intensification of the deviations of actual demand from the average:

$$\sigma = \sqrt{\frac{\sum_{i=1}^{n} (d_i - \bar{d})^2}{n-1}} \quad [7.4]$$

where:
d_i is the actual demand in period i;
$\bar{d}$ is the average demand calculated on n periods;
n is the number of periods for which data on demand is available.

In the case of demand subject to some forms of variability and trends, it can be effective to calculate the **moving average**, by which demand forecasts for a future period are determined by the simple mean of the sales observed during *n* prior periods (for example, months). The average is called "moving" because it is constantly updated by regularly substituting the oldest data with the most recent available.

Analytically, the calculation of the moving average applied to the forecast of a future period t+1 is shown by the following expression:

$$f_{t+1} = \frac{\sum_{i=t-n}^{t} d_i}{n} \qquad [7.5]$$

where:
f_{t+1} is the forecast sought, set as the moving average of *n* previous values;
d_i is the actual demand in the period *i*;
n is the number of periods for which the sales (or consumption) data is acquired.

The number *n* is generally established empirically, such as the number of months or weeks indicating a certain season, for example. If it is necessary to estimate the values for hundreds or thousands of articles with different possible values of *n*, we suggest searching for the value that minimizes the sum of squares of the difference between the forecast and the actual demand for each instant observed.

From the formula shown above it follows that for high values of *n*, the weight of the more recent demand data will be less; for low values, this weight will be emphasized. The number of periods will preferably be large during phases of slow evolution, while it will be small in periods of intense fluctuations, in order to make the average more sensitive to the latter.

The procedure for calculating the average associates a constant weight (the coefficient 1/n) with all of the data taken into consideration: thus, each figure contributes with the same proportion to the informational content of the forecast. If it is believed that the data referring to specific periods is more indicative of future evolution of demand (for example, the most recent ones), the **weighted moving average** can be used. In this case, the calculation is performed as follows:

$$f_{t+1} = \frac{\sum_{i=t-n}^{t} d_i \times w_i}{n} \qquad [7.6]$$

where:
f_{t+1} is the forecast sought for the period t+1, set as the weighted moving average of the *n* previous values;
d_i is the actual demand measured in the period *i*;
w_i is the weight assigned to the demand measured in the period *i*;
n is the number of periods in which the sales (or consumption) data is acquired.

A more sophisticated, reliable, and easily automizable version is the **exponential smoothing** of the weight of the farthest data. This is a particular type of moving average, where the factors are weighted based on a parameter α subject to continuous revision, that decreases geometrically as a function of their age, thus offering the way to act on the informational contribution of each, altering their weight as a function of how recent, and thus how significant, they are. As that parameter approaches 1, the influence of the more recent terms in the series increases strongly, while as α approaches zero, the contribution of the more recent terms decreases in favor of those farthest away in time. In the first case, the forecast is promptly adjusted to the evolution of the series, guaranteeing responsiveness in the forecast, but introducing the risk of being distorted by random fluctuations in demand; in the second case, the forecast is more inertial, giving higher value to the farther terms, reducing the effects of disturbance and offering forecasts which are more stable, but slow to react to systemic changes in the series. This parameter also allows for correcting the forecast based on the divergence between the estimate made in the immediately prior period and the data actually detected.

Analytically, the estimated demand for a future period $t+1$ is expressed by the following function:

$$f_{t+1} = f_t + \alpha \times (d_t - f_t) \qquad [7.7]$$

where:
f_{t+1} is the forecast sought for the period t+1
α is constant between 0 and 1
d_t is the actual demand measured in the period t
f_t is the forecast formulated for the period t

The formula above can be rewritten as follows:

$$f_{t+1} = (1 - \alpha) \times f_t + \alpha \times d_t \qquad [7.8]$$

In this equation, substituting f_t with the same forecast formula for the period t and then proceeding with the same substitution in each prior period, we obtain the following:

$$f_{t+1} = \alpha \times d_0 + \alpha \times (1 - \alpha) \times d_1 + \alpha \times (1 - \alpha)^2 \times d_2 + \ldots + \alpha \times (1 - \alpha)^t \times d_t \qquad [7.9]$$

In this equation, the sum of the constants, i.e. α, $\alpha \times (1 - \alpha)$, $\alpha \times (1 - \alpha)^2$ etc., is equal to 1.

The simplified formula, if rewritten highlighting α, shows that the forecast for a future period is equal to the previous forecast, increased or decreased of a portion (equal to α) of the error that emerged in the previous period. As already stressed, this indicates the critical nature of the choice of the coefficient α which, if high, tends to assign considerable weight to variations in demand (including random variations), guaranteeing extreme sensitivity in the model; if it is low, it attenuates the incidence, guaranteeing reliability in the presence of stable trends.

As in the case of the number n in the moving average method, in the exponential smoothing technique the optimal α coefficient is computed determining the value that minimizes the summation of the squares of the absolute forecasting errors.

For an example involving the calculation of a simple, moving, and weighted average and of the exponential smoothing, see Box 7.1.

Box 7.1 Time series methods: an example

Consider the actual sales data for the products A and B, shown in Table 7.2 and relating to the last 15 weeks.

Table 7.2 Weekly sales of A and B (in pieces)

	Sales A	Sales B
W.1	100	100
W.2	104	98
W.3	108	101
W.4	96	100
W.5	101	99
W.6	102	101
W.7	96	103
W.8	94	105
W.9	99	106
W.10	103	107
W.11	101	112
W.12	103	115
W.13	97	117
W.14	96	120
W.15	99	122

As can easily be seen by reading the data, in week 1 sales of both products were 100 pieces. However, while the demand profile for A follows a rather stable trend, for product B, after the first 5 weeks in which oscillations are seen around the value of 100, an evident growth trend begins. Imagine wanting to estimate the sale of the two products for week 15, using the simple average, moving average, and weighted average methods. The forecast will then be compared with the actual sales in week 15 in order to assess the appropriateness of the various techniques with respect to the two products.

The simple average of the demand for products A and B, respectively, is equal to:

$$f_{15(A)} = \frac{\sum_{i=1}^{14} d_i}{14}$$

$$f_{15(A)} = \frac{100 + 104 + 108 + 96 + 101 + 102 + 96 + 94 + 99 + 103 + 101 + 103 + 97 + 96}{14} = 100$$

$$f_{15(B)} = \frac{\sum_{i=1}^{14} d_i}{14}$$

$$f_{15(B)} = \frac{100 + 98 + 101 + 100 + 99 + 101 + 103 + 105 + 106 + 107 + 112 + 115 + 117 + 120}{14} = 106$$

In calculating the moving average, it is necessary to establish the value of the parameter n, that is, the number of more recent periods to be considered in the calculation. Suppose having established that a reasonable value can be 5 weeks. In this case, the moving average of A and B, respectively, will be:

$$f_{15(A)} = \frac{\sum_{i=14-5}^{14} d_i}{5}$$

$$f_{15(A)} = \frac{103 + 101 + 103 + 97 + 96}{5} = 100$$

$$f_{15(B)} = \frac{\sum_{i=14-5}^{14} d_i}{5}$$

$$f_{15(B)} = \frac{107 + 112 + 115 + 117 + 120}{5} = 114.2$$

Since, especially in the case of B, the demand growth rate is rather evident and accentuated, it can be considered more appropriate to assign a greater weight to the sales recorded in the more recent periods. This implies assigning weights, as shown in **Table 7.3**.

Table 7.3 Weights assigned to the sales recorded in the last 5 weeks

W.1	5%
W.2	5%
W.3	10%
W.4	30%
W.5	50%

It follows that the weighted averages of A and B, respectively, are equal to:

$$f_{15(A)} = \frac{\sum_{i=14-5}^{14} d_i \times w_i}{5}$$

$$f_{15(A)} = \frac{\sum_{i=14-5}^{14} 103 \times 5\% + 101 \times 5\% + 103 \times 10\% + 97 \times 30\% + 96 \times 50\%}{5} = 97.6$$

$$f_{15(B)} = \frac{\sum_{i=14-5}^{14} d_i \times w_i}{5}$$

$$f_{15(B)} = \frac{\sum_{i=14-5}^{14} 107 \times 5\% + 112 \times 5\% + 115 \times 10\% + 117 \times 30\% + 120 \times 50\%}{5} = 117.55$$

Summing up, we obtain what is presented in **Table 7.4**.

Table 7.4 Simple, moving, and weighted average of products A and B

	Product A	Product B
Simple Average	100	106
Moving Average	100	114.2
Weighted Average	97.6	117.55

Comparing the values of the averages in **Table 7.4** with the actual sales data for week 15 (**Table 7.2**), we can make some interesting observations. For product A, characterized by a stable demand profile, the results obtained with the simple and moving averages are aligned with each other, since the absence of trends or seasonality make the performance in the longer interval of 14 weeks and in the shorter and more recent interval of 5 weeks overlap. The values of the two averages are also consistent with the actual demand in period 15, equal to 99 pieces. The value obtained with the weighted average, on the other hand, deviates more from the actual sales data, since it is obtained by attributing an arbitrarily higher weight to the more recent sales, that only due to random variations can be higher or lower than the simple (or moving) average.

Different results emerge with reference to Product B, whose sales show an evident growth trend. In this case, the values of the three averages differ considerably. In particular, the lowest value is found with the simple average, that is based on the longest time series. A significantly higher value is obtained with the moving average, calculated on the last 5 periods, in which the sales have already reached higher levels than in the first weeks. Lastly, the weighted average produces the highest absolute value, since in this case it is based on a system of weights that assigns greater importance to demand in the more recent periods. Here, the predictive capacity of the three different averages changes considerably. In particular, as can be reasonably expected for this product, the weighted average is that closest to the actual sales in week 15, equal to 122 pieces.

Now, using the same data, imagine applying the exponential smoothing technique with specific reference to product B, for which, as we have seen, it can be appropriate to use a forecasting system that gives more importance to more recent data. In this case, it will be necessary to assign a value to the parameter α, that must be included between 0 and 1. Suppose that an appropriate value can be 0.6, due to the significant sales growth rate in recent weeks. Suppose also that the forecast for week 14 was 121 (f_{14}), compared to actual sales of 120 (d_{14}). In this case, the forecast for week 15 will be as follows:

$$f_{15(B)} = (1 - \alpha) \times f_{14} + \alpha \times d_{14}$$

$$f_{15(B)} = (1 - 0.6) \times 121 + 0.6 \times 120 = 120.4$$

When demand is not stationary and is characterized by phenomena such as trends and seasonality, it is necessary to use more sophisticated techniques. In particular, in the presence of trends, the trend-adjusted exponential smoothing method can be used, in which a trend component is added to the forecasting equation already illustrated and linked to the stationary part of the demand, improving the overall accuracy of the forecast. The best-known trend-adjusted exponential smoothing model is the Holt-Winters forecasting model, for which we refer to the specialist literature on the subject (Chatfield, 1978).

Analogously, when forms of seasonal demand appear, it is necessary to estimate the entity of the increase (or decrease) of demand with respect to the value it would assume in stationary conditions. In this case, the adjustment is based on the calculation of a "seasonal index," that expresses the entity of the demand expected in a specific "season" as a percentage of the overall average. Various techniques exist for the calculation of the seasonal index, for which we refer to the specialist literature on the subject.

7.4.2 Causal methods

Causal methods are used when it is expected that the quantity to be forecasted is influenced by some environmental variables. For example, it can reasonably be imagined that the performance of the real estate market in a future time interval will depend on the level of interest rates, the growth rate of Gross Domestic Product, and expected employment levels. In this case, a statistical method able to identify the relationships between these variables and the number of new homes requested, for example, will be more reliable than a technique based on the analysis of only time series. However, causal methods require a significant effort in terms of data collection and analysis, and thus are not generally used extensively, especially in situations where it is necessary to develop forecasts for a high number of products, as happens in large-scale retail.

The simplest causal method is represented by the **linear regression**, in which it is assumed that the dependent variable is determined by a single phenomenon, expressed by the independent variable. By way of example, we can imagine that the sales of an article can depend linearly on advertising investment. This type of relationship is expressed as follows:

$$Y = a + b \times X$$

where:
Y is the dependent variable to be forecast;
a is the intercept;
b is the slope, that expresses the degree of dependency of Y with respect to X;
X is the independent variable.

To estimate the values of the parameters of the linear regression, it is necessary to use appropriate statistical procedures to analyze the historical data for the dependent and independent variables. This way, we obtain a regression equation that for each value of X will allow for estimating the related value of the dependent variable Y.

In some situations, though, a single independent variable does not appear sufficient, since from a logical standpoint there can be numerous factors that, together, determine the phenomenon subject to the forecast. This is what happens in the real estate sector, as mentioned above. In situations like this one, it is necessary to use the **multiple regression** method, characterized by the simultaneous presence of multiple independent variables, each weighted for a specific regression coefficient. The equation is as follows:

$$Y = \beta_0 + \beta_1 \times X_1 + \beta_2 \times X_2 + \ldots + \beta_k \times X_k \qquad [7.10]$$

where:
Y is the dependent variable to be forecast;
β_0 is the intercept;
$\beta_t \ldots \beta_k$ are the regression coefficients of the independent variables;

k is the overall number of independent variables considered in the model;
$X_1 \ldots X_k$ are the independent variables.

In this case as well, in order to estimate the parameters of the model, it is necessary to analyze a dataset of observations, that allows for describing the relationships between the variables considered.

An essential aspect of regression models is represented by the identification of the set of independent variables (or of the single independent variable, in the case of linear regression), to be carried out on a logical level even before a statistical level. Subsequently, to verify the soundness of the hypotheses formulated on the possible predictors of a dependent variable, it is suggested to perform a check through the calculation of the correlation coefficient (r) between the dependent variable and each single independent variable. The closer that indicator is to its extreme values (–1; +1), the greater the presence of a strong relationship, that justifies the inclusion of the predictor in the regression equation. If, on the other hand, the correlation coefficient is close to 0, the model should not take into account the independent variable considered, that evidently does not have any significant influence on the quantity to be forecast.

By way of example, consider the example shown in **Box 7.2**.

Box 7.2 Linear regression: an example

Consider the data shown in **Table 7.5**, that describes the quantity of pieces of product A sold in the last 15 weeks and the advertising investment made by the company for this item.

Table 7.5 **Weekly sales of A (in pieces) and advertising investment (€)**

Week	Advertising investment (€)	Sales (pieces)
1	1500	2000
2	1600	1900
3	1590	1950
4	1630	2050
5	1690	2100
6	1700	2150
7	1690	2200
8	1740	2100
9	1760	2100
10	1790	2150
11	1800	2300
12	1780	2350
13	1810	2400
14	1840	2350
15	1870	2450

Observing these values, it seems evident that there is a certain correlation between the two quantities, that is confirmed by the graph in Figure 7.3.

Figure 7.3 **Advertising investment and pieces of A sold**

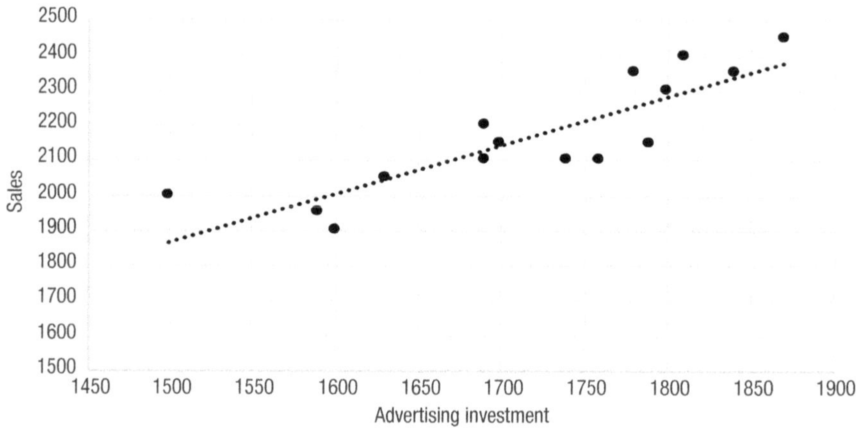

To predict the level of sales in week 16, we can use the linear regression approach, that implies the estimation of the parameters shown in the following formula:

$$Y = a + b \times X$$

In this case, Y will be the level of sales expected in week 16 and X the advertising investment planned for this week, that is equal to € 2,000.
Analyzing the data, we obtain the following linear equation that describes the relationship between sales and the advertising investment:

$$Y = -214.99 + 1.3872 \times X$$

Therefore, the expected level of sales of A in week 16 will be as follows:

$$Y = -214.99 + 1.3872 \times 2000 = 2559.41$$

7.5 The measures of forecasting accuracy

However accurate forecasting techniques may be, forecasting errors cannot be completely avoided. Each forecast must therefore be accompanied by an estimate of the error, which can be defined as the difference between the estimated data and the actual data, measured over the same reference period of time:

$$E_t = d_t - f_t \qquad [7.11]$$

where:
E_t is the forecasting error in the period t
d_t is the actual value in the period t
f_t is the forecast for the period t

The measure of the forecasting error is the basis for calculation of additional indicators aimed at evaluating the reliability of the forecast. These are generally divided into two categories: **standard error measures** and **relative error measures** (Siemsen, 2015; Sanders, 2017).

In the first category, the most common indicator is the **mean absolute deviation** (**MAD**), calculated as follows:

$$MAD = \frac{\sum_{i=1}^{n} |actual_i - forecast_i|}{n} \quad [7.12]$$

where:
$actual_i$ is the actual demand recorded in the period i
$forecast_i$ is the forecast for the period i
n the number of periods for which a forecast is available

This indicator, that expresses the size of the average error in absolute terms, is expressed by the unit of measure of the variable to be forecast, and thus does not allow for easy comparison with the same accuracy indicators calculated for other types of products. An additional limit of this tool is the fact that it treats positive and negative forecasting errors in the same manner. However, in business practice, it is rare for the two situations to be equivalent for a company, since from a financial standpoint they have different implications. A positive error can give rise to problems of stock-out, and thus losses due to lack of sales. A negative error is associated with too much stock, that leads to holding costs. The level of these two different types of costs varies depending on the specific characteristics of each case, but is unlikely to be equal. As a consequence, the same MAD can correspond to higher or lower costs for the company.

A second indicator of accuracy belonging to the category of standard error measures is the **Mean Squared Error** (**MSE**), calculated as follows:

$$MSE = \frac{\sum_{i=1}^{n} \left(actual_i - forecast_i \right)^2}{n} \quad [7.13]$$

This indicator, like the previous one, also aims to avoid mechanisms of compensation between errors of opposite signs. However, it also has some significant limits. First of all, it is not very intuitive, because it is expressed in a unit of measure that is the square of the unit of measure of the variable considered. Furthermore, squaring the error ends up assigning excessive weight to large (positive or negative) errors. Lastly, as happens for the MAD, this indicator does not allow for comparison between the levels of accuracy of the forecasts relating to variables expressed in different units of measure.

To overcome this last limit, we use relative error measures. The first indicator to consider in this category is the **Mean Percentage Error** (**MPE**), that is calculated as follows:

$$MPE = 100\% \times \frac{\sum_{i=1}^{n} \left[(actual_i - forecast_i) / actual_i \right]}{n} \quad [7.14]$$

The MPE expresses the average forecasting error in percentage terms, and thus can be used for comparative purposes. However, since forecasting errors can be both positive and negative, forms of compensation often take place that lead to overestimating the accuracy of the forecasting model. Nevertheless, this indicator can be particularly useful in situations in which the goal is to determine the existence of biases in forecasts, that can lead to the prevalence of positive or negative forecasting errors. As already stressed above, this has different financial implications, related in the first case to a greater incidence of stock-out costs, and in the second to the overall amount of holding costs. If one of these types of costs is significantly higher, the positive or negative sign of the MPE constitutes an important piece of information to recognize the need to improve the demand forecasting process in order to enhance its accuracy and reduce the financial impact for the company.

A second indicator belonging to the category of relative error measures is the **Mean Absolute Percentage Error** (**MAPE**), calculated with the following formula:

$$MAPE = 100\% \times \frac{\sum_{i=1}^{n}\left[\left|actual_i - forecast_i\right|/actual_i\right]}{n} \qquad [7.15]$$

The difference with respect to MPE consists of the fact that the MAPE takes absolute forecasting error into account, and thus is not subject to phenomena of compensation that can lead to overestimating the accuracy of the forecast. On the other hand, compared to MPE, this indicator does not allow for identifying forecasting biases that could lead to underestimating or overestimating demand.

Both of these two indicators can be used to perform comparisons between different products, even when the relative demand is measured in different units of measure. This is due to the measurement in percentage terms, a characteristic of the MPE and the MAPE.

For an example of the calculation of forecasting accuracy indicators see **Box 7.3**.

Box 7.3 Measure of forecasting accuracy: an example

Consider the sales data for product A shown in **Table 7.6** and assume that the company uses the moving average technique on the data for the last 5 weeks to formulate sales forecasts.

Table 7.6 Product A – actual sales and moving average as a forecast

	Actual sales (pieces)	Forecast (pieces)
W.1	100	
W.2	104	
W.3	108	
W.4	96	
W.5	101	

	Actual sales (pieces)	Forecast (pieces)
W.6	102	101.8
W.7	96	102.2
W.8	94	100.6
W.9	99	97.8
W.10	103	98.4
W.11	101	98.8
W.12	103	98.6
W.13	97	100.0
W.14	96	100.6
W.15	99	100.0

Based on the available data, the following forecasting accuracy indicators are calculated:

$$MAD = \frac{\sum_{i=1}^{n} |actual_i - forecast_i|}{n}$$

$$MAD = \frac{0.2 + 6.2 + 6.6 + 1.2 + 4.6 + 2.2 + 4.4 + 3 + 4.6 + 1}{10} = 3.4$$

$$MSE = \frac{\sum_{i=1}^{n} (actual_i - forecast_i)^2}{n}$$

$$MSE = \frac{0.04 + 38.44 + 43.56 + 1.44 + 21.16 + 4.84 + 19.36 + 9}{10} = 16$$

$$MPE = 100\% \times \frac{\sum_{i=1}^{n} [(actual_i - forecast_i)/actual_i]}{n}$$

$$MPE = 100\% \times \frac{0.00 - 0.06 - 0.07 + 0.01 + 0.04 + 0.02 + 0.04 - 0.03 - 0.05 - 0.01}{10} = -1.0\%$$

$$MAPE = 100\% \times \frac{0.00 + 0.06 + 0.07 + 0.01 + 0.04 + 0.02 + 0.04 + 0.03 + 0.05 + 0.01}{10} = 3.5\%$$

7.6 The choice of forecasting technique

The choice of which forecasting model to adopt is based, first of all, on the observation of the demand for finished products. The use of quantitative models becomes possible in the event that demand can be forecasted on a statistical basis, while it would make no sense in a situation of an entirely random performance.

The degree of forecastability can be determined by calculating the "**coefficient of variation**," equal to the standard deviation of the forecasting error (σ) divided by the simple average of the demand ($\bar{d}$):

$$CV = \frac{\sigma}{\bar{d}} = \frac{\sqrt{\frac{\sum_{i=1}^{n} (f_i - d_i)^2}{n-1}}}{\frac{\sum_{i=1}^{n} d_i}{n}} \qquad [7.16]$$

where:
f_i represents the value of the demand forecast for the period i
d_i represents the value actually measured in the period i
n is the number of periods in which the sales (or consumption) data is gathered

The smaller the ratio, the more statistically forecastable demand is, and the more accurate the forecasts. In general, the degree of forecastability of demand depends on a series of circumstances, including:

- **the size of the volumes**: according to the law of large numbers, the accuracy of forecasts tends to improve with the increase of sales volumes, due to the reduction of the phenomena of unexplained variance. For example, if we estimate a demand of 5,000 pieces and the actual figure is 4,500 sales, the absolute forecasting error is 500 units, with an accuracy of the forecast of nearly 90%. On the other hand, if the forecast is 10,000 and the sales are 9,500, the absolute forecasting error is still 500 pieces, but the accuracy rises to nearly 95%;
- **the level of aggregation of the data**: the degree of accuracy of forecasts increases with the growth of the level of aggregation of the data. For example, the estimate of demand for a certain product, regarding a specific geographic area, will tend to be less precise than the estimate made in relation to a broader area. Likewise, the forecasts relating to a single product will be less accurate than those concerning an entire product family. The analysis of data with a higher level of aggregation thus allows for better recognition of trends and seasonality;
- **the stability of demand**: if demand is stable in time, sales forecasts will be accurate. This condition occurs especially with essential goods, such as milk or bread, whose consumption has little elasticity with respect to the unit price or income of individuals. In these conditions, the sales forecast is simpler and less subject to significant errors. This does not happen with luxury goods, whose purchase is generally secondary with respect to the satisfaction of the need for essential goods, and is higher during periods of economic growth. This greater variability of sales implies greater risk of forecasting error;
- **intensity of competition**: when the dynamics of competition in a sector are very intense, reaching reliable forecasts is more complex. This is due to the fact that the actions carried out by competitors to orient customer decisions in their favor are numerous and potentially effective. This increases the variability of demand, that is per se a source of less accuracy. In addition, with the increase of circumstances aimed at influencing the demand for a specific company's products, the complexity of the forecasting process increases, along with the risk of low accuracy;
- **difficulty in identifying cause-effect relationships**: for some products, especially those introduced to the market recently, it is difficult to have sufficiently long time series to conduct statistical analysis. Moreover, the factors that determine the level of demand may be numerous and difficult to identify. In those cases, the company takes the risk of developing unreliable forecasting models,

because they are based on incorrect hypotheses regarding causal relationships. This risk is higher in the presence of high correlation indexes, whose level and direction, however, do not always correspond to an actual relationship of dependence between the quantities observed. For example, it may happen that a statistical correlation is linked to mere random factors and coincidences, and that the two numbers are actually linked to a third phenomenon not included in the model. In these cases, it is possible that good levels of accuracy are seen in the short-term, but that the model's weakness will subsequently become apparent. In these situations, it is necessary to make use of the knowledge and skills of experts in the sector, who can contribute to the development of the forecasting model by identifying the determinants of the sales on a logical level, even before using statistics.

Due to the numerous forecasting techniques available and the considerable number of factors that can affect the forecastability of demand for a product, it is necessary for companies to understand which approach is most appropriate in each situation. For this purpose, as illustrated in **Figure 7.4**, it is useful to classify the products offered by a company based on two dimensions:

1. their **value for the company**, understood as the relevance that the specific product has for the company's financial performance, above all prospectively; to evaluate this dimension, it is possible to concentrate on the sales growth rate;
2. the **degree of forecastability of sales**, understood as the company's ability to formulate reliable sales forecasts.

Based on this classification, the following four types of products can be identified.

Figure 7.4 Portfolio matrix for choice of forecasting model

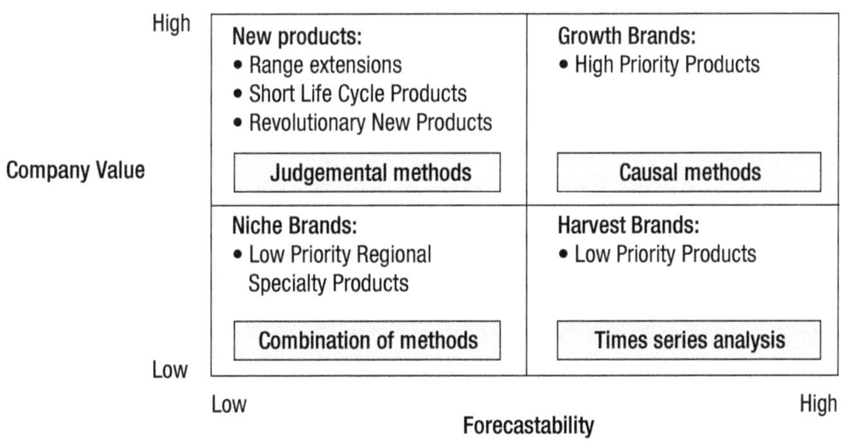

Source: adapted from Chase, 2016.

1. **New products**: these are in turn divided into three sub-categories: *range extensions, short life cycle articles,* and *revolutionary new products.* Range extensions consist of the use of an already consolidated brand for the launch of an article with different characteristics than existing articles, or belonging to other product categories. In this case, despite sales data not being available, it is possible to identify similar products with a consolidated sales history, allowing to develop forecasts by analogy. The intervention of a brand or product expert is necessary here to identify the article that most represents an adequate benchmark. Short life cycle articles are very common in some sectors, including clothing and consumer electronics, and are characterized by remaining on the market for an average of 6 months, and in any case for no more than one year. In this case as well, it is possible to formulate sales forecasts by analogy, identifying similar products launched in the past. However, unlike range extensions, the time series available relating to the benchmark article are by definition short and unlikely to help identify stable and reliable patterns. Therefore, even more than in the previous case, the role of experts in the sector or on the product is important to reach a reasonable forecast. Lastly, radically new products, often developed thanks to "breakthrough" technological innovations, pose greater challenges for forecasting, since they are able to profoundly transform the market and the competitive context. Sales data does not exist to support the forecasting process, nor are similar articles available with which to work by analogy. Therefore, the only techniques possible are those aimed at leveraging the knowledge and skills of experts within the company, and if necessary, of outside specialists in the relevant technology or sector, able to formulate long-term forecasts. In this case, all of the qualitative techniques already illustrated can be used.
2. **Niche brands**: these brands are generally developed to satisfy the specific needs of limited groups of customers, generally belonging to specific geographic areas, or identifiable based on variables such as age, ethnicity, or social status. They are often products with relatively long time series, but whose sales values remain at rather low levels. This factor, as already illustrated, can influence the effectiveness of quantitative demand forecasting techniques. This is further burdened by the fact that the moderate turnover generated with these products does not justify an investment by the company to gather detailed data, that can support the forecasting process. To overcome this limit, it is suggested starting from the time series to then improve the forecast through an intervention by market experts, able to recognize important dynamics and phenomena in the context, that can influence future sales. Concerning the operational method that allows for combining quantitative techniques based on statistics with the evaluation of experts (such as the sales force, for example), consider **Box 7.4**.
3. **Growth brands**: these are brands and articles developed to generate most of the company's sales, which have high rates of growth. In these cases, statistical data is generally available, but is in part contaminated by the numerous initiatives aimed at stimulating sales, such as promotions, merchandising

operations, and incentives for the sales force. Here, although patterns can be identified regarding the demand, it is also necessary to understand the impact these initiatives can have on sales. Causal methods appear particularly useful for this purpose.

4. **Harvest brands**: these are products with a long sales history and consolidated patterns of demand, that allow for an effective application of demand forecasting techniques linked to the analysis of time series. Although the sale of those articles can be very profitable, significant growth rates of sales are not expected, as happens instead with growth brands. One implication is a lesser intensity of promotional efforts, that makes the impact of these initiatives on sales levels less important in the medium to long-term. This justifies the fact that causal techniques are not used for harvest brands, as instead happens for growth brands.

Box 7.4 Combination of methods: an example

The Alpha company has two product lines, A and B, each divided into two Stock Keeping Units (A1, A2, B1, B2), and operates in a business characterized by a growth trend. However, the products in the portfolio have different market dynamics. Three are in a phase of rapid growth, induced in part by promotional initiatives aimed towards specific trade clients. One, on the other hand, is at the end of its life cycle and will soon be eliminated from the product range. To take these circumstances into account, the Demand Planner has decided to organize the forecasting process as illustrated in Figure 7.5.

Figure 7.5 The forecasting process at Alpha

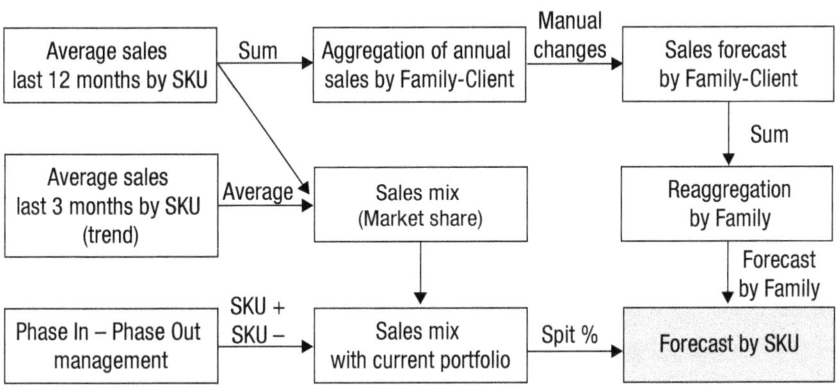

The Demand Planner starts from the analysis of the sales data for the last year and that for the last quarter. The latter is used to formulate forecasts based on the hypothesis that in the coming year, the quantity of sales will be 4 times the sales in the last quarter. These two elements of information are used by the Demand Planner to reach an estimate of the demand given by the arithmetic mean of the two input data, as seen in Figure 7.6. Furthermore, knowing that product A2 is in phase-out, it is possible to update the forecast, bringing the estimate of sales of this article to zero. The key information that this procedure makes it possible to obtain is a reasonable forecast of how sales will be divided within the two product families.

To estimate sales in terms of the number of pieces, it is appropriate to make use of the experience and information of the sales force, which being aware of the dynamics in the sector and the specific promotional initiatives aimed at three of the company's clients (Jay, Key, and Mae), represents a crucial point of reference for the forecasting process. The sales force thus reaggregates the sales data for the product families for each single client, and then produces an estimate of future demand. This estimate is subsequently reaggregated by product family. (See **Figure 7.7**)
At this point the Demand Planner, based on the percentage division among SKUs obtained previously and the subjective estimates formulated by the sales force, determines the forecasts for each single SKU, as shown in **Figure 7.8**.

Figure 7.6 Computation of the sales mix

Product Family	SKU	Pieces sold in the previous year
A	A1	90
	A2	10
B	B1	50
	B2	50
TOTAL		**200**

Product Family	SKU	Forecasted annual sales estimated on the basis of the last quarter
A	A1	100
	A2	0
B	B1	60
	B2	70
TOTAL		**230**

Product Family	SKU	Simple average	Sales mix
A	A1	95	95%
	A2	5	5%
B	B1	55	48%
	B2	60	52%
TOTAL		**215**	

A2: phase out

Product Family	SKU	Simple average (new product range)	Sales mix
A	A1	95	100%
	A2	0	0%
B	B1	55	48%
	B2	60	52%
TOTAL		**210**	

Figure 7.7 Computation of annual forecast for each product family

Product Family	SKU	Pieces sold in the previous year
A	A1	90
	A2	10
B	B1	50
	B2	50
TOTAL		**200**

Client	Product Family	Pieces sold in the previous year
Jay	A	30
	B	20
Key	A	40
	B	30
Mae	A	30
	B	50
TOTAL		**200**

Client	Product Family	Judgemental forecast for the next year (pieces)
Jay	A	30
	B	30
Key	A	40
	B	40
Mae	A	40
	B	60
TOTAL		**240**

Product Family	Judgemental forecast for the next year (pieces)
A	110
B	130
TOTAL	**240**

Figure 7.8 Computation of annual forecast for each SKU

Product Family	SKU	Simple average (new product range)	Sales mix
A	A1	95	100%
	A2	0	0%
B	B1	55	48%
	B2	60	52%
TOTAL		**210**	

Product Family	Judgemental forecast for the next year (pieces)
A	110
B	130
TOTAL	**240**

Product Family	SKU	Forecasted sales for the next year
A	A1	110
	A2	0
B	B1	62
	B2	68
TOTAL		**240**

8 Production Plan: Sales and Operations Planning

by *Alberto Grando*

8.1 The goal of Sales and Operations Planning

The expression **Sales and Operations Planning – S&OP**, in the past more frequently called **Aggregate Planning**, was born in the context of information systems and expressed well the **goal of balancing supply and demand** of planned volumes: the Required Production Capacity to satisfy sales forecast, and the Available Production Capacity of the company. They can refer to finished products (in the case of MTS production) or assemble-modules and components (in the case of ATO and MTO production). In the mid to long-term horizon, companies must in fact decide how to set the necessary resources (machines, people, external capacity, etc.) to effectively and efficiently respond to sales forecasts, appropriately sizing the production capacity needed to satisfy requirements of finished products (MTS) or of articles placed in decoupling points located further upstream (ATO and MTO). The terms effective and efficient imply that the final goal of this phase is not the involvement only of production, but also of other functions: sales, logistics, administrative and financial functions, and the units responsible for new product development.

The S&OP phase thus draws on demand planning and takes into account the system of constraints linked to critical capacity resources, and the company's economic-financial goals. The S&OP takes shape generating possible medium to long-term scenarios, based on **what-if simulations aimed at optimizing feasibility** (production capacity) **and economic performance** (typically the minimization of costs and investments or the maximization of the service level guaranteed by inventory goals). Therefore, the S&OP plays the role of interface between:

- sales, generally organized by product lines, channels, geographic markets, or other aggregates;
- and production capacity, defined by production units dedicates to the production of homogeneous product families from the standpoint of the morphology and production cycle.

So the aim is to define the appropriate metrics to allow for a "dialogue" between sales and production functions.

As we have stated, the demand forecasting (Chapter 7) represents one of the critical inputs for the Sales and Operations Planning phase. The requests resulting from the demand forecast and from orders book, netted based on available stock, are transformed into production requirements and Required Production Capacity, which means into the translation of the market requests in terms of Production Capacity units. Making this conversion means reading the market language, that is typically expressed by revenues or volumes of product lines, with the language most useful for Production. In other words, it means transforming the requests, expressed in units of output, into production capacity needs, in the units most suited to the technical/operational characteristics of the plant or the department, that are generally hours or physical units (pieces, tons, liters, cubic meters, etc.). In this sense, we refer to a defined production unit (production system or single machine) or a critical resource that constrains the others, with the goal of reconciling the needs expressed by the sales budget with those of production. If, for example, the same plant is used to process different types of products which have different throughput rates, it is necessary to define and use **equivalent units**. Consider the typical case of a shoe factory that assembles boots, shoes and sandals on the same conveyor belt, absorbing different production times (Cycle Times or Throughput Times). If we want to determine a first rough measure of the Required Production Capacity, expressed in pieces, it is necessary to compare the quantities expressed in product mixes through the construction and use of appropriate equivalency coefficients. These can be built, for example, by comparing the standard times associated to each different types of product, with the standard time of the real or virtual article, taken as a specific unit of measure, as exemplified in **Box 8.1**.

Box 8.1 Capacity analysis and equivalent units

Imagine a machine that produces a composite mix of 8 products, for which the sales functions have drawn up sales forecasts for the next semester, expressed in volumes, and shown in the following table.

1 Part Number	2 Sales Forecast (units/sem)	3 Standard Time (std h/unit)	4 = 2 * 3 RPCh (std h/sem)	5 Equiv unit Coeff. (T_i/T_A)	6 = 2 * 5 RPC_A (A Equiv Units/sem)
A	400	0.4	160	1	400
B	350	0.6	210	1.5	525
C	175	0.4	70	1	175
D	250	0.8	200	2	500
E	150	1.1	165	2.75	412.5
F	200	1.6	320	4	800
G	125	0.6	75	1.5	187.5
H	300	0.8	240	2	600
TOTAL			1440		3600

Knowing the standard times, and for reasons of simplicity not considering the set-up times, it is possible to determine the Required Production Capacity (RPC_h), equal to 1,440 standard hours. As illustrated by the table, it is likewise possible to calculate the Required Production Capacity (RPC_{eu}) in equivalent units. For that purpose, it is necessary to:

- choose the product to be used as a unit of reference. Although the calculation can be carried out indifferently choosing any product, in general the choice is a product that best represents the company's offering, or the product that is best-known or sold the most, or that has been produced for the longest time;
- calculate the equivalency indices, as a ratio between the standard times of each single product (ST_i) and the standard time of the reference product (ST_x), such that: Equivalent Index = ST_i / ST_x;
- determine the RPC_{eu}, as a product of the expected sales volumes and the respective equivalency index, i.e. equal to 3,600 units, expressed in equivalent units of product A.

Knowing also that:

- The Plant Calendar Time expected in the semester is equal to 120 days;
- It is planned that work will take place in 2 shifts of 7.5 hours each;
- The monitoring of the inactivity of the machines shows losses of:
  - 18 hours due to Lack of Materials Time (LMT)
  - 15 hours due to Maintenance Down Time (MDL)
  - 12 hours due to Failure Down Time (FDL)

it is possible to calculate, respectively, the Theoretical Production Capacity (TPC), the Available Production Capacity (APC), and a rough Utilization index (U), that can be expressed indifferently in hours or equivalent units. In fact:

- TPC = 120 days/semester x 2 shifts/day × 7.5 hours/shift = 1,800 hours/semester;
- APC = TPC – ∑losses = 1,800 – 45 = 1,755 hours/semester
- $U_h = RPC_h / APC_h = 1{,}440 / 1{,}755 = 82\%$
- $U_{eu} = RPC_{eu} / APC_{eu} = 3{,}600 / (1{,}755 / 0.4) = 82\%$.

The calculation shown here is an initial rough measurement of production capacity, and does not take into account other parameters such as Efficiency and Quality, whose utilization will be illustrated below.
In the event *labor intensive production* that results in very different types of products – this happens with many light assemblies – it can be convenient to use the hours of direct labor available, as a parameter for measurement of production capacity. Likewise, machine hours are used in *capital intensive companies*; in other situations, physical or conventional measurements of both input and output are used (pieces, tons, liters, barrels, etc.).

- *Input measures* are used more frequently in cases where a single input material corresponds to multiple output products, as in the case of cracking processes in the oil industry.
- *Output measures*, on the other hand, are suggested in situations where multiple factors contribute to the composition of the product, such as in the automobile sector.

Those measurements, whatever they are, must in any event be parameterized to the unit of time; that way, measures of *Throughput Rate* (a potential flow measure) can be defined that, if they refer to set time periods (shift, week, month, etc.), are commonly included in the notion of production capacity.

8.2 Sizing the Available Production Capacity

Previously (Chapter 5), we already introduced the issue of **Production Throughput Rate** (or Pace) and **Throughput Time** (or Unit Cycle Time), whether standard or effective, referring to a generic product x and a generic interval of time i, according to the ratio [5.3]:

$$TTx\ (time/unit) = 1/TRx_i\ (units/time)$$

If we abandon the simplifying hypothesis of the generic product x and we refer to a set of products (n), produced based on an assigned mix on the same machinery or department, we can calculate the **Mix Throughput Rate ($\mathbf{TR_{mix}}$)** as:

$$TR_{mix} = \frac{\sum_{in}(CQ_{in} + DQ_{in})}{\sum_{in}(PTcq_{in} + PTdq_{in}) + TST}\ (units/time) \qquad [8.1]$$

where, using the notations already illustrated:

CQ_{in} = quantity of good (or conforming) products;
DQ_{in} = quantity of reject (or defective) products;
$PTcq_{in}$ = production time used to produce quantity CQ;
$PTdq_{in}$ = production time used to produce quantity DQ;
TST = total actual setup time.

Therefore, $\sum_{in}(CQ_{in} + DQ_{in})$ expresses the number of units (or quantities) belonging to a family of products, whether good or defective, produced in a set of periods *i* in which the plant was in operation. And $\sum_{in}(PTcq_{in} + PTdq_{in}) + TST$ expresses the time dedicated for the production of the same set of products (good or defective) summed with the total actual set-up time, in multiproduct cases, for the retooling of the machine in the alternation between different products. In other words, the final calculation of the TR_{mix} is equal to:

$$TR_{mix} = \frac{total\ quantity\ produced\ relating\ to\ an\ assigned\ mix}{hours\ used\ to\ produce\ the\ assigned\ mix}$$

By substituting the actual measures of the Throughput Rate and the Unit Cycle Time with the related **standard values**, we can obtain estimated values. For example, if $\sum_{in}(CQ_{in} + DQ_{in})$ is equal to $\sum_{in}\left[\overline{TR}_{in}(PTcq_{in} + PTdq_{in})\right]$ and $\sum_{in}(PTcq_{in} + PTdq_{in})$ is equal to $\sum_{i}\left[(CQ_{in} + DQ_{in})/\overline{TR}_{in}\right]$, it is possible to estimate the standard $\overline{TR}_{mix}$, based on an assigned mix, obtaining the following expression:

$$\overline{TR}_{mix} = \frac{\sum_{in}(CQ_{in} + DQ_{in})}{\sum_{in}\left[(CQ_{in} + DQ_{in})/\overline{TR}_{in}\right] + \overline{TST}}\ (units/time) \qquad [8.2]$$

knowing the quantity to produce, the average incidence of defective volumes and the articles standard Mix Throughput Rate, transformable into

$$\overline{TR}_{mix} = \frac{\sum_{in}\left[\overline{TR}_{in}(PTcq_{in} + PTdq_{in})\right]}{\sum_{in}(\overline{PT}cq_{in} + \overline{PT}dq_{in}) + \overline{TST}}\ (units/time) \qquad [8.3]$$

where $\overline{TST}$ represents the total Standard Setup time.

Table 8.1 **Example of calculation of standard $\overline{TR}_{mix}$**

Product 1	Product 2
CQ_1 = 120,000 (units)	CQ_2 = 100,000 (units)
DQ_1 = 4,000 (units)	DQ_2 = 1,500 (units)
$\overline{TR}_1$ = 800 (units/hour)	$\overline{TR}_2$ = 500 (units/hour)
$\overline{PT}cq_1$= 150 (hours)	$\overline{PT}cq_2$ = 200 (hours)
$\overline{PT}dq_1$= 5 (hours)	$\overline{PT}dq_2$ = 3 (hours)
$\overline{TST}$ = 20 (hours)	

Table 8.1 shows an example of calculation of the Standard Mix Throughput Rate ($\overline{TR}_{mix}$) relating to a plant that produces two products, where all of the times are expressed in standard values.

Applying [8.2]

$$\overline{TR}_{mix} = \frac{120{,}000 + 4{,}000 + 100{,}000 + 1{,}500}{\frac{120{,}000 + 4{,}000}{800} + \frac{100{,}000 + 1{,}500}{500} + 20} = 596.56\ (units/hour)$$

or, being expressed in standard measures [8.3]:

$$\overline{TR}_{mix} = \frac{800 \cdot (150 + 5) + 500 \cdot (200 + 3)}{150 + 5 + 200 + 3 + 20} = 596.56\ (units/hour)$$

It is evident that in this case, the Standard Mix Throughput Rate is expressed in units of mix (with a given, constant mix). If, on the other hand, we wish to express the throughput in specific units of product, it becomes necessary to perform a calculation in equivalent (or comparable) units, as illustrated below (in equivalent units of product 1).

$$\overline{TR}_{mix} = \frac{120{,}000 + 4{,}000 + (100{,}000 + 1{,}500) \times 800/500}{\frac{120{,}000 + 4{,}000}{800} + \frac{100{,}000 + 1{,}500}{500} + 20} = 757.67\ (equivalent\ units\ 1/hour)$$

or:

$$\overline{TR}_{mix} = \frac{800 \cdot (150 + 5) \times 500 \cdot [(200 + 3) \times 800/500]}{150 + 5 + 200 + 3 + 20} = 757.67\ (equiv.\ units\ 1/hour)$$

From an operational standpoint, what most concerns production management is the **determination of producible quantities** of defined product mixes, on specific machineries or departments, **in a certain period of time**. To obtain the calculation of expected production volumes in a given time (or the time to be used for an assigned production volume), it is necessary to refer to the concepts of **Plant Opening Time** and **Possible Status of plant**, already illustrated in Chapter 5, and to introduce the indicator of ***Availability***.

With reference to a generic plant, thus given T and the different times connected to its status, it is possible to calculate its expected **Theoretical Throughput Rate (TTR)** as the product of the **Nominal Throughput Rate (NTR)** (defined elsewhere as peak), for the **Availability coefficient (*A*)**, representing the probability that a resource is able to operate at the time it is called on to do so.

$$TTR = NTR \times A \; (units/time) \qquad [8.4]$$

where:

$$A = \frac{T - LDT - LMT - LWT - OTL - FDL - MDL}{T - LDT - LMT - LWT - OTL} \qquad [8.5]$$

Deducting from the numerator and the denominator the stoppages for causes not due to the machine, the Availability coefficient is equal to the ratio of the **Actual Operating Time** (or the theoretical time of utilization net of the unavailability due to breakdowns and maintenance downtimes), with respect to the **Theoretical Utilization Time.**[1] In determining the Availability index, we deduct the times of non-functioning, LDT, LMT, LWT and OTL (lack of demand, materials, workers time and organizational time losses – see Chapter 5), with the proper consideration that the incidence of breakdowns and stoppages must be compared to the time the plant is operational, or the probability that a machine will break down is proportionate to the time for which it operated.

Knowing the Standard Mix Throughput Rate ($\overline{TR}_{mix}$) and the Availability coefficient A, it is possible to estimate for a time T the expected throughput of the period, or the **Standard Available Production Capacity**:

$$\overline{APC} = \overline{TR}_{mix} \times A \times T' \; (units) \qquad [8.6]$$

where $T' = T - TTT - LWT - OTL$, since, in the case of estimation of production capacity, we consider the same incidence of technical tests, strikes, and organizational breaks deduced from the past, while the elements LDT and LMT remain imponderable, and not desired during predictions.

Here we see the distinction between the concepts of Throughput Rate and Production Capacity. The first refers to a **measure of flow**, expressed in units/time, with short time values, such as pieces/minute or tons/hour, while the second is a **measure of volume** (capacity), expressed in units, and equal to the application of

---

[1] From a more strictly technical standpoint, in maintenance practice the measure of availability of a machine is defined as A = MTTF/ MTTF + MTTR, where the notations MTTF and MTTR represent the temporal magnitudes *Mean Time To Failure* and *Mean Time To Repair*, respectively. On this subject, see Levitt (2011). If MTTR is considered analogous to the times FDL and MDL, the technical availability index is entirely comparable to that illustrated in [8.5].

the measure of Throughput Rate for a fixed temporal interval (T) such as week, month, year, etc.

It should be pointed out that in multiproduct situations, the measures of Utilization and Efficiency, that in the case of monoproduct production (Chapter 5) are illustrated with reference to an Actual or Standard Throughput Rate (TRx_i) is replaced with Mix Throughput Rate (TR_{mix}).

In fact, if we multiply the Utilization coefficient calculated in [5.4] by the $\overline{TR}_{mix}$ we obtain:

$$U = \frac{\sum_{in}\left[(PTcq_{in} + PTdq_{in}) + TST\right] \times \overline{TR}_{mix}}{T \times \overline{TR}_{-mix}} \quad [8.7]$$

or, the *standard production volumes obtainable based on the hours actually worked*, compared to the *theoretical production volumes obtainable in time T*, or in other words, the ratio between Used Production Capacity – UPC and Theoretical Production Capacity – TPC (gross of any potential unavailability).[2] On this point, see the example in **Table 8.2**.

In that case, the times used for conformant and defective production, and the set-up time, differ from the times measured in the previous example, relating to the throughput mix, since they are actual, not standard.

Table 8.2 Example of calculation of utilization and yield indexes in a multiproduct case

Product 1		Product 2
CQ_1 = 120,000 (units)		CQ_2 = 100,000 (units)
DQ_1 = 4,000 (units)		DQ_2 = 1,500 (units)
TR_1 = 750 (units/hour)		TR_2 = 400 (units/hour)
$PTcq_1$ = 160 (hours)		$PTcq_2$ = 250 (hours)
$PTdq_1$ = 5.3 (hours)		$PTdq_2$ = 4.7 (hours)
TTT = 8 (hours)		TTT = 12 (hours)
	LWT = 2 (hours)	
	LMT = 4 (hours)	
	FDL = 10 (hours)	
	MDL = 12 (hours)	
	TST = 22 (hours)	
	T = 490 (hours)	

[2] If, on the other hand, the Availability coefficient A were introduced in the denominator of equation [8.5], and time T were substituted with time T′, we would obtain a Utilization index related to the Available Production Capacity – APC, where the latter is equal to $\overline{TR}_{mix} \times A \times T'$ [8.6].

Therefore, applying equation [5.1]:

$U = (490 - 48) / 490 = 0.90$

or applying equation [8.7], with Standard Throughput Rate in equivalent units of product 1, calculated in [8.2].

$$U = \frac{(160 + 250 + 5.3 + 4.7 + 22) \times 757.67}{490 \times 757.67} = 0.90$$

Similarly, the calculation of the **Efficiency**, in the case of multiple products, is equal to the actual production compared to the standard production producible in the actual hours, or the actual production measured in standard hours compared to the hours actually utilized. Returning to the equations illustrated in Chapter 5, and substituting Mix Throughput Rate ($\overline{TR}_{mix}$) for Throughput Rate ($\overline{TR}_x$) we obtain the following:

$$\eta = \frac{actual\ production}{std\ prod\ based\ on\ act\ h} = \frac{\sum_{in}(CQ_{in} + DQ_{in})}{\left[\sum_{in}(PTcq_{in} + PTdq_{in}) + TST\right] \times \overline{TR}_{mix}} =$$

$$\frac{\sum_{in}(CQ_{in} + DQ_{in})}{\left[\sum_{in}(PTcq_{in} + PTdq_{in}) + TST\right] \times \dfrac{\sum_{in}(CQ + DQ)}{\sum_{in}\left[(CQ_{in} + DQ_{in})/\overline{TR}_{in}\right] + \overline{TST}}} \quad [8.8]$$

$$\text{that is } \eta = \frac{act\ prod\ valued\ in\ std\ h}{actual\ production\ hours} = \frac{\sum_{in}\left[(CQ_{in} + DQ_{in})/\overline{TR}_{in}\right] + \overline{TST}}{\sum_{in}(PTcq_{in} + PTdq_{in}) + TST} \quad [8.9]$$

$$\text{or also } \eta = \frac{standard\ hours\ produced}{actual\ production\ hours} = \frac{\sum_{in}(\overline{PT}cq_{in} + \overline{PT}dq_{in}) + \overline{TST}}{\sum_{in}(PTcq_{in} + PTdq_{in}) + TST} \quad [8.10]$$

where the elements in the numerator (marked) assume standard values.

Using the data shown in **Tables 8.1** and **8.2**, we thus obtain:

$$\eta = \frac{120{,}000 + 4{,}000 + (100{,}000 + 1{,}500) \times 800/500}{(160 + 5.3 + 250 + 4.7 + 22) \times 757.67} = 0.8552, \text{ applying [8.8]}$$

if expressed in equivalent units 1, or also,

$$\eta = \frac{(120{,}000 + 4{,}000)/800 + (100{,}000 + 1{,}500)/500 + 20}{160 + 5.3 + 250 + 4.7 + 22} = 0.8552, \text{ applying [8.9]}$$

$$\text{or finally, } \eta = \frac{150 + 5 + 200 + 3 + 20}{160 + 5.3 + 250 + 4{,}7 + 22} = 0.8552, \text{ applying [8.10].}$$

We previously [8.6] defined Available Production Capacity as $APC = \overline{TR}_{mix} \times A \times T'$, that is, expressing it in terms of expected standard production, based on estimated available time ($T' \times A$). It is now possible to add a second measure of expected production in an interval of time T that takes into account the historically measured productivity of the plant. By using the coefficients of Utilization and Efficiency, i.e. by introducing the relative deviation between actual and standard production into the calculation, and possibly a Quality index, we obtain a further estimate of the expected **Available Production Capacity**. This represents the expected production volumes in the time interval T, presumably close to the future actual production, because it is adjusted with values of Utilization, Efficiency and Conformity (Quality index) taken from the past, in this case considered a valid approximation of future trends.

$$ADC' = \overline{TR}_{mix} \times T \times U \times \eta \times Q \; (units) \qquad [8.11]$$

8.3 The Available Production Capacity from the standpoint of Total Productive Maintenance

It is possible to obtain similar results by adopting the approach known as **OEE – Overall Equipment Effectiveness**, used by companies that operate following the principles of **World Class Manufacturing** and **Lean Management** (Chapter 17), starting with the work of Nakajima (1988) on *TPM – Total Productive Maintenance*. Returning to the diagram of plant status presented in Chapter 5, we can highlight the method of calculation of the indicators developed in the context of TPM, and the points of contact with what we have illustrated above,[3] using the notations already introduced.

Consider **Figure 8.1** that again presents graph 5.2, showing the status of a machine and the causes of losses, reviewed from the OEE perspective.

As already mentioned (Chapter 5), with reference to a generic resource, by deducting **Unscheduled Time** (because of holidays, vacations, reduced hours, etc.) from **Solar Time (ST)**, we obtain **Plant Calendar Time (T)**, also called **Gross Available Time.**[4] Subtracting from that time the **Planned Shutdowns** for preventive maintenance (MDL'), workers training plans (LWT'), technical test (TTT) and organizational interventions (OTL) planned for the observed time horizon, such as stoppages for relayout or scheduled adjustments of the plants, scheduled labor union meetings, etc., we obtain the **Net (or Planned) Available Time**. As already described, that time contains a series of non productive times linked to the loss causes just cited, such as setup time (TST), machines breakdown (FDL), unscheduled maintenance (MDL), lack of workers for strikes (LWT), lack of materials (LMT) or lack of orders (LDT), the latter only if not predictable, to the contrary of trends linked to seasonal cycles. By deducting those times, defined as **Unplanned Downtime** (or

[3] For a detailed examination of the differences between methodologies, see Grando and Turco (2005).

[4] See De Toni and Panizzolo (2018).

Figure 8.1 Plant status and causes of losses from OEE perspective

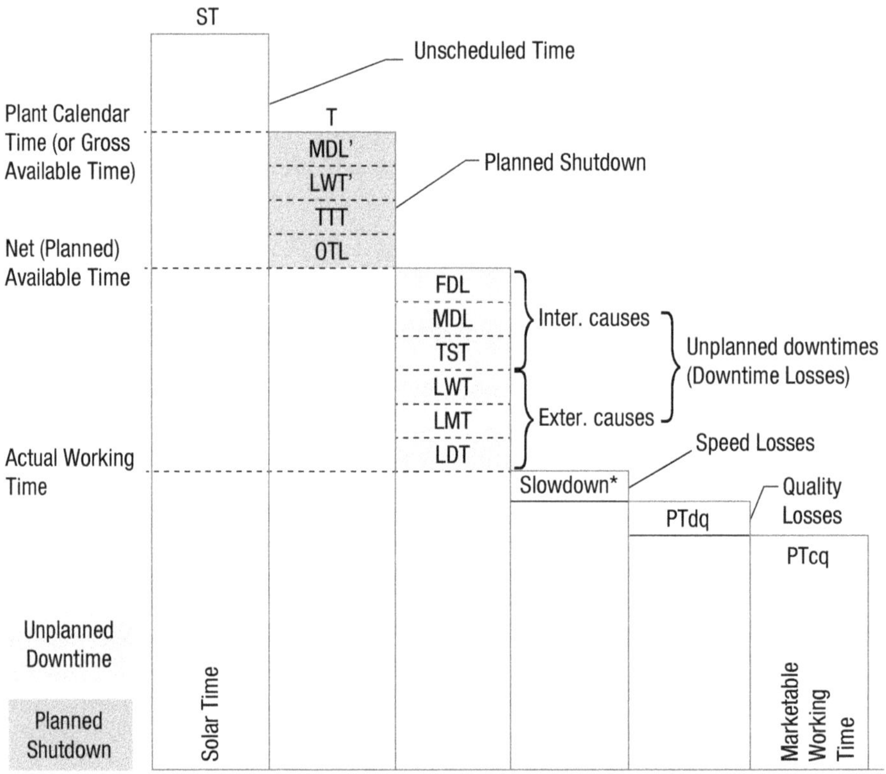

* Speed losses, here represented in terms of working time reduction.

Downtime Losses), we obtain **Actual Working Time** ($PTcq_i + PTdq_i$ + Slowdown). The main difference with respect to the convention previously illustrated lies in the fact that according to the TPM approach, which is currently widespread in practice, the set-up times (TST) represent one of the elements of waste to pursue, and as such to be considered as lost time to be minimized or eliminated. In the alternative approach, they are considered physiologic (and already optimized by engineering departments) in multiproduct production and thus included in Actual Operating Time. The treatment of time lost due to strikes (LWT) is more debatable. Since it regards labor not paid for, that time does not contribute to determining the amount of hours paid, but can, however, impact Net Available Time, and thus the Available Production Capacity depending on whether those stoppages are announced with suitable advance notice or are called without warning. To generalize, it is important to note that in different situations, the different causes adopted for the calculation of the indexes described here depend on the specific operational context observed and the conventions adopted, as well as their extent and the possibility of reliable measurements and metrics.

Table 8.3 **Analysis of times and calculation of indexes**

Index	Calculation	It measures	Aim
Utilization (U)	$\frac{\text{Gross available time}}{\text{Solar time}}$	Unscheduled time	Measures the percentage of non-utilization of plants versus the solar time, useful in capital intensive processes
Shutdown (S)	$\frac{\text{Net available time}}{\text{Gross available time}}$	Planned shutdown	Measures scheduled shutdown due to planned maintenance or other planned stoppages
Availability (A)	$\frac{\text{Actual working time}}{\text{Net available time}}$	Downtime losses	Measures unplanned stoppages (time losses) such as setup time (TST), machines breakdown (FDL), unscheduled maintenance (MDL), lack of materials (LMT), etc.
Efficiency (or yield) (η)	$\frac{\text{Volumes produced expressed in standard times}}{\text{Actual working time}}$	Speed losses	Measures losses of efficiency or yield
Quality (Q)	$\frac{\text{Conforming volumes expressed in standard times}}{\text{Volumes produced expressed in standard times}}$	Quality losses	Measures losses due to non-conformities

Source: adapted from De Toni and Panizzolo, 2018.

In the calculation of OEE and the other related indicators, efficiency losses due to slowdowns of machines or labor with respect to standard times, previously measured with the Efficiency (or Yield) indicator, are entered as lost time and defined as **Speed Losses**. Those losses can also significantly affect the performance of a production system, especially if the dominant resource is labor that, by its nature, is more subject to deviations from standard values, justified by differences in abilities, motivation, and other natural causes like seniority, or environmental operations conditions (light, fatigue, etc.). Lastly, as already observed, the product output can have non-conformities with respect to quality specifications, previously measured by the indicator Q, defined here as **Quality Losses.**[5]

Table 8.3 shows the adimensional indexes obtainable from the relations between the times shown above.

The indexes described above can be used to calculate the main indicators used to monitor the production performances: P – Productivity, OEE – Overall Equipment Effectiveness, LF – Load Factor and TEEP – Total Effective Equipment Performance.

[5] According to the OEE approach, there are six principal causes of losses, grouped into the three families. In particular, downtimes losses include the losses due to set-up & adjustment and breakdown; speed losses include those due to small stops and reduced speeds; quality losses include those due to startup rejects and production rejects.

In particular, using the previous notations:

- ***Productivity*** **(P)** is equal to the product of the Availability and Efficiency (or Yield) indexes and measures Downtime and Speed Losses:[6]

$$P = A \times \eta = \frac{Actual\ working\ time}{Net\ available\ time} \times$$

$$\times \frac{Volumes\ produced\ expressed\ in\ standard\ times}{Actual\ working\ time}$$

$$P = \frac{T - LDT - LMT - LWT - OTL - FDL - MDL - TTT - TST}{T - MDL' - OTL - TTT} \times$$

$$\times \frac{\sum_{in} \left[(CQ_{in} + DQ_{in}) / \overline{TR}_{in} \right]}{T - LDT - LMT - LWT - OTL - FDL - MDL - TTT - TST} =$$

$$= \frac{\sum_{in} \left[(CQ_{in} + DQ_{in}) / \overline{TR}_{in} \right]}{T - MDL' - OTL - TTT} \text{ or } \frac{Volumes\ produced\ expressed\ in\ standard\ times}{Net\ available\ time} \quad [8.12]$$

- ***Overall Equipment Effectiveness*** **(OEE)**, an index usually applied in the Lean Manufacturing projects, is equal to the product of the Availability, Efficiency and Quality indexes, and measures Downtime, Speed and Quality Losses:

$$OEE = A \times \eta \times Q =$$

$$= \frac{Actual\ working\ time}{Net\ available\ time} \times \frac{Volumes\ produced\ expressed\ in\ standard\ times}{Actual\ working\ time} \times$$

$$\times \frac{Conforming\ volumes\ expressed\ in\ standard\ times}{Volumes\ produced\ expressed\ in\ standard\ times}$$

$$OEE = \frac{T - LDT - LMT - LWT - OTL - FDL - MDL - TTT - TST}{T - MDL' - OTL - TTT} \times$$

$$\times \frac{\sum_{i} \left[(CQ_{in} + DQ_{in}) / \overline{TR}_{in} \right]}{T - LDT - LMT - LWT - OTL - FDL - MDL - TTT - TST} \times$$

$$\times \frac{\sum_{in} \left[CQ_{in} / \overline{TR}_{in} \right]}{\sum_{in} \left[(CQ_{in} + DQ_{in}) / \overline{TR}_{in} \right]} = \frac{\sum_{in} \left[CQ_{in} / \overline{TR}_{in} \right]}{T - MDL' - OTL - TTT} \quad [8.13]$$

[6] There is an evident analogy with the conventions illustrated in Chapter 5. In particular, in the TPM logic, Productivity P is equal to the Productivity previously calculated as P = U × η, with the difference that Net Available Time, not Gross Available Time (Plant Calendar Time) is used as the denominator; the indicator OEE is equal to U × η × Q with the same notation regarding the denominator used; the indicator of Load Factor coincides with the productivity illustrated in Chapter 5, where P = U × η × Q, using Plant Calendar Time as the denominator, while the TEEP uses the same numerator referred to the entire Solar Time.

$$\text{or } \frac{\textit{Conforming volumes expressed in standard times}}{\textit{Net available time}}$$

- ***Load Factor* (LF)** is equal to the product of the OEE and Shutdown indexes, or the product of Availability, Efficiency, Quality and Shutdown, and measures the ratio between actual conforming production expressed in standard hours and gross available time:

$$LF = \mathrm{A} \times \eta \times \mathrm{Q} \times \mathrm{S} = \mathrm{OEE} \times \mathrm{S} = \frac{\textit{Conform. vols. in std t}}{\textit{Net avail. time}} \times \frac{\textit{Net avail. t}}{\textit{Gross avail. t}}$$

$$LF = \frac{\sum_{in}\left[(CQ_{in})/\overline{TR}_{in}\right]}{T - MDL' - OTL - TTT} \times \frac{T - MDL' - OTL - TTT}{T} =$$

$$= \frac{\sum_{in}\left[(CQ_{in})/\overline{TR}_{in}\right]}{T} \quad [8.14]$$

$$\text{or } \frac{\textit{Conforming volumes expressed in standard times}}{\textit{Gross available time}}$$

- ***Total Equipment Effective Performance* (TEEP)** is equal to the product of Load Factor and the Utilization of the resource, or the product of Availability, Efficiency, Quality, Shutdown and Utilization. It measures the ratio between conforming volumes expressed in standard times and Solar Time:

$$TEEP = \mathrm{A} \times \eta \times \mathrm{Q} \times \mathrm{S} \times \mathrm{U} = \mathrm{LF} \times \mathrm{U} =$$

$$= \frac{\textit{Conforming volumes expressed in standard times}}{\textit{Gross available time}} \times \frac{\textit{Gross available time}}{\textit{Solar time}}$$

$$TEEP = \frac{\sum_{in}\left[CQ_{in}/\overline{TR}_{in}\right]}{T} \times \frac{T}{ST} = \frac{\sum_{i}\left[CQ_{in}/\overline{TR}_{in}\right]}{ST} \quad [8.15]$$

$$\text{or } \frac{\textit{Conforming volumes expressed in standard times}}{\textit{Solar time}}$$

In light of the analysis conducted to this point, we should present some summary considerations.

- The indicators presented, precisely because they are the result of deployments, have high **diagnostic value**, since they allow for detecting potential losses of productivity and production capacity, attributing them to specific causes.
- In principle, all of the indicators presented **can be calculated for machines and labor**, with the exception of TEEP, that can be calculated only for the latters.
- The possibility to determine the indexes presented depends on the system of data collection from the field, such that each company will shape its own reporting system finding the proper balance between the costs linked to the implementation of a measurement system and the value provided by the indexes that can be calculated. The development and growing diffusion of **MES – Manufacturing Execution Systems** currently facilities the acquisition of data from the field, through

the automatic measurement of states of functioning of the machines and standard causes for stoppages, reducing their cost and increasing their reliability.

For the purpose of calculating **Available Production Capacity** (APC), the indicator of **Load Factor (LF)** becomes key. Starting from Gross Available Time – known because it has been calculated from Solar Time net of Unscheduled Time – and from the verification of reliability of the Load Factors of critical resources deducible from the past, it allows us to determine the available time for the production of conforming products.[7] In fact, the Load Factor being the product of OEE and Shutdown index, considers the three main losses (downtime, speed and quality losses) related to the OEE as well as the unscheduled shutdowns related to the Gross Available Times. Therefore, the expected APC can be referred to the Expected Planned Available Time equal to the product of Load Factor and Gross Available Time (or Plant Calendar Time) (De Toni and Panizzolo, 2018).

It is important to note that the problems of measuring production Throughput Rate, and of the indexes presented and consequently of capacity, gradually increase when going from the case of a **single machine** or a single stage of the production process, to a **system of machines** or to the production system as a whole. This is because the measurements of the status of the plant are significant if applied to each machine belonging to a department, and not if dealt with in terms of cumulative values (Cigolini and Grando, 2009). Limiting ourselves to the distinctive characteristics of the two production areas – job shop and flow lines – we note that in the **job shop**, the Throughput Rate, and the derived indexes, strictly depend on the product mix, which is in fact very broad, and on the criteria on which the scheduling of production is based. The typical ability of a job shop to process a broad range of work processes, according to machine load criteria, on machines that are significantly unsaturated, or by the presence of significant bottlenecks, considerably limits any possible estimate of expected throughput, which de facto is measurable only with probabilistic criteria, linked to particular cases of mixes and availability. In this regard, we note that in the job shop we generally see a strong prevalence of machine hours compared to man-hours, in terms of availability; given the scarcity of resources, the expected throughput must be parameterized in terms of man hours. This difficulty of measurement and estimation is countered by the relative ease with which the throughput of a job shop structure can be modified if needed. Indeed, given its characteristics, it is possible to gradually increase the system's throughput, through investments for expansion in single machines and the number of workers. Such a process is not feasible in situations characterized by **flow lines**, in which changes in Throughput Rate and Capacity require long times and very costly investments, that depend on true restructuring

[7] Since the Load Factor is equal to the ratio between conforming volumes expressed in standard times and Gross Available Time, the latter equal to Solar Time net of Unscheduled Time, it is evident that there is an analogy with the convention for calculation of APC illustrated in [8.6], which however doesn't consider efficiency and quality.

operations or expansion with complete parallel lines or through significant portions of the plant. On the other hand, in such cases the expected capacity and indexes presented can be calculated with less difficulty, conducting measurements and estimates for independent lines that, due to their "monolithic" characteristics and essentially monoproduct (or mono-family) nature, can be considered as a single machine. Lastly, the performance of a line depends strictly on the care taken in the process design phase, from the method of scheduling production and the criticalities assigned to the maintenance activities. The importance of economies of scale requires high volumes in such cases, and availability and utilization levels close to saturation. In conclusion, we stress that regardless of the type of process observed, in the presence of obvious bottlenecks in the physical flow, useful operational indications can be drawn by operating on critical points, or limiting activities to measurement and estimation to particular machines or critical phases, as they constrain the rest of the production flow.

8.4 S&OP process: the comparison between Required Production Capacity and Available Production Capacity

Once a uniform unit of measurement has been defined, it is necessary to **compare Required Production Capacity**, the expression of forecasted demand or a consolidated orders book, net of stock, and **Available Production Capacity**, equal to Theoretical Production Capacity (based on Gross Available Time) adjusted by the Unscheduled Time of the resource multiplied by its Load Factor.

As we have seen, those instances of unavailability can be **physiological** (stoppages due to planned maintenance, breaks required by labor contracts, time dedicated to training, technical test, etc.) or **pathological** (breakdowns, absenteeism above normal values, lack of materials, etc.). The former can be calculated with reasonable certainty, and thus planned, while the latter can only be measured *ex post,* or at the most estimated based on past track records, as they are difficult to forecast. We see this, for example, in the case of a unit in which production capacity is closely linked to labor; the attendance coefficient (the one's complement to the rate of absenteeism) is to be considered normal if kept within average values; otherwise, it takes on pathological characteristics.

It is interesting to note how, especially in the mid-term horizon, it exists a strict connection between the outcomes of the Production Planning process and the determination of Available Production Capacity. In fact, one of the results of the medium-term planning process is lot sizing, to which the number and, at equal technology and operational logics, the entity of the set-up times correspond. Furthermore, those times can in turn considerably reduce Available Production Capacity, imposing additional rescheduling cycles. This phenomenon, that we could call the "**scheduling loop**," is often underestimated. In fact, many production managers who are faced with saturated capacity try to seek to reduce high utilization through costly and difficult to reverse investments in expansion of Theoretical Production Capacity (purchase of machinery and hiring of labor), when significant gains in Available Pro-

Table 8.4 **The nature of interventions aimed at increasing production capacity**

Intervention	Timeliness	Structurality
Overtime	Very high	Very low
2nd and 3rd shift	High	Mid
Outsourcing of Capacity	Mid-high	Mid
Outsourcing of Specialty	Mid	Mid-high
Investment in machines	Low	High

duction Capacity could be obtained with more careful scheduling processes, aimed at optimizing set-up values, or as we will see below (Chapter 17), through the systematic elimination of waste. On the other hand, if that is not possible, production capacity can be increased by different degrees of structural interventions: from the use of overtime, second or third shifts – or through the reduction of Unscheduled Time – to outsourcing, or to investments in new machineries. It should be noted that the **timeliness and structurality of those interventions are inversely correlated**, as illustrated in Table 8.4. While the use of overtime can be implemented almost immediately, the use of reliable subcontracting and incremental investment in machines and labor (doubling of shifts) entail longer times for reaching full operation or pursuing learning curves and achieving skills that translate into essentially permanent expansions of productive capacity. As concerns the choice of outsourcing, a distinction is generally made between outsourcing of capacity and outsourcing of specialty. The first creates capacity increases in both the level of any possible stage of a production process, supplementing internal production; the second, which operates on a specific process phase or for the overall process of a product line, usually entails internal reorganizations of a relatively permanent nature. This is what happens, for instance, when available labor resources are concentrated on final assembly activities, removing them from non-core upstream operations that can be assigned to outsourcers. The opposite is true for interventions that aim to reduce supply (divestments, temporary unemployment benefits, etc.).

The verification of congruence between Required Production Capacity and Available Production Capacity at the aggregate level becomes a necessary, but not sufficient, condition to develop the **Master Production Schedule**. Even in a situation where Available Production Capacity exceeds Required Production Capacity in aggregate terms, it is necessary to carefully assess the entity of the request along two lines of deployment, as illustrated in Figure 8.2:

- **time**, observing the division by period;
- **mix**, observing the division by product.

The representation in Figure 8.2 shows that, in the presence of imbalances in terms of cycles or mix of demand, once again an alternative arises between the set-up of ex-

Figure 8.2 **Comparison between required and available production capacity**

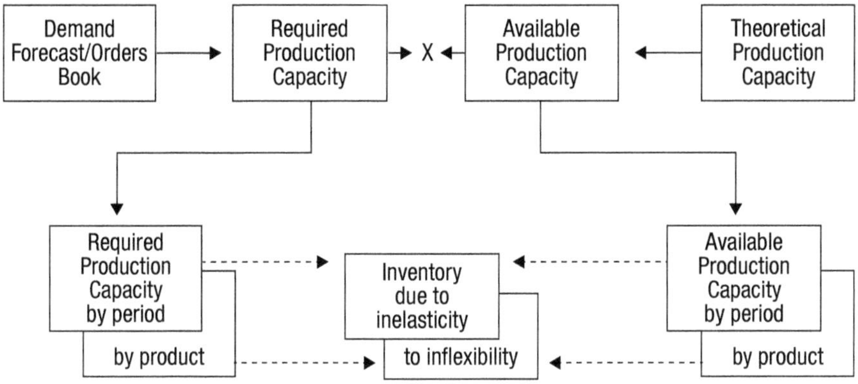

cess capacity – generally to a limited extent in the short period – and performing advance work with the consequent accumulation of stock. This phenomenon at times is driven by inelasticity (in the presence of variable demand and rigid supply) and inflexibility (in the presence of demand divided into a mix that cannot be guaranteed due to restrictions on the plant's versatility).

As we have already noted, the **Production Plan or S&OP refers to aggregate outputs** (groups of products) and pursues the most appropriate reconciliation between the aggregate productive resource needs (Required Production Capacity) and availability of aggregate productive resources (Available Production Capacity), over medium to long-term term horizons, equal to the period covered by the production budget. The horizon covered by the plan is divided into monthly or quarterly cycles, within which there is a comparison between representative technical data, concerning the key company resources. This is to stress that, due to its "aggregate" nature, the figures covered by the plan are always presented in aggregate terms relating to (Vollman et al., 1984; Da Villa, 1985; Sianesi, 2011; De Toni and Panizzolo, 2018):

- **standard data** for consumption of materials and commitment of resources (machine hours or man hours);
- **product families**, that generally refer to only critical aspects: machine hours of the work centers that are potentially saturated or critical in terms of quality of work;
- the **capacity of certain suppliers**, as concerns external resources;
- **costly or difficult to obtain components**, or those characterized by **long lead times**, as concerns materials.

It is here that, upon detecting gross differences between needs and availability of resources, it is possible to intervene on commitments with suppliers and contractors, adjusting internal capacity with interventions on machines, the introduction of overtime and doubling shifts, and hiring personnel. Such interventions are aimed at verifying general feasibility, a precondition for authorizing the Aggregate Plan and launching the second phase of the more general process of Production Planning and Control, the definition of the Master Production Schedule. Here, multiple softwares and IT modules are used that produce particular reports called Resources Requirement Planning – RRP, with reference to all of the company resources involved in the process, or Rough Cut Capacity Planning – RCCP modules that verify the availability only of productive resources, processing the standard data representing aggregations by family. If the result of that capacity/load check is positive, the plan is feasible and generally takes the name of Authorized or Approved Aggregate Plan.

8.5 S&OP: the integration of different perspectives

As already mentioned, the S&OP, offering a holistic vision of the principal measures affected by the planning processes, allows the different company's functions to enter into dialogue and make decisions, based on specific goals and information integrated into a single logical view. Below we provide some examples of S&OP calculations, referring to two main production situations (MTS and MTO) and their typical objectives (Wallace and Stahl, 2008).

The first case is that of a company operating within the **Make to Stock (MTS) logic**. The main goal is the **service level** guaranteed to the market through an optimal size of the **Finished Products Inventory**.

As illustrated in Table 8.5, in the month of October, having the actual data for the previous three months, monthly forecasts are projected for an additional 6 months, to then continue by quarters until completing a demand forecasting with a rolling horizon of 12 months. That forecast data is compared with the final and planned production data for the same horizon, and with the related Finished Goods Inventory data.

In this example, we see that actual sales have exceeded the forecast by 800 units, resulting in a reduction of the finished products inventory and in an actual service level far from the 99% target. The **integrated view** of sales performance, the delta between planned and actual production, and the impact on days on hand of Inventory (and as a consequence on the service level presumably offered) appears to be very useful. Indeed, management, using an integrated depiction of the different perspectives, can make the most appropriate decisions to bring the system back into balance in the coming months; for example, by increasing the sales forecast and thus introducing overtime or additional production shifts.

In the second case, imagine that the same company operates with a **Make to Order (MTO) system**, and thus starts production once orders are booked.

In that case, shown in Table 8.6, the target linked to the finished products inventory is substituted by the **customer order backlog**.

The **backlog** is represented by orders received, but not yet shipped, regardless of the Promised Delivery Date (PDD). In other words, since it is MTO production, and thus there is no inventory, there are orders received and still to be produced and shipped. **The size of the backlog directly influences the Customer Delivery Time**,

Table 8.5 S&OP for October in an MTS operations system

TARGET LINE FILL %	99%											
TARGET DAYS ON HAND	**10**	**days**										
	JUL	AGO	SEPT	OCT	NOV	DEC	JAN	FEB	MAR	QII	QIII	TOTAL 12 months
SALES												
SALES FORECAST	4000	4000	4000	4200	4200	4400	4400	4400	4400	13800	13800	65600
ACTUAL SALES	4400	3800	4600									
DELTA SALES	400	-200	600									
CUMULATIVE DELTA		200	800									
OPERATIONS												
PRODUCTION PLAN	4000	4000	4000	4200	4400	4500	4600	4600	4500	13800	13800	66400
ACTUAL PRODUCTION	4000	4060	3980									
DELTA PRODUCTION	0	60	-20									
CUMULATIVE DELTA	0	60	40									
FINISHED GOODS INVENTORY												
PLANNED INVENTORY	2000	2000	2000	1240	1440	1540	1740	1940	2040	2040	2040	
ACTUAL INVENTORY	1600	1860	1240									
DAYS ON HAND	8	9	6	6	7	8	9	10	10	10	10	
ACTUAL LINE FILL %	97%	98%	89%									

Source: adapted from Wallace and Stahl, 2008.

Table 8.6 S&OP for October in a MTO operations system

TARGET LINE FILL %	99%											
TARGET ORDER BACKLOG	**6**	**weeks**										
	JUL	AGO	SEPT	OCT	NOV	DEC	JAN	FEB	MAR	QII	QIII	TOTAL 12 months
BOOKINGS												
FORECAST BOOKING	400	400	400	400	400	400	400	400	400	1200	1200	6000
ACTUAL BOOKING	440	400	420	400	200							
DELTA BOOKING	40	0	20									
CUMULATIVE DELTA		40	60									
PRODUCTION/SHIPMENT												
PRODUCTION PLAN	400	400	400	400	400	400	400	400	400	1200	1200	6000
ACTUAL PRODUCTION	400	420	400									
DELTA PRODUCTION	0	20	0									
CUMULATIVE DELTA	0	20	20									
ORDER BACKLOG												
PLANNED BACKLOG	600	600	600	600	600	600	600	600	600			
ACTUAL BACKLOG	640	620	640									
BACKLOG (weeks)	6	6	6	6	6	6	6	6	6			
ACTUAL ORDER FILL %	97%	100%	100%									

Source: adapted from Wallace and Stahl, 2008.

Table 8.7 **S&OP for October in an MTO operations system – Revised plan**

TARGET LINE FILL %	99%											
TARGET ORDER BACKLOG	4	weeks										
	JUL	AGO	SEPT	OCT	NOV	DEC	JAN	FEB	MAR	QII	QIII	TOTAL 12 months
BOOKINGS												
FORECAST BOOKING	400	400	400	400	400	400	400	400	400	1200	1200	6000
ACTUAL BOOKING	440	400	420	400	200							
DELTA BOOKING	40	0	20									
CUMULATIVE DELTA		40	60									
PRODUCTION/SHIPMENT												
PRODUCTION PLAN	400	400	400	400	400	400	400	400	400	1200	1200	6000
REVISED PRODUCTION PLAN				**400**	**420**	**440**	**460**	**460**	**420**	**1200**	**1200**	
ACTUAL PRODUCTION	400	420	400									
DELTA PRODUCTION	0	20	0									
CUMLATIVE DELTA	0	20	20									
ORDER BACKLOG												
PLANNED BACKLOG	600	600	600	600	600	600	600	600	600	600	600	
REVISED PLANNED BACKLOG				**640**	**620**	**580**	**520**	**460**	**440**	**440**	**440**	
ACTUAL BACKLOG	640	620	640									
BACKLOG (weeks)	**6**	**6**	**6**	**6**	**6**	**6**	**5**	**5**	**4**	**4**	**4**	
ACTUAL ORDER FILL %	97%	100%	100%									

Source: adapted from Wallace and Stahl, 2008.

such that it is necessary to keep it under control. If it grows excessively, it can penalize sales actions; on the other hand, if it decreases too much, it could impact the efficiency of production departments. With reference to Table 8.6, imagine that the Sales Department, considering the 6 weeks of backlog to be excessive, sets the goal of reducing it by one-third to sustain its sales actions. **Table 8.7** shows the effects of a Revised Production Plan aimed at temporarily increasing the volumes by up to a maximum of 15% and as a consequence reabsorbing part of the backlog, considered to be excessive.

The simultaneous depiction of the main variables at play offered by S&OP systems guarantees an integrated analysis perspective, and thus the possibility to simulate the effects of possible corrective actions on the main elements, such as sales, production, inventory, and backlog, that may be completed by economic evaluations linked to the related dynamics in terms of turnover, costs, margins, and investments in both inventory and capacity.

9 Inventory Functions and Control

by *Alberto Grando*

9.1 Introduction

The stock management, whether materials and components, semi-finished or finished products, and the related physical supply and distribution flows, constitutes a fundamental component of Operations and Supply Chain Management processes. While we will conduct a detailed examination of the different inventory management techniques in later chapters, here we must stress that two common intents can be recognized in each method used:

- the rationalization of investment in stock, tending to gradually eliminate imbalances and waste and to limit the related **working capital**;
- the integration between stock management, production planning, and the other components of the operations system aimed at optimizing the provided **service level**.

As already stressed, the management of physical flows takes on importance not only at a specifically operational level – the subject of this book – but also on an economic level, due to the impact on operating income and the related expenses, and on a strictly financial management level of the cash flows requirements. The proper Stock Management, therefore, must be aimed on the one hand at guaranteeing continuity and fluidity in the purchasing, production, and sale processes typical of core operations, and on the other, at limiting the related economic and financial needs. Further below, after outlining the relevant conceptual framework and the links with the other components of the operations system, we will analyze the elements that influence the proper sizing of inventory and the related costs; in Chapters 10 and 13 the different approaches most frequently used in stock management will be introduced.

9.2 Type and functions of inventory

Inventory can be defined as the sum of materials, semi-finished products (or Work-in-Process – WIP), and finished products that at a given time are waiting to participate or participate in the transformation or distribution process. The **essential function of stock is to offset the external and internal variability** of the operations system that is expressed in the various phases of purchase, transformation, and distribution. The **ex-**

ternal variability is between the discontinuous performance of purchase, upstream, and target markets, downstream, and the uniformity required by the production process; the **internal variability** regards harmonizing the possible differences (in terms of times and volumes produced) detectable in the adjacent stages of the transformation process. Thus, some types of stock ensure flexibility in purchases, allowing for optimization of procurement policies independent of production requirements; others make production, aimed at the search for normalization of work phases and cycles, compatible with the requirements of distribution systems and the variability of market demand.

An initial classification that emerges from these brief notes is that of the **position of inventory**, along the various phases of the production process: raw materials and components, semi-finished and finished products.

- **Raw materials and components (RM & C)** consist of incoming factors of production, destined for transformation, that supply the production process.
- **Semi-finished products or Work in Process (WIP)**, are materials that have undergone some transformations (semi-finished materials, sub-assemblies, groups, assemblies, etc.) but have not yet been finished. They perform the function of connection and balancing between production phases characterized by different levels of Throughput Rate and Capacity or different management approaches: for instance, it is the case of a process managed according to MTS – Make to Stock logic in the upstream stages and to ATO – Assembly to Order in the final phases: the two sub-processes are connected by a **decoupling stock** to balance the different flow management perspectives.
- **Finished products (FP)** are goods that, once the transformation process at the company has been completed, are ready for sale (despite not necessarily being ready for final consumption).
- **Spare Parts (SP) and Maintenance, Repair & Operations (MRO)**, generally considered *auxiliary materials*, characterized by a merely accessory and ancillary functions. In turn, these are distinguished into **spare parts and consumables**; the former are purchased or manufactured components or products usually required to guarantee an appropriate service level to the after sales activities; the latter are materials and components (fuel, lubricants, etc.) needed to keep working machines and spare parts for internal use.

It should be mentioned that this classification must be distinguished each time by the destination that the stock in question has in the sequence of the transformation and sale process of a particular company. It is indeed evident that what represents a component or semi-finished product for one, can be a finished product for another company upstream in the overall supply chain. Thus, stock first of all has the function of separating operations in the sequence of typical phases of a production process, making the management of each phase efficient in conditions of relative autonomy and cost-effectiveness of operations.

From the observation of the reasons that justify the accumulation of stock and the **functions performed** by it, it is possible to make a second distinction, between:

Operational Stock, Organizational Stock, Safety Stock, Speculative Stock and **Strategic stock** (Magee and Boodman, 1967; Max and Candes, 1984). We will now examine the different types and sub-categories of stocks, as shown in **Figure 9.1**.

Operational stock means inventory that is accumulated to cover the needs of a period of time necessary for transport or production of a good or to achieve the already mentioned function of decoupling of two or more phase in the purchase, production, and sale process. In this regard, a distinction is generally made between:

- **Transit Stock** (being transferred or processed) is proportionate to the lead time required to transfer a good from one storage or processing point to another. The average amount of that inventory – also called Pipeline Inventory – is generally configured by the following expression:

$$I = \overline{d} \times \overline{LT} \qquad [9.1]$$

  where:
  I = Transit stock necessary in a certain stage (units);
  $\overline{d}$ = average sales (or consumption) in units per time (units/period);
  $\overline{LT}$ = average lead time required to go from one stage to the next (periods);

- **Decoupling Stock**, that, as mentioned, makes different stages of an internal process independent each other because of dissimilar flow management logic, Throughput Rates or to face specific bottlenecks;

Figure 9.1 **Types of stocks according their function**

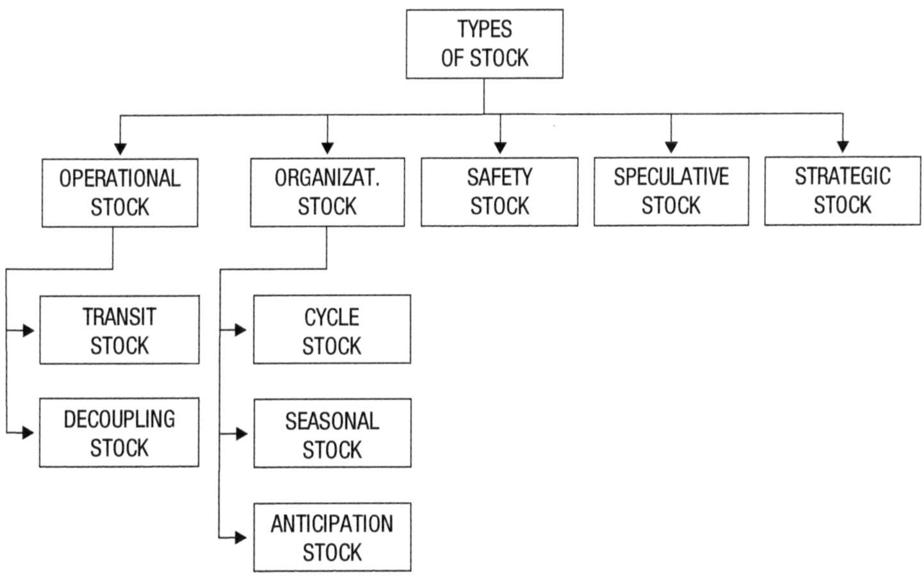

Transit stock are particularly relevant in case of great distance between the sourcing point and the consumption point; for examples in global supply chains characterized by suppliers in the Far East and manufacturers in Europe or in the case of distribution chains serving customers all around the world. It is clear that, to modify the level of transit stock, without affecting the efficiency of the production process, it is necessary to principally intervene on the lead times, since the sales (or consumption) rhythms are less controllable or are set by the planning of downstream departments.

Organizational stock performing the function of a cost optimization or of a shock absorber aimed to mitigate an expected or **planned** flow variability for the company. Organizational stock is in turn divided into three large categories, in relation to the general functions it must perform.

- **Cycle Stock** (or *lot size inventory*), accumulated to benefit from possible discount or set-up or transportation optimization. It is manifest at the time of purchases (or productions) in quantities exceeding immediate needs, justified by potential price discounts or optimization of operational (set-up) and transport costs.
- **Seasonal stock**, or stock accumulated to deal with strong seasonality of the demand or other fluctuations, linked to the need to compensate for possible oscillations in internal consumption and external demand, for examples related to the consumption during specific periods of the year (Christmas, Easter, etc.).
- **Anticipation stock**, generally accumulated to protect the company from potential procurement difficulties, or to anticipate temporary planned stoppages of plants and to provide business continuity to customers (sometime called *transition stock*).

In the proper management of a production system, functional and organizational stock must be accompanied by additional stock.

Safety Stock, built to deal with **uncertainty** in demand or consumption in internal departments, and potential delays in supply; it ensures the balanced and uninterrupted conduct of operations. Phenomena such as a delay in delivery from suppliers, abnormal functioning of the logistical system, and unexpected machine stoppages or products quality issues, make it necessary to hold higher levels of inventory than those normally kept in situations of certainty. However, since uncertainty, that is the source of these variations, avoids any absolute quantification, it is necessary to identify a safety standard (and stochastic models) that takes into account the expected volatility of demand (or supply) and lead times, and the service level to be ensured, that is, the company's intention to guarantee the prompt availability of materials in the different phases of the transformation and distribution cycle. The level of the safety stock, as will be shown later, is linked first to the forecast accuracy concerning future trends of consumption and demand and the service level desired.

Speculative Stock, represented by inventory established (independent of the operational function it performs) in order to draw benefits from an anticipated variation in prices over a specific time horizon. It regards both the **purchasing prices and the sale prices**, and can be determined by an expected variation in both directions (increase or reduction) of purchase or sale prices. Generally, advance speculative purchase is combined with functional purchase (although in that case operational needs are only of secondary importance) to determine the inventory level of materials, such as seasonal use, to be held in the presence of considerable price instability in the short-term. A merely speculative purchase, on the other hand, ultimately finds its justification only in the comparison of the related risks and costs, and revenues and margins that are earned. Thus this type of stock, unlike the others cited, can have an episodic nature. By way of example, consider the case of producers of cables that purchase copper for their operations. The volumes purchased will be influenced not only by planned production requirements, but also by the expected fluctuations in the material's price on the commodity stock exchange or raw materials markets. The same can be said for those who purchase oil, grain, cocoa, or other goods.

Lastly, **Strategic Stock** refers to intentional accumulation of raw materials, semi-finished and finished products to meet "lumpy" consumption or demand, meaning demand characterized by sudden, significant increases over average values (Sianesi, 2011). Those situations cannot be faced with safety stock, such that statistical models are adopted aiming to estimate possible oscillations in demand and lead-time. In this case, we refer to **unforeseen events**, not present in historical statistics on consumption or procurement times, such as the huge logistics problems emerged during the Covid-19 pandemic. Other cases can be, for example, the need to accumulate stock for the launch of a new product, when there may be a peak of requests in a short time, incompatible with the company's production capacity; or the case of the opening of a new geographic market that makes it necessary to prepare stock allowing to cover the entire distribution network.

9.3 The traditional methods for Stock Control

The most frequently used ratios adopted in the inventory control are the **Average Inventory Coverage Period (IC)** and the **Average Inventory Turnover (IT)** (Grando, 1995).

These ratios can be calculated in physical terms, or by value. Generally, from the perspective of Operations Management, physical quantities are used, operating by article part-numbers or reasonably uniform families of articles. In cases where synthetic evaluations are sought within a set of non-homogeneous articles or using large aggregates, such as the overall trend of company stock, they must inevitably be expressed as values. First, we illustrate the calculation of indicators in physical quantities.

- The **Average Inventory Coverage Index (IC)** – also called Flow Time – represents the number of periods (days, weeks, etc.) of average demand (or consumption) that is "covered," i.e. satisfied, by average stock. In other words, it expresses the average time spent by the article in stock.

Assuming:
- Q/2 = average stock in a period (units)
- SS = Safety Stock in the same period (units)
- D = overall demand in the same period (units)
- $\bar{d}$ = average daily demand in the same period (units/day)
- n = number of days in the same period (days)

the coverage (in days) is calculated as:

$$IC = \frac{\textit{Average stock in a period}}{\textit{Demand in the same period}} \times \textit{days in the period}$$

$$IC^1 = \frac{\frac{Q}{2}+SS}{D} \times n = \frac{\frac{Q}{2}+SS}{\bar{d}} \text{ (days)} \qquad [9.2]$$

- The **Inventory Turnover Index (IT)** expresses the number of times this stock "rotates" or is renewed, as a result of sales (or consumption). The Inventory Turnover Index is thus equal to $\frac{n}{IC}$:

$$IT = \frac{\textit{Demand in a period}}{\textit{average stock in the same period}}$$

$$IT = \frac{D}{\frac{Q}{2}+SS} = \frac{n}{IC} \text{ (times)} \qquad [9.3]$$

The calculation of the two indicators and the relationships between them are exemplified in **Box 9.1**.

Box 9.1 Example of Inventory Turnover and Coverage indices

Assuming 240 working days in a year, an yearly demand (D) of 12,000 units, an average stock of 1,600 units and a safety stock of 400 units, applying [9.2] and [9.3]:

$$IC = \frac{1{,}600 + 400 \ (\textit{units})}{12{,}000 \ (\textit{units})} \times 240 \ (\textit{days}) = \frac{2{,}000 \ (\textit{units})}{50 \ (\textit{units/day})} = 40 \text{ (days)}$$

$$IT = \frac{12{,}000 \ (\textit{units})}{1{,}600 + 400 \ (\textit{units})} = 6 \text{ (times or rotations)}$$

$$IC = \frac{240 \ (\textit{days/year})}{IT \ (\textit{times/year})} = \frac{240 \ (\textit{days/year})}{6 \ (\textit{times/year})} = 40 \text{ (days)}$$

$$IT = \frac{240 \ (\textit{days/year})}{IC \ (\textit{days})} = \frac{240 \ (\textit{days/year})}{40 \ (\textit{days})} = 6 \text{ (times)}$$

---

[1] For the details relating to the calculation of average stock and safety stock in the different stock management methods, see Chapter 10.

9.4 The problems in determining measurements by value

As mentioned, for the purpose of an aggregate analysis, quantities are expressed in values. In that case, it is necessary **to carefully choose the values** that express demand and consumption, and those at which to value the stock (Grando, 1995).

Those parameters, for which examples are shown in **Table 9.1**, are calculated as a function of the different evaluation needs, distinguishing the inventory by homogeneous categories in terms of destination (raw materials, semi-finished and finished products), and in terms of product category (by families, classes, groups, and part numbers or SKU – Stock Keeping Unit).

With reference to the indicators reported, it is necessary to recall that the first expresses the number of periods (days, weeks, months) of coverage guaranteed by the inventory, and as a consequence, the duration of the investment in the stock and the related economic cycle. The second, that is essentially the inverse of the first, expresses the number of times in which the average inventory is replenished in the

Table 9.1 Stock management ratios

		(Value)
Net Sales Revenues	(SR)	5.200
1 RM Beginning Inventory	(BRM)	400
2 WIP Beginning Inventory	(BWIP)	200
3 FP Beginning Inventory	(BFP)	500
4 Purchases	(P)	1.700
5 Manufacturing Cost	(MC)	1.100
6 RM Ending Inventory	(ERM)	350
7 WIP Ending Inventory	(EWIP)	180
8 FP Ending Inventory	(EFP)	550
Cost of Good Sold (4 + 5 + 2 - 7)	(COGS)	2.820
RM Consumption (1 + 4 – 6)	(RMC)	1.750
Full Industrial Cost (COGS - 3 + 8)	(FIC)	2.870

AVERAGE INVENTORY COVERAGE (Days)

- RM = RM x 360/COGS = 48 → RM = RM x 360/RMC = 77
- WIP = WIP x 360/COGS = 24 → WIP = WIP x 360/FIC = 24
- FP = FP x 360/COGS = 67 → FP = FP x 360/COGS = 67

AVERAGE INVENTORY TURNOVER (Times)

- RM = COGS/RM = 7.5 → RM = RMC/RM = 4.7
- WIP = COGS/WIP = 14.8 → WIP = FIC/WIP = 15.1
- FP = COGS/FP = 5.4 → FP = COGS/FP = 5.4

period under observation, relating to the level of sales or consumption. In calculating those ratios, the values generally used are **average stock value** (usually used gross of the respective depreciation provision), compared to the **Cost of Goods Sold**, or in the absence of this ratio, and less correctly, Net Revenues. With the goal of refining the analysis, it however seems appropriate to introduce an initial adjustment, differentiating the denominator of the ratios in question, with the intention to obtain greater uniformity with the respective numerators. In that sense, the **raw materials** can be compared to the **Consumption of materials** (beginning inventory + purchases – ending inventory, in the case of purchase of only raw materials). **Semi-finished products**, evaluated based on the cumulative value added in the process stage subject to evaluation, are compared with the **Full Cost of the Product** or the **Industrial cost**, thus cleaning the ratio of the inventory movements of finished products, as they are governed by choices in the commercial sphere; for the latter, we can refer to the entire **Cost of Goods Sold**.

Concerning the choice of the specific value to adopt, we must remember the general principle of valuation according to which inventory must be valued at the **lower between the historical cost and the market value**, with the intention to determine the actual utility or functionality of an original stock value. The most frequent criteria for determination of the cost are the specific identification of the cost, F.I.F.O. (first in first out), the average weighted cost, L.I.F.O. (last in first out), the last cost in the invoice, etc. These are **conventional valuation measures** that every company can define from an operational standpoint, while respecting the tax and legal constraints of the country in which they operate.

Concerning the valuation of the semi-finished products, work in process, and finished products, the costs of transformation that can be derived from management financial control tools are considered (by job, cost center, or standard costs). The method of the lesser between historical cost and market value aims at eliminating those inventory values that are thought not to be recoverable in the future due to deterioration, technical or economic obsolescence, etc. Therefore, the task is to estimate the **presumed realisable value of the inventory**, in accordance with the general criterion of prudence. Concerning the market value to be applied, the replacement cost and the net realisable value are generally taken, net of completion for semi-finished products and sales expenses, such as sales commissions, transport, and packaging, for finished products. The need to contain costs within the limits of the market value becomes fully manifest in the determination of the net realisable value of obsolete or slow-moving materials. In the first case, the depreciation is justified by the need to take into due account the reduction of the degree of liquidity of the investment in stock, related to an objective obsolescence of inventory; in the second, on the other hand, the reduction expresses the well-grounded worry of potential obsolescence. The incidence of that decrease, that can be extended to the point of accepting its value only as scrap, is to be estimated taking into account certain evidence such as the trend of market demand, the relationship between sales in the most recent period and inventory, and certain future use, as in the case of orders already acquired or foreseen.

On this point, consider the case shown in **Table 9.2**, where in order to reach a correct valuation, different finished products are classified as a function of product uniformity and analyzed from the standpoint of **liquidation times**. In the particular example shown, the finished products with liquidation times of less than nine months were written up by 20% in order to take into account the expected prices, based on a consolidated orders book. The products with liquidation times of between nine and twelve months were valued at their book value, not having particular risks of obsolesce. The products with liquidation times exceeding twelve months were written down by 20% and 40% for the values with liquidation times of between one and two years and for those exceeding two years. The decision to introduce those adjustments must also take into account the company policies on stock. If, for example, the existence in the inventory of slow-moving SKUs justifies the need for completion of the offer range, with the certainty to place the products on the market, even in the medium-term, it does not seem advisable to modify the values entered in the financial statement. The case of spare parts seems to be particularly significant in that regard.

Those indicators are generally compared with the relevant average data or analyzed in terms of historical statistics. An initial limit of that practice can be found in the diagnostic potential of the indicators, useful only for the particular scope of financial statement analysis. It indeed seems rough and partial to evaluate the correctness of an investment in stock from a management and economic-financial perspective with reference only to its degree of liquidity or only the speed of rotation. Those assertions are justified only "all other conditions being equal," which from both the standpoint of analysis and even more so from that of operations management, does not seem to be particularly useful.

In fact, it is evident that an overly limited coverage period can mean inefficient procurement choices or exposure to high risks of stock-out. On the other hand, a high index can imply accumulation justified by speculative decisions. Different rates of rotation can come from the diverse logistics flow management approaches used

Table 9.2 Analysis of liquidation times of finished products in warehouse (in thousands of euros)

Finished Products Inventory Class	Liquidation time (months) < 6	6-9	9-12	12-24	>24	Total
AAA	472	503	606	491	376	
BBB	–	251	506	–	–	
CCC	180	215	210	–	–	
DDD	–	–	100	50	70	
Total	652	969	1.422	541	446	4.030
Adjustment	1.2	1.2	1	0.8	0.6	
Total	782	1,163	1,422	433	268	4,068

(think of Just in Time logics, for example), and may not be a benchmark expressing the true performance of the inventory for all companies. A second limit is found in the difficulties of determining the values used in the indicators. Without dwelling on the problems linked to the calculation of the Cost of Goods Sold, we will limit our observations to the **typical defects of the determination of average inventory**. Average inventory of a product means the quantity (or value) of goods that is present on average in the inventory at any time, calculated as the arithmetic mean of the weighted levels of inventory for the respective durations, and thus (Urgeletti Tinarelli, 1981):

$$\bar{I} = \frac{I_1 N_1 + I_2 N_2 + \ldots + I_n N_n}{T} \qquad [9.4]$$

where:
$\bar{I}$ = average inventory (units);
I_i = stock levels (units);
N_i = days of stock for each level (periods);
T = temporal interval of observation (overall period).

The difficulties of obtaining the actual data useful for the calculation led the analysts to simplify the operation, assuming that the stock varies linearly between the maximum level and the minimum level reached, respectively, immediately after the arrival of one batch and immediately before the next one. In that case, represented in **Figure 9.2**, the actual average inventory is equal to the arithmetic mean between those values, weighted according to the duration (t_i) of the respective intervals:

$$\bar{I} = \frac{\frac{I_1 + I_{1'}}{2} t_1 + \frac{I_2 + I_{2'}}{2} t_2 + \ldots + \frac{I_n + I_{n'}}{2} t_n}{T} \qquad [9.5]$$

Figure 9.2 Trend of inventory and calculation of average values

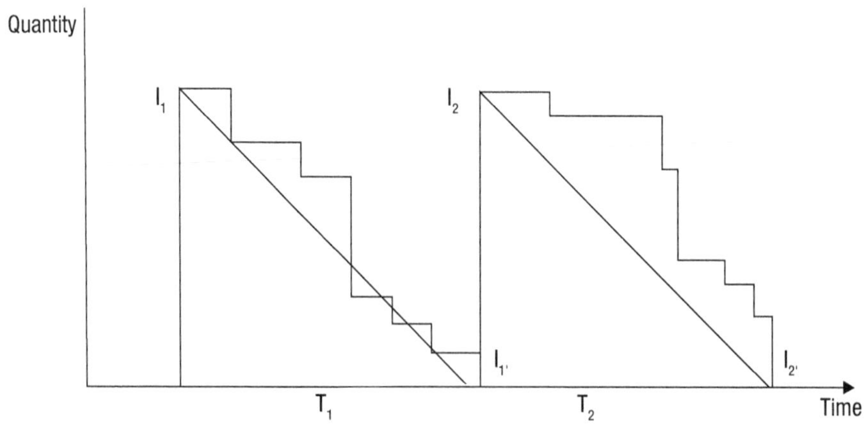

While in that simplification the distortion of the valuation is limited, in practice the analyses are generally conducted by calculating the **non-weighted arithmetic mean between** two subsequent values of annual inventory, with obvious distortions in estimation.

$$\bar{I} = \frac{I_0 + I_n}{2} \qquad [9.6]$$

Additional limits can then be found in the calculation of value, that as is known, is strongly influenced by the **valuation criterion** adopted (LIFO, FIFO etc.) in the presence of variable price regimes. Looking carefully, we see that the average inventory depends on other elements – specifically, the safety stock and the lots size – determined based on the service level offered and the optimization of the stock management variables (lead times, demand and costs sustained). It is therefore possible to reduce the scope of the indicators in question to mere **ex post synthetic control tools**, that clarify little from the standpoint of the analysis, and almost nothing from that of diagnosis and management. The pure containment of the level of inventory, and consequently its measure expressed by the rate of turnover or coverage, cannot but represent a strongly mediated goal, because it is the result of a sum of management choices that are located principally in the operational sphere of the company's activity.

9.5 The use of ABC curves and Inventory-Consumption matrix

ABC curves (or 20-80 or Pareto curves) are one of the most effective and simple tools that can be used by management in different contexts: evaluation of suppliers with respect to quantities or values supplied, analysis of clients with respect to turnover generated, or of products sold with respect to profit margin, and so on.

In order to construct the **ABC curve**, that expresses the connection between an independent variable (e.g. clients, articles, inventory SKUs) and a dependent variable (e.g. turnover, profit margin, inventory value), the cumulative values of the latter are generally segmented and classified as a function of cut-offs widely used in operational practice. In this regard, see **Table 9.3**.

Table 9.3 The parameters empirically used in the ABC curve

	Link between variables			
Class	Variable x		Variable y	Cut-off on Cumulative Value Variable y
A	20%	→	80%	80%
B	30%	→	15%	95%
C	50%	→	5%	100%

Table 9.4 An example of ABC classes

Channel	Class ABC	No. of customers	% of customers	No. of pairs	% of pairs
Wholesale	XA	23	10%	103,793	50%
	A	40	18%	63,610	30%
	B	62	28%	31,619	15%
	C	99	44%	9,698	5%
	Total	**224**	**100%**	**208,720**	**100%**
Retail	XA	30	5%	83,978	50%
	A	67	11%	50,521	30%
	B	153	24%	25,227	15%
	C	376	60%	8,951	5%
	Total	**626**	**100%**	**168,677**	**100%**

An additional class is often identified within Class A, defined as super A (XA), that contains a limited percentage of elements of the independent variable, which corresponds to the first 50% of the cumulative value of the dependent variable.[2] For example, **Table 9.4** provides an example of ABC classification relating to the number of clients and the number of pairs of shoes sold by a footwear company through Wholesale and Retail channels.

The tool can be useful also in the inventory management, through the calculation of the items' usage value. The **Usage Value (UV)** of a material is the product of the unit value by the quantity demanded (or consumed) over the time period considered. On this point, see **Box 9.2**, which shows that, as often happens in reality, the concentration of the inventory value is very high and follows the above-mentioned classes.

Box 9.2 Example of Usage Value – Pareto Curve

With reference to the simplified case shown in the following table, if we set:

- UV_i = Usage Value (euros)
- i = different items
- v_i = unit value of the i items
- D_i = demand (or consumption) of i items in the observed period (units)

[2] The classification shown here, despite being widely used, represents a convention. In many companies, we can find slightly different versions, such as A=70, B=20, and C=10. What counts is maintaining the same convention over time in order to drive useful indications for management.

to build the ABC curve we must proceed as follows:

- calculate $UV_i = v_i \times D_i$ for each of the 20 SKU_i held on stock;
- calculate the % value of UV_i with respect of the total $\Sigma_i UV$;
- reorder the SKU_i by decreasing values of $\%UV_i / \Sigma_i UV$;
- calculate the cumulative value relating to the $\%UV_i / \Sigma_i UV$;
- classify the different articles (SKU_i) using the conventional values referenced above (**Table 9.4**).

SKU_i	Unit value (v_i) (€)	Demand (D_i) (units)	Usage Value ($Di \times v_i$) (€)	Usage Value (%)	SKU	Usage Value (%)	Cumulative Usage Value (%)	ABC Classes
X1	9	850	7650	27.30	X1	27.30	27.3	XA
K6	5	200	1000	3.57	V1	22.84	50.1	XA
F3	7	700	4900	17.49	F3	17.49	67.6	A
D5	8	200	1600	5.71	M3	12.85	80.5	A
G2	4	130	520	1.86	D5	5.71	86.2	B
A3	6	80	480	1.71	K6	3.57	89.8	B
B4	3	55	165	0.59	G2	1.86	91.6	B
L8	5	70	350	1.25	A3	1.71	93.3	B
M3	8	450	3600	12.85	L8	1.25	94.6	B
F4	3	70	210	0.75	F4	0.75	95.3	B
V1	8	800	6400	22.84	B6	0.73	96.1	C
F9	9	10	90	0.32	Z2	0.68	96.7	C
Z2	5	38	190	0.68	B4	0.59	97.3	C
A7	4	40	160	0.57	A7	0.57	97.9	C
B6	6	34	204	0.73	Y2	0.43	98.3	C
H2	3	35	105	0.37	J8	0.40	98.7	C
J8	4	28	112	0.40	H2	0.37	99.1	C
Y2	4	30	120	0.43	F9	0.32	99.4	C
07	7	12	84	0.30	07	0.30	99.7	C
W6	2	40	80	0.29	W6	0.29	100.0	C
20			Σ 28020	100.00			100.0	

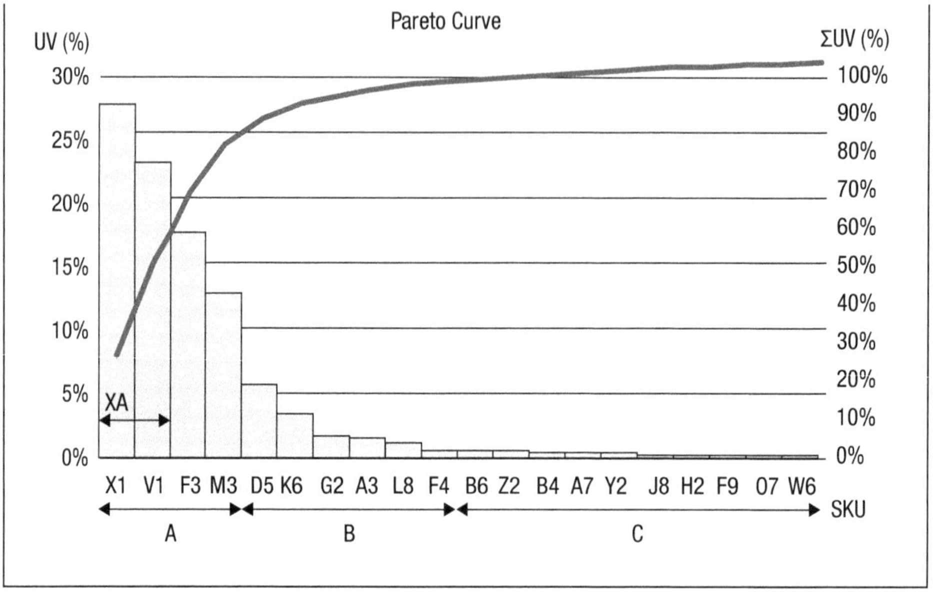

The classification shows that a very small number of products have a very significant influence on the UV, while many SKUs have very little impact. In the example, we in fact see that the first 20% of the SKUs influence 80% of the overall UV (and thus are denominated Class A); just half of these, denominated XA, has an impact of approximately 50% of the total UV; the next 30% generates an additional 15% of the total UV (Class B) and a large set of SKUs (Class C) influences only 5% of the total UV. The ABC curve thus shows how a few articles are "worth" a lot, and on the other hand, many are "worth" very little.

A useful tool for the inventory check-up, developed through ABC curves, is found in the construction of the **Inventory-Consumption Matrix**, or **ABC Cross Analysis**.

This tool is based on the construction of two ABC curves, using for both the articles held in stock, as an independent variable, and as dependent variables, respectively:

- the Average Consumption Value;
- the Average Inventory Value.

Both set as a function of the specific company practices (for example, using values of LIFO, FIFO, Average cost, last cost in invoice, etc.) and depending on whether it is stock of Finished Products, Semi-finished products, or Raw materials.

Once the articles in stock in ABC clusters are classified, respectively in terms of Average Inventory Value and Average Consumption (or Demand) Value, it is possible to build the grid, as illustrated in **Figure 9.3**.

Figure 9.3 The Inventory-Consumption matrix

		INVENTORY			
		A	B	C	
CONSUMPTION	A	AA	AB	AC	80%
	B	BA	BB	BC	15%
	C	CA	CB	CC	5%
		80%	15%	5%	

In the case illustrated above, using ABC curves, we come to a representation with 9 quadrants, respectively named AA, AB, AC, BA, BB, BC, CA, CB, and CC. In each quadrant, articles are placed with comparable behavior in terms of relevance from the standpoint of investment in stock or from that of contribution to sales or internal consumption.

Observing the matrix, it is possible to express some considerations linked to the criteria underlying the construction of the ABC curves.

- The first column, corresponding to Class A of Inventory, represents the sum of the articles that have the most weight in terms of working capital invested and Holding Cost, and thus must be carefully monitored.
- The first line, corresponding to Class A of Consumption (or Demand), represents the sum of articles that have the most weight in terms of sales, if they are finished products, or internal consumption, if components and materials.
- It is also possible to express some considerations on the main quadrants, those placed at the angles of the grid. In particular:
  - **quadrant AA** represents an area to investigate. It is necessary to determine if the high average stock is adequate to sustain the high corresponding consumption or if there is the opportunity to rationalize the stock. An intervention in this quadrant evidently has a high amount of "**leverage**": a limited number of articles will be analyzed, for which a "shift to the right" of the most important part numbers can be translated into significant savings from the standpoint of capital invested in stock. On the other hand, it is necessary to avoid the risk of stock-out for these articles, that are subject to heavy market demand or internal consumption;
  - **quadrant AC** represents an area from which to draw any lessons learned. It refers to articles in high demand, for which it has been possible to maintain

very low stock levels and therefore high rotation indexes (fast moving articles). It is possible to investigate which management methods are used and assess whether to draw indications for AA articles. In this case, though, the **risk of stock-out** is particularly significant, precisely due to the limited level of stock and it can push to higher safety stock;
- **quadrant CA** represents another area to analyze. This represents slow moving articles for which there is high stock but low sales or consumption. In this case, there is a high probability of identifying **obsolete** products. Before intervening with write-downs or elimination of stock, we suggest isolating any articles that can be used as spare parts to sell in after-sales processes or for internal use;
- **lastly, quadrant CC** represents an area to analyze with a different type of attention than the others. This is a quadrant in which we will presumably find a very high number of articles, that are not very demanding in terms of stock, and for which demand or consumption is very limited or that often end up in stock-out. We should not be deceived by their low importance in terms of investment in stock, though. This is a quadrant which contains high **complexity costs**, linked to the high number of part numbers, the low frequency of use, and the related operating costs.

Therefore, although the articles placed along the diagonal have characteristics of greater coherence in the relationship between investment in stock and consumption, as pointed out, it is necessary to carefully observe each quadrant. In regard to the articles placed in the other quadrants – AB, BA, CB, BC – it is a good rule to see if there is the opportunity to "push them" towards more coherent positions, aimed at limiting the investment in stock, but without reducing the service level offered to the market or internal consumption.

Lastly, as a final observation, we note that the dynamic nature of this tool means that it must be **used periodically** to assess the impact of improvement actions implemented and any changes in demand or consumption.

Table 9.5 shows an example in which, in addition to the values of average stock and consumption, there is the number of SKUs for each quadrant and the Average Inventory Coverage Index (IC). It is also evident that Class D represents the cases in which consumption is generated in the apparent absence of stock; this is a symptom of misalignment between physical stock and accounting stock; or the case of stock not moved in the period taken into observation for consumption.

9.6 The factors that influence stock levels

For management purposes, the optimal sizing of inventory is represented by that ideal level, that in normal operating conditions, ensures the correct performance of purchase-production-sale operations. The differences that can be observed in the comparison between the levels of actual stock and ideal stock are linked to circumstances

Table 9.5 **The Inventory-Consumption Matrix**

CONSUMPTION			INVENTORY									
			A	%	B	%	C	%	D	%	Total	%
CONSUMPTION	A	Consumption (€)	€ 6.840.157	36,1%	€ 1.854.019	9,8%	€ 608.390	3,2%	€ 5.834.382	30,8%	€ 15.136.948	80,0%
		Avg Inventory (€)	€ 635.131	42,3%	€ 42.539	2,8%	€ 3.215	0,2%	€ -	0,0%	€ 680.886	45,4%
		Number of articles (#)	547	4,2%	328	2,5%	125	1,0%	719	5,6%	1719	13,3%
		Coverage Index (months)	1,1		0,3		0,1		0,0		0,5	
	B	Consumption (€)	€ 292.802	1,5%	€ 518.103	2,7%	€ 610.747	3,2%	€ 1.418.097	7,5%	€ 2.839.749	15,0%
		Avg Inventory (€)	€ 144.416	9,6%	€ 54.546	3,6%	€ 14.849	1,0%	€ -	0,0%	€ 213.811	14,3%
		Number of articles (#)	248	1,9%	464	3,6%	624	4,8%	1382	10,7%	2718	21,1%
		Coverage Index (months)	5,9		1,3		0,3		0,0		0,9	
	C	Consumption (€)	€ 34.408	0,2%	€ 91.262	0,5%	€ 273.808	1,4%	€ 546.921	2,9%	€ 946.400	5,0%
		Avg Inventory (€)	€ 119.808	8,0%	€ 65.847	4,4%	€ 34.277	2,3%	€ -	0,0%	€ 219.931	14,7%
		Number of articles (#)	213	1,7%	583	4,5%	1987	15,4%	3837	29,8%	6620	51,4%
		Coverage Index (months)	41,8		8,7		1,5		0,0		2,8	
	D	Consumption (€)	€ -	0,0%	€ -	0,0%	€ -	0,0%	€ -	0,0%	€ -	0,0%
		Avg Inventory (€)	€ 275.333	18,4%	€ 62.078	4,1%	€ 22.579	1,5%	€ 25.610	1,7%	€ 385.601	25,7%
		Number of articles (#)	250	1,9%	557	4,3%	993	7,7%	21	0,2%	1821	14,1%
		Coverage Index (months)										
Total Consumption (€)			€ 7.167.368	37,9%	€ 2.463.384	13,0%	€ 1.492.944	7,9%	€ 7.799.400	41,2%	€ 18.923.096	100,0%
Total avg inventory (€)			€ 1.174.687	78,3%	€ 225.010	15,0%	€ 74.920	5,0%	€ 25.610	1,7%	€ 1.500.228	100,0%
Total number of articles (#)			€ 1.258	9,8%	€ 1.932	15,0%	€ 3.729	29,0%	5959	46,3%	12878	100,0%
Coverage Index (months)			2,0		1,1		0,6		0,0		1,0	

of variability (both inside and/or outside the company) that translate into a higher or lower accumulation at the operationally desirable level, justified, as already mentioned, by the presence of anticipation, speculative, safety stock, etc, With reference to the operational stock, the **factors that have the most influence on the size** refer to the following aspects:

1. the characteristics of the production system in question;
2. the characteristics of the product, the market, and the distribution systems;
3. the costs linked to inventory management;
4. the control method used in the choices regarding the management of materials flows.

The last point will be analyzed further in two specific chapters (Chapters 10 and 13); here we immediately address the first three.

9.6.1 The characteristics of the production system

The type of production process (to order or to stock, intermittent flow, continuous flow, etc.), the number of production stages, the technology used, and the related degree of flexibility of the work cycles, are all factors able to influence the sizing of the inventory. By their nature, **Make to Order – MTO** and **Assemble to Order – ATO** entail considerable investments in stock of raw materials and semi-finished products (or component parts), while they generally do not require significant accumulation of finished products. Those accumulations take place in the forms discussed previously: speculative stock, anticipation stock, or more simply safety stock, while they are not necessary in the **Purchase to Order – PTO** and **Engineering to Order – ETO** production systems. This does not happen, however, for line or flow productions (continuous or intermittent), that since they usually work on forecast, based on **Make to Stock – MTS** logics, often base their competitiveness on high levels of inventory of finished products and on the uninterrupted procurement of incoming materials (especially for continuous flow productions). The greater the number of manufacturing stages (and the throughput time), the greater the stock of semi-finished products of work in process will be.[3] The adoption of dedicated or flexible technologies once again translates into different inventory problems, in quantitative terms (for example, determined by the need to ensure the uninterrupted flow of the process, avoiding machine downtime) and qualitative terms (linked to the need to have the "right" component at the "right" time in the "right" place").

[3] The issue is represented by the Little rule, that given the throughput of a process, establishes a relationship between its lead time and the volume of WIP.

9.6.2 The characteristics of the product, the market, and the distribution systems

The sizing of the inventory is also affected by the distinctive characteristics of the product, its susceptibility to deterioration and obsolescence, and the degree of differentiation along the various stages of the transformation process. The predictability of market trends, the size and frequency of orders, and the service level required are also important. As concerns the composition of the product, we can see, for example, that in modular productions, based on **postponement policies**, it is possible to achieve savings, keeping adequate levels of stock of semi-finished modules, while in the case of products differentiated since from the point of raw materials, significant inventories are formed from the initial phases of manufacturing or assembly. Products subject to deterioration (like many foodstuffs) or subject to rapid obsolescence (such as fashion products or consumer electronics), require the maximum compression of stock, while goods belonging to particular product sectors (typically agro-industrial goods) can see seasonal fluctuations of the level of inventory within the year, with high accumulations of raw materials in the harvesting period, and subsequently, high investments in finished products, after manufacture them.

As we have already highlighted, the **predictability of the demand** plays an important role in the choices relating to the accumulation of stock, since the clearer future trends are, the more it will be possible to limit inventories of materials and products. Large orders with uncertain frequency entail a rise in inventory, while small and frequently scheduled orders lead to considerable stock optimization. With reference to distribution systems, we can say that the stronger downstream integration is, the more sensitive the investments in finished products will be, and the more proportionally lower the uses of raw material and semi-finished products will be as well.

The rapid survey we have just conducted leads to the fact that investments in stock depend on the **objective conditions in which the company is operating** at a given time; when those conditions change, the level of inventory tends to change as well. From a management standpoint, evaluating the efficiency of a stock control system thus means verifying the adequacy of the relationship between the actual inventory and the operational conditions in which the purchase-production-sale cycle takes place. To do this, though, the first step is represented by the objective measurement of those conditions. Supposed inefficiencies in the inventory system often have their roots in inefficient purchasing, management, production, or distribution systems, that lead to seeking flexibility and recovering competitiveness with high investments in stock, neglecting the specific measures needed in the areas where the problems originate. The **optimal level of stock**, that which minimizes investment despite satisfying the technical and economic constraints of core operations, is in any event linked to the overall efficiency of the production system. For example, the lower the purchase lead time, the lower investment in raw materials and components stock will be, all other conditions being equal. Likewise, the lower the throughput time of a process, the lower the inventory of WIP products will be. Moving from productions for large batches to small lots production (a choice made possible to the increased flexibility of plants), it is possible to considerably reduce the level of inventory. Ultimately, raising the level of efficiency of part or all of the production system leads to significant reductions in the level of inventory.

9.6.3 The relevant costs in the stock management

The fundamental decisions relating to inventory management policy are linked, first of all, to two well-known questions:

- **when to order**, i.e. on what date the procurement order (or the production order) is to be issued;
- **how much to order**, i.e. the number of units to procure (or process).

The answers to those questions are to be determined in light of the operational and economic aspects of the transformation processes observed and of the risks and costs related to exceeding high or low levels of inventory. From a management standpoint, the decisions linked to **reordering times** and the **lot size** thus entail careful evaluation of the **relevant differential cost** components. For the purpose of proper stock management, it is necessary to consider only those costs susceptible to variation due to the different choices that can be implemented, while ignoring those elements that are in no way influenced by those decisions. For example, as long as the level of inventory does not exceed the capacity of a storage area, the relevant rental fee is irrelevant as regards decisions on the frequency of reordering or sizing of the batch, since it does not vary with changes in the type of decision made. On the other hand, as regards the cost classes to be considered for sizing the order (or production) batch, the following elements are generally considered relevant:

a. the ordering cost in the case of a purchase batch or the set-up cost in the case of a production batch;
b. the stock holding cost;
c. the stock-out cost.

Ordering costs refer to the costs sustained by the company to replenish stock, that vary depending on the **method of purchase or internal production** (set up or production change over cost). Those costs are formed by two principal components:

1. the costs linked to purchasing operations, that do not depend on the quantities procured; that item generally includes the costs of shortage assessment, search, contact, and selection of suppliers. In case of internal production the relevant cost is the machine set up cost;
2. the cost due to the activity of issuing the order to the supplier, receiving and controlling the goods (shipment, registration, quality control, etc.), in the case of pur-

chases, or the activities entailed by planning, scheduling, and launch, in the case of production.

The ordering costs are usually considered fixed costs, independent of the size of the batch and proportionate to the frequency of the orders issued.

Holding costs (or carrying costs) are formed by:

- the costs linked to the **handling and transport** of materials (the part that varies with the variation of the quantities);
- the costs directly related to the **operations of warehouses and storage areas** (rent, heating, cooling, lighting, personnel, maintenance, insurance, and taxes); that is, the part influenced by the decision relating to the size of the procurement or production batch; this essentially means the variable component of those costs, since the fixed component is irrelevant for the decision;
- the costs linked to the **obsolescence and deterioration** of the stored products, and to any thefts, losses, or waste;
- the cost of the **financial resources** absorbed by the stock. This element evidently depends on the cost of capital for the company, understood as the average cost of the various sources of funding, also known as **WACC – Weighted Average Cost of Capital**, on the volume of financial resources used and the time the stock is held. From a differential viewpoint, the investment must be assessed carefully, referring only to the monetary outlays (cash out) actually sustained for the factors absorbed by the inventory (materials, labor, and general expenses). Since the stock holding cost is proportionate to the value of the goods in stock at any point in a given interval of time, that is equivalent to saying that it is proportionate to the value of the average stock in that interval.

Stock-out costs (or shortage costs) are generated when the company is unable to promptly face (or is not able to respond at all to) market requests or internal requirements. In the case of stock of raw materials or semi-finished products, an inventory shortage situation can entail costly stoppages of plants, retooling costs to convert production, wait times, etc. In the case of finished products, the costs in question are linked to the temporary (back orders) or definitive loss of the unsatisfied client and the contribution margin associated, the implementation of emergency measures, urgent purchases from more expensive suppliers, and so on. The exact quantification of those costs is difficult when attempting to determine the strategic component due to the loss of a client and the deterioration of the company's image and reputation, while it seems much simpler to determine the cost connected to the exhaustion of raw materials and semi-finished products.

The issue of Stock Management is inseparably linked to Production Planning. The intensity of that link varies with the changes in the techniques used. Some approaches consider the choices relating to the size and the assortment of the inventory as a consequence of scheduling decisions; others, to the contrary, treat the two phenomena jointly, making the two aspects mutually dependent. In the choice between alternative production plans, it is necessary to consider not only the costs linked to scheduling alternatives in a strict sense (costs for retooling, overtime work, subcontractors, late deliveries, etc.), but also those generated by the different impact on stock management. The close integration observed between stock management and production planning is only a single link in a larger chain that includes the entire physical flow, from the procurement of raw materials to the distribution of the finished products.

9.7 Stock management systems

The considerations presented above do not exhaust the analysis of the elements that influence the inventory sizing. It is necessary to also consider the type of models that can be used in the stock management and control activities. The literature and business practice have developed numerous management methods and tools: some that are relatively simple in their application, and others that imply the support of complex information systems and databases, and still others that assume precise design choices (in terms of layout and physical flows), that influence the very characteristics of the production systems involved. Each method has its own peculiar characteristics, advantages and disadvantages, specific fields of application, and different levels of complexity. As will be illustrated below (Chapters 10 and 13), there exist multiple techniques for stock management.

However, in order to offer a synthetic, but to the extent possible complete, view of those models, we note that ultimately, all of them refer to one of **two basic approaches** (Da Villa, 1985; 1991):[4]

a. the **look-back** approach, based on the historical trend of the variables considered, that includes the methods defined in practice as to stock management methods;
b. the **look-ahead** approach, that is based on demand forecast, that includes the methods defined in practice as on requirements management methods.

The **look-back** logic assumes that the procurement (or production) order is issued when that stock is lower than a predefined quantity. That criterion is based on the concept of replenishing stock that is running out; each time the stock reaches a preset reordering level (or interval), it is replenished.

The **look-ahead** logic, on the other hand, assumes knowledge of the production plan of finished products, or the elements subject to planning. Its validity is based on the possibility to calculate the requirements of semi-finished products, components,

[4] The two approaches described, are at times classified based on the underlying logic of flow management and defined, respectively, as the *pull* approach, because in principle the overall physical flow is "pulled" by downstream orders, and the *push* approach, because the physical flow is essentially "pushed" from upstream to downstream as a function of planning process, based on demand forecast.

and materials used, in a cascading manner for dependent demand articles.

The use of look-back or look-ahead approaches descends from multiple variables, among which the nature of demand for stock goods is important. As already mentioned, a distinction is generally made between:

- **Independent demand goods**;
- **Dependent demand goods**.

The former refer to the articles whose demand depends on the market requests served, i.e. the finished products, or in any event what can be considered a finished product in the observed production system. A company that produces electronic components, for example, considers the component that is produced and sold to be a finished product, even though that product, in turn, will be considered an input for a downstream company that produces PlayStations, amplifiers, or cooling systems. To this we must add the requirements generated by specific purposes, such as spare parts or MRO. Independent demand goods management systems will be illustrated in Chapter 10.

The second category refers to articles whose demand, or to be more precise, their requirements, depend on higher-level articles in a BOM. This is the case with the cylinders of an internal combustion engine, for example, for which the need is not estimated, but is calculated as a function of the demand for engines that use them, or the case of quantities of durum wheat flour needed to produce pasta, or square meters of cotton to produce shirts, etc. Dependent demand goods management systems will be illustrated in Chapter 13.

For **independent demand goods**, the only criteria that can be adopted are **look-back criteria**, since there are no higher-level goods from which to derive their requirements deterministically. For **dependent demand goods**, on the other hand, **both look-back and look-ahead criteria** are possible. A look-back or look-ahead approach can be adopted for them, even though especially in the presence of high management complexity, the latter are preferable and as a result increasingly used.

An analysis of the relevant variables to take a choice between the two approaches is presented at the end of Chapter 13, after having clarified the characteristics of the most used inventory management techniques.

10 Inventory Control Methods for Independent Demand Goods

by *Alberto Grando*

10.1 Setting stock levels in conditions of certainty

In regard to the best method for managing inventory, the cost analysis conducted in Chapter 9 was aimed at introducing two of the fundamental questions: **how much** and **when** to purchase or produce. Below, we first present the basic elements for the parameterization of the principal Inventory Control Systems, to then analyze the most important ones.

10.1.1 How much to produce or purchase: the Economic Order Quantity

With reference to the **size** (how-much) of procurement (or production) lots, in the situations where it is possible to purchase greater quantities than those immediately necessary for transformation (or sale), **two conflicting tendencies are generated, related to stock ordering and holding cost**.

The first, that favors the use of larger batches, is due to those cost components (procurement or production **ordering costs**) that tend to grow with the increase of the number of purchase orders, or with the number of set-up or production launches.

The second, which favors smaller batches, is exercised by those cost components (stock **holding or carrying costs**) that tend to increase with the growth of the average quantity on stock.

As already mentioned, costs for administration, orders and retooling of machines (set-up) are generally all or mostly independent of the size of the batch, but proportionate to the frequency of orders. If those costs are high, the decisions regarding stock will be oriented towards increasingly large (and less frequent) purchase or production lots.

On the other hand, holding costs, risks of obsolescence, and specifically, the overall cost of invested capital, increase with the growth of average inventory, which, as we see in **Figure 10.1**, is proportionate to the lot size.

If we place **time** on the x-axis and the **quantity** to be procured (or produced) on the y-axis, it will be clear that for purchases (or manufactured lots) equal to Q the average inventory in each procurement (or production) interval is equal to $\frac{Q}{2}$. The problem of the choice of the optimal batch, is thus reduced to the search for the point of equilibrium between these conflicting needs, that is, the quantity for which the

Figure 10.1 Consumption trend and average stock

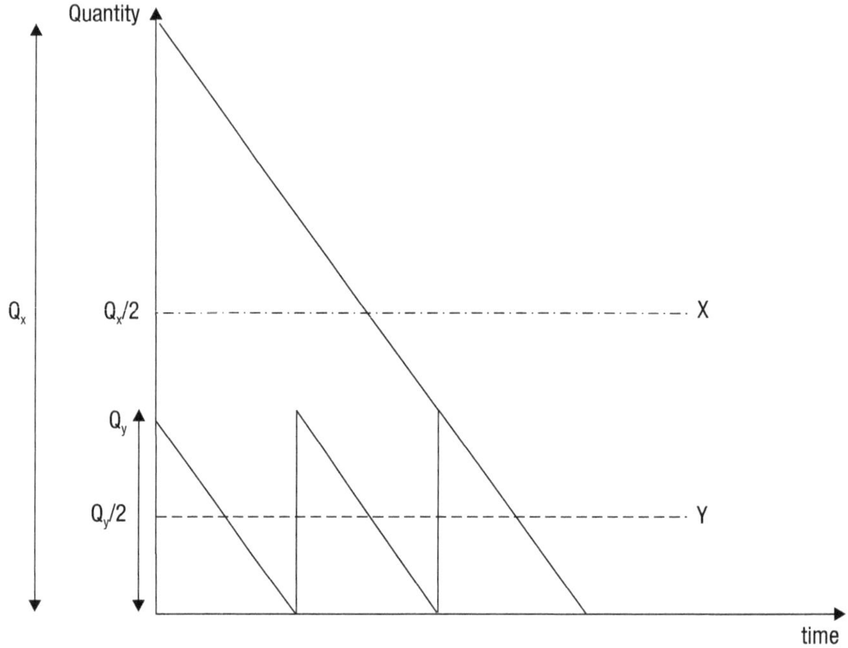

differential total costs function associated with the decision is minimized. The best-known approach to solve the problem, although it is quite simplified, is the **Economic Order Quantity model (EOQ)**.[1]

EOQ and batch replenishment of inventory

Over the period of time considered, the relevant costs for determining the Economic Order Quantity, represented in **Figure 10.2**, are:

- the stock **Holding Cost** (or Carrying Cost), expressed in euros/period, proportionate to the average stock ($\frac{Q}{2}$), represented graphically by a straight line passing through the origin, increasing along with the quantity ordered (or produced);
- the **Ordering Cost** (or Set-up Cost), expressed in euros/period, proportionate to the number of orders issued $n = \frac{D}{Q}$, represented graphically by an equilateral hyperbola, decreasing along with the quantity ordered (or produced).

[1] The model originally developed by Harris (1913) and then by Wilson (1934) is based on the following simplifying hypotheses: a) the demand for the period (d) is predictable and constant in time; b) the unit purchase price is known, constant in time, and independent of the quantities ordered; c) the quantities purchased in each order are all delivered at the same time; d) the lead time is known and constant, or sometimes, assumed to be null; e) the goods are not subject to deterioration and the goal is to satisfy the demand in full.

Figure 10.2 Inventory management costs and the Economic Order Quantity

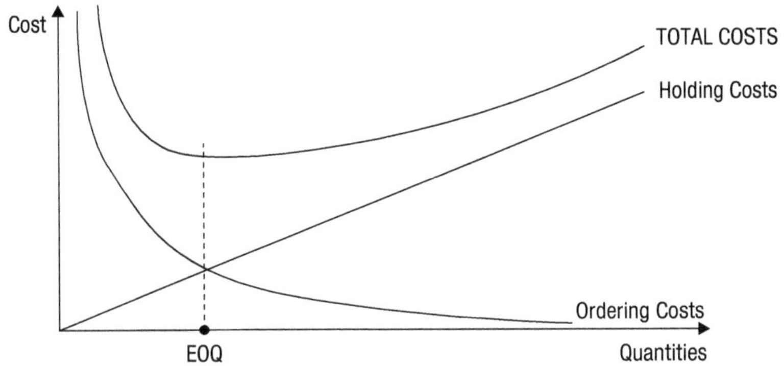

So if we set:

EOQ = the batch to order or produce each time (unknown) (units)
D = overall demand (or requirement for internal consumption) in the period taken into consideration, generally one year (units/period)
Hc = holding cost of a unit of stock for one year, expressed applying a percentage cost (*i*)[2] to the value of a stock unit (*v*) (euros/unit-period)
Oc = unit ordering cost (euros/order)

we can easily deduce that the stock **Holding Cost** is obtained by multiplying the Unit Holding Cost (Hc) by the average inventory $\frac{Q}{2}$; the **Ordering Cost**, on the other hand, is given by the Unit Ordering Cost (Oc) for the number *n* of orders issued in the period, equal to $\frac{D}{Q}$. The **Total Differential Cost** (TC) of managing stock, represented by the sum of Holding Costs (*HC*) and Ordering Costs (*OC*), is:

$$TC = OC + HC = Oc \times \frac{D}{Q} + Hc \times \frac{Q}{2} = Oc \times \frac{D}{Q} + v \times i \times \frac{Q}{2} \qquad [10.1]$$

The expression indicates that with the increase of the quantity ordered *Q*, the Holding Cost increases and the Ordering Cost decreases, and vice versa. The minimum point of the Total Cost function – expressed graphically by **Figure 10.2** – is obtained for those values of *Q* for which the first derivative is equal to zero and the second derivative is positive:

$$TC' = -Oc \times \frac{D}{Q^2} + v \times i \times \frac{1}{2} \qquad [10.2]$$

$$TC' = 0 \qquad EOQ = \sqrt{\frac{2DOc}{vi}}(units) \qquad [10.3]$$

[2] Note that the term i is composed by two sub-elements: i' which represents the company's weighted average cost of capital (WACC) or the opportunity cost measured by the IRR (Internal rate of Return) of an alternative investment; and i'' represented by all differential holding costs, such as insurance, obsolescence, people, energy, space, etc., as a percentage of the total value invested in stock.

where *EOQ* represents the optimal purchase (or production) batch.[3]

As already observed, the difficulties inherent in the calculation of the EOQ refer principally to the exact determination of the terms included in the expression for each material.

If, by way of example:

- the Unit Ordering Cost of an article is 200 €/order;
- the Value of each stock unit is equal to 600 €/unit;
- the Holding Cost (including the cost of invested capital) is set at 20% of the annual value of the stock;
- Demand for the article is equal to 12,000 units; then

$$EOQ = \sqrt{\frac{2 \times 12{,}000 \times 200}{0.20 \times 600}} = \sqrt{\frac{4{,}800{,}000}{120}} = \sqrt{40{,}000} = 200 \text{ (units)}$$

EOQ expressed in value

The principal limit to the application of the method illustrated above lies in the fact that the determination of the EOQ must inevitably be carried out for each article independently. When numerous SKUs are present, though, simplified methods can be applied to reduce the burden of the calculation.

A path that can be taken consists of the identification of the **Economic Order Quantity, expressed in value (EOQ_v)** (Magee and Boodman, 1967; Da Villa, 1991); by this we mean referring to the concept of **Usage Value (UV)** of each article subject to analysis, or of the Value of a unit v (euros/unit) for its predicted consumption *D* (units/period) in the period of time considered, such that:

$$UV = D \times v \qquad [10.4]$$

Therefore, the EOQ_v is:

$$EOQ_v = EOQ \times v$$

or:

$$EOQ_v = v \times \sqrt{\frac{2DOc}{vi}} = \sqrt{\frac{2DOcv}{i}} = \sqrt{\frac{2UVOc}{i}} = \sqrt{\frac{2Oc}{i}} \times \sqrt{UV} \qquad [10.5]$$

[3] The base formulation of the Wilson model is thus:

$$\frac{\delta CT}{\delta Q} = -Oc\left(\frac{D}{Q^2}\right) + \frac{Vi}{2}$$

$$Oc \times \frac{D}{Q^2} = \frac{Vi}{2}$$

$$Q^2 = 2Oc \times \frac{D}{Vi}$$

$$Q = \sqrt{2Oc \times \frac{D}{Vi}}$$

In the literature, models can be found that take into account some particular cases that can complicate the determination of the economic order, such as the presence of discounts and rebates based on quantity, minimum transport quantities, production times, and so on.

As we can see, the parameters contained in the expression underlying the first square root are independent of the characteristics of the material observed – value and consumption –, but related to elements common to all materials and mostly connected to the ordering cost (assumed the same for the all articles considered) and the company's holding expenses and financial policy: the ordering cost (Oc) and interest rate (i), related to the company's WACC and holding cost percentage. We can therefore consider the equation in question a **constant K**, equal for all materials purchased, such that:

$$EOQv = K \times \sqrt{UV} \qquad [10.6]$$

from which we pragmatically deduce that the EOQ by value of an article is proportionate to the square root of its **Usage Value**.

Think of the case of a wholesaler who, in handling a high quantity of articles, must periodically order from suppliers the quantities ideally defined by the EOQ.

It is reasonable to suppose that, in the medium-term, the Ordering Cost (Oc) and the rate (i) used to calculate the Holding Cost are the same for all of the articles (or families made up of articles) with similar characteristics in terms of method of procurement. In that case, it will be possible to adopt the following simplified empirical approaches.

If the periodic order is expressed in value and we assume a monthly consumption also expressed in value ($dm_v = \overline{dm} \times v$), the purchase batch ($EOQ_v$), expressed in value, will be a constant (K) times the square root of the monthly sales (dm_v):

$$EOQv = v \times \sqrt{\frac{24\ dm_v\ Oc}{vi}} = \sqrt{\frac{24\ Oc}{i}}\ \sqrt{dm_v} = K\ \sqrt{dm_v}\ (euros) \qquad [10.7]$$

Table 10.1 shows the **simplified method** of application of [10.7] relating to 5 different articles having unit values (v) and monthly demand (dm) that are different, but share the Oc and i% values.

Table 10.1 Example of calculation of lot by value

Article	v (€/unit)	dm (units/month)	Oc (€/order)	i (%)	K	$\sqrt{dm_v}$ (€/month)	EOQ_v (€/order)
A	30	1,000	20	5%	97.980	173.2	16,971
B	25	400	20	5%	97.980	100.0	9,798
C	5	2,000	20	5%	97.980	100.0	9,798
D	4	10,000	20	5%	97.980	200.0	19,596
E	16	900	20	5%	97.980	120.0	11,758

It is also possible to adopt an analogous method, operating with physical quantities and not values. In that case, assuming Ordering Cost (Oc) and Holding Cost ($Hc = v \times i$) as uniform among different variants (x) of the same article A, each with its own annual demand ($D_x = dm_x \times 12$), it is possible to express EOQ as:

Table 10.2 **Example of simplified calculation of lot by quantity**

Variants Article A	v (€/unit)	dm (units/month)	Oc (€/order)	i (%)	K	√D (units/year)	EOQ (units/order)
A1	30	200	20	5%	5.164	49,0	253
A2	30	100	20	5%	5.164	34,6	179
A3	30	400	20	5%	5.164	69,3	358
A4	30	50	20	5%	5.164	24,5	126
A5	30	250	20	5%	5.164	54,8	283
Total Month Demand		1000					

$$EOQ_x = \sqrt{\frac{24\ dm_x\ Oc}{vi}} = \sqrt{\frac{2\ Oc}{vi}}\ \sqrt{12\ dm_x} = K\ \sqrt{D_x}\ (units) \qquad [10.8]$$

Applying [10.6] to the data shown in Table 10.2, we come to a calculation of EOQ for the different variants x of article A, equal to a constant (K) times the square root of yearly demand (D_x).

The calculation of the **Economic Production Quantity (EPQ)** is similar to that of the economic purchase order, in the event that we see instant (in batch) supply of the machine observed. This is the case in which, for example, the batch consists of a fixed quantity of green coffee that after roasting, is instantly and fully stored in silos for decantation; or in the case of chemical products obtained by synthesis or mixing, generally performed in batch. The only difference is related to the Ordering Cost that in the **EPQ** is represented by the **Set-up Cost**, as shown in Figure 10.3.

EOQ and linear replenishment of inventory

In the more frequent cases in which the **supply is linear**, due to the production released gradually from the departments or machines upstream, such as a production line, it becomes necessary to modify the calculation. See Figure 10.4 and consider:

Figure 10.3 **Ordering and holding cost in purchasing and production operations**

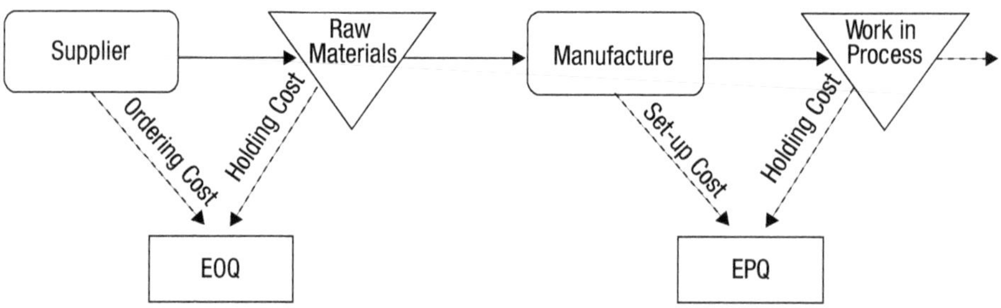

- p = daily Production or **Supply Rate**, equal to $\frac{P}{t}$ (units/period), where P represents the production in the period (level B in **Figure 10.4**), and t the number of days in that period;
- d = daily **Demand Rate** or daily consumption or withdrawal (units/period), equal to $\frac{D}{t}$, where D represents the demand or consumption in the period (level A in **Figure 10.4**) and t the number of days in the period;
- T = **time required** for the production (replenishment) of a lot Q at the production rate p (periods);
- p-d = **Rate of Inventory change** (increase, given a Production Rate by definition greater than the Consumption Rate) in the period T, also equal to $\frac{P}{t} - \frac{D}{t}$ (units/period);
- $T \times (p - d)$ = amount of **inventory at the end of period** T (units).

Assuming an annual time horizon, in that case:

- the production batch $Q = T \times p$ represents the **Economic Production Quantity (EPQ)** sought, equal to the Production Rate for the interval of time sufficient to accumulate it (level B in **Figure 10.4**);
- the **average inventory** $Q' / 2 = T \times (p - d) / 2$ is lower than the average inventory held in the case of instant supply (or batch production), since the quantity Q' (level C in **Figure 10.4**), the maximum inventory in the period, is increased in time T at a rate $(p - d)$. Clearly, this entails lower need for space and less absorption of costs than in the case of instant supply;
- the annual total **Ordering Cost** (or of *set-up cost*) is $OC = n \times Oc = \frac{D}{Q} \times Oc = \frac{D}{(T \times p)} \times Oc$ [10.9]

Figure 10.4 EPQ with linear supply or production

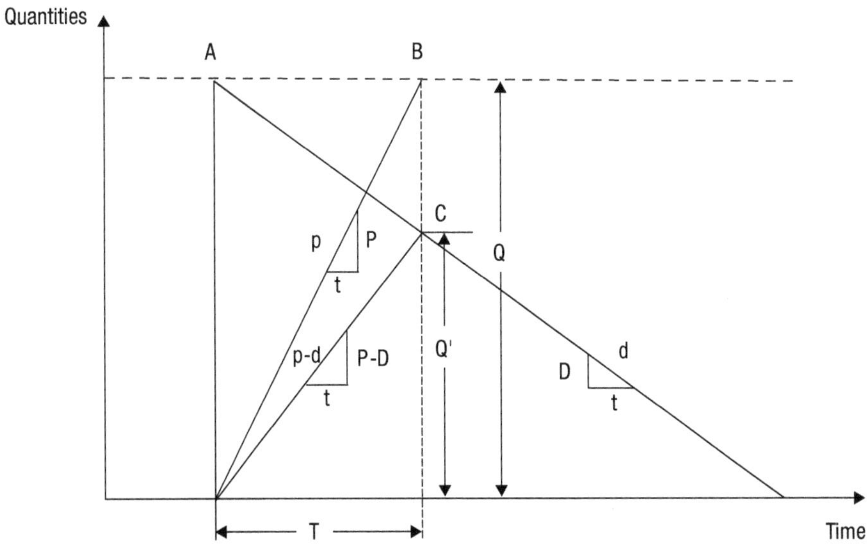

- the annual total **Holding Cost** is $HC = \frac{Q'}{2} \times Hc = \frac{T \times (p-d)}{2} \times v \times i$ [10.10]

and thus the annual **Total Cost** (TC) comes out as:

$$TC = OC + HC = \frac{D}{(T \times P)} \times Oc + \frac{T \times (p-d)}{2} \times v \times i \quad [10.11]$$

Deriving TC with respect to Q ($Q = T \times p$) and zeroing the first derivative, we obtain the formula of the EPQ with linear supply or production:

$$EPQ = \sqrt{\frac{2D\,Oc}{vi\left(1 - \frac{d}{p}\right)}} \quad \text{or} \quad EPQ = \sqrt{\frac{2D\,Oc}{vi\left(1 - \frac{D}{pT}\right)}} \quad [10.12]$$

That formula is actually entirely similar to the EOQ /EPQ in batch, except for the coefficient $(1 - \frac{d}{p})$. Indeed, if p is significantly larger than d, the coefficient is close to 1 and the outcome of the formulas in the two cases is the same. Note that the purchase or production order in batches, the p goes to infinity, since it is an entry "in bloc" of a certain quantity (Da Villa, 1991). In that case, as expected, the [10.12] becomes similar to the [10.3].

In fact, in that case the rhythm of purchase or production (p) is assumed to be infinite, leading to a purchase or production by the supplier or department upstream in batches (P), produced instantaneously (T close to 0), as in the case cited of the transfer of a batch of roasted coffee from the roasting department to the grounding and packaging department.

10.1.2 When to purchase or produce: the Reorder Point

After having determined "how much" to order or produce, we need to answer the second question: "**when**" to do it. In the inventory management, that response is not given by searching for an instant in time, but by determining a definite **level of stock**. That level represents the point in which the need becomes manifest to issue the order to replenish the inventory needed for transformation processes or sale, and is called the **Reorder Point** or Reorder level. On this point, see **Figure 10.5**.

The **Reorder Point (ROP)** coincides with the inventory sufficient to cover consumption (or sale) during the order lead time (assumed to be known and constant). In this model, in light of the simplifying hypotheses introduced, the ROP is exactly equal to the average consumption during the average lead time, and thus:

$$ROP = \overline{LT} \times \overline{d} \text{ (units)} \quad [10.13]$$

where:
ROP = Reorder Point (units);
$\overline{LT}$ = Average Lead Time in the case of purchase or production and transfer from one department to another (periods);
$\overline{d}$ = Average Demand Rate or consumption rate (units/period).

Figure 10.5 **Determination of reorder point**

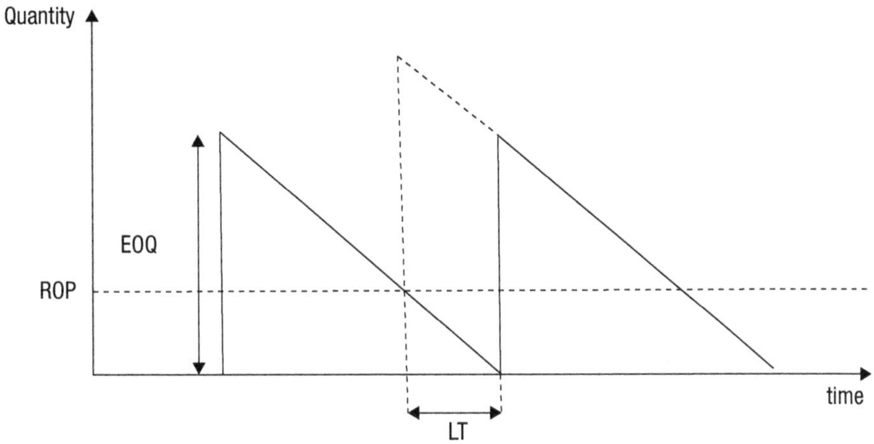

Table 10.3 **The determination of the Reorder Point**

Periods	**1**	**2**	**3**	**4**	**avg**
Purchase Lead Time (days)	20	22	18	20	20
Consumption (units/day)	20	10	15	15	15

For example, suppose we have the data expressed in **Table 10.3**.

As we see, the average daily consumption is equal to 15 units and the average supply lead time 20 days. We can then easily calculate the ROP, that is equal to 15 units/day × 20 days = 300 units. At the time the stock level reaches these quantities, it becomes necessary to launch a purchase or production order to replenish the stock in time.

10.2 EOQ insights

The EOQ model, despite being an extremely simplified construct, offers a series of **critical insights** that is worthy to consider.

- An initial significant consideration is based on **the nature of the algebraic link between the dynamics of demand (D) and the EOQ**; the latter does not change with the same proportion as the demand, but by its square root. Two products, *x* and *y*, that differ by a factor of 25 in the request quantities, differ by a factor of 5 in the size of the purchase or production order. This shows the presence of obvious economies of scale in the inventory management.
- In addition, the **Total Cost**, in proximity to its minimum point, is relatively insensitive to the variations of the batch size for low value products, as illustrated in

Figure 10.6 **EOQ and product value**

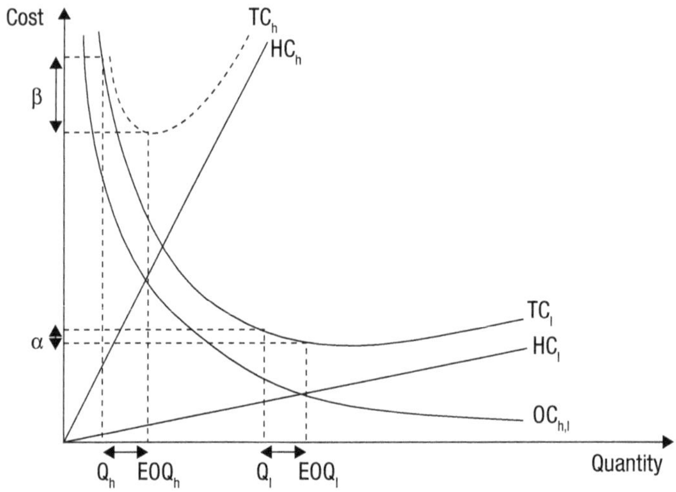

Figure 10.6, which shows the determination of the batch for two products characterized by the same Ordering Cost (OC) and by different Holding Costs (HC), high (h) and low (l). In that case, consistent with the considerations presented, the batch of product (h) is lower and the economic impact of potential deviations from the ideal quantity (Q – EOQ) is proportionally higher, as shown by the interval β, larger than α. In consideration also of the riskiness of calculation of certain cost items, it is thus possible to obtain satisfactory results also close to the optimal point, especially for low value articles. In that way, the search for that interval, that at times is very burdensome in the presence of numerous articles in stock, is significantly simplified.

- Lastly, the **direct and inverse relationships observed** between Unit Ordering Costs, Unit Holding Costs, and Economic Order Quantity (EOQ), provide business managers with important decision-making leverage in view of an efficient management of the uses of financial resources. For example, with the growth of cost of capital (WACC), the optimal size of the order and the average stock level are reduced. The economic need to rationalize the management of stock can lead to interventions aimed at stimulating greater flexibility in production or supply (that allows for purchase or production in small batches) and, when possible, lower lead times.

10.3 Setting stock levels in conditions of uncertainty

In the previous section, we considered the problem of stock control in conditions of absolute certainty. We assumed that demand (or internal consumption) is constant and known, and that lead time is constant and known as well. Now it is necessary to abandon those unrealistic simplifying assumptions and take into account the **uncertainty** that characterizes any business and company's operations. In observing

Table 10.3, we in fact see that only in the fourth period does the actual value of lead time and consumption reflect the average values used for the sizing of the Reorder Point, while in the second and third periods, there is excess stock at the time of replenishment, and in the first, there is a situation of stock-out. Making the situation expressed in **Figure 10.5** more realistic, it is possible to graphically represent the different hypotheses of inventory levels, at the time they reach the reorder point, if demand (or consumption levels) are not in fact known with certainty, but they are random variables. On this point, see **Figure 10.7**.

Figure 10.7 The case of aleatory consumption

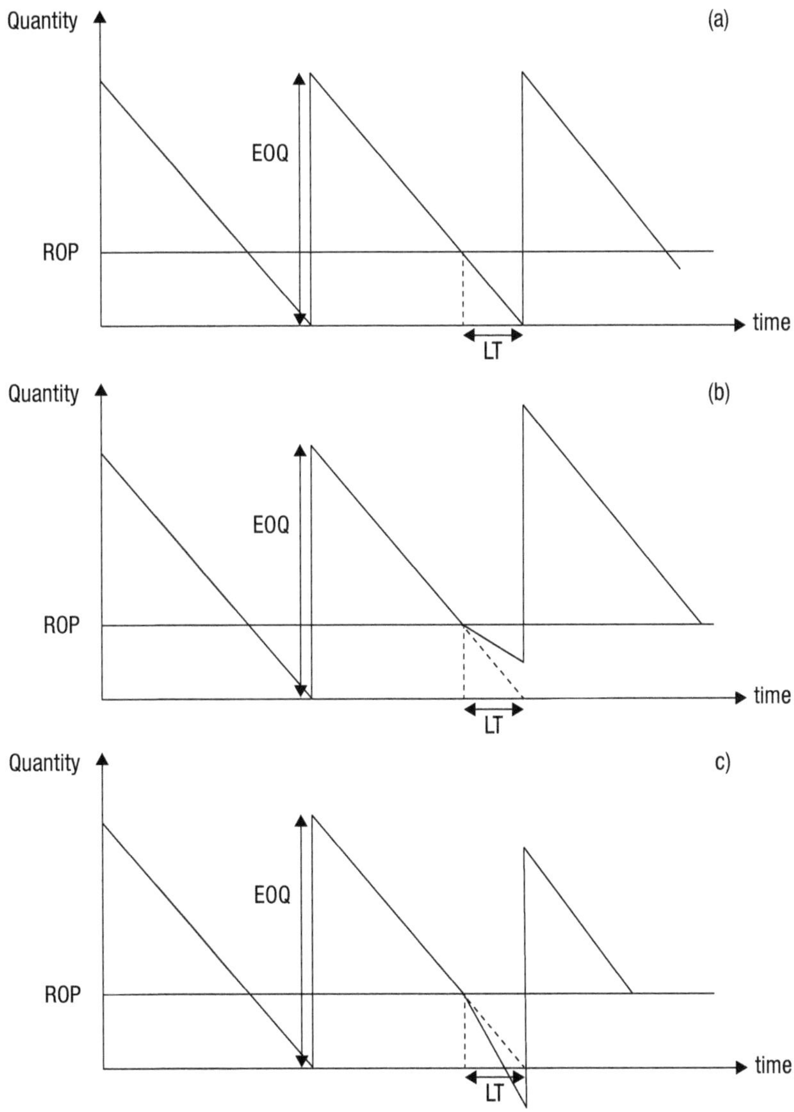

In case (a), that again corresponds to Figure 10.5, consumption in the procurement or production lead-time (expressed by the slope of the curve) is **considered constant**; in that case, inventory is exhausted at the same time that a new batch of goods arrives.

In case (b), during the lead time, **consumption is lower** than the estimated level; in such a situation, when new quantities of goods arrive, the stock is not completely exhausted and the stock quantities are summed, translating into an excess of accumulation, or pro-tempore **over-stock**.

The case shown in (c) can lead to much more serious consequences, as **consumption is higher** than expected in the lead-time. That situation translates into insufficient stock levels, generally defined as **stock-out**.

Examples with entirely analogous consequences can be built by considering **anticipations or delays** with respect to the expected average **lead time** (random variable), even in the presence of constant and known levels of consumption; the two possible variations are represented in **Figure 10.8**. In the extreme case, that in which the two unforeseen events combine – deviation of consumption and of lead-time from the average value – cumulative or compensatory effects occur in terms of excess or insufficiency of stock. The uncertainty in consumption (or demand in the case of finished products) and/or in lead times, can be dealt with using particular, appropriately sized volumes of inventory, previously (Chapter 9) called **Safety Stock**. The level of those volumes is thus **proportionate to the uncertainty** of demand, or of internal consumption, to the reliability of the supply system and the service level desired. In theory, its optimal size derives from a comparison between the holding costs that the business must sustain to maintain additional quantities of inventory with precautionary intent, and the cost generated by stock-out situations. As we have already said, the estimate of this value, especially in situations of insufficiency of finished products, is very arduous and risky. This has led to the use of another parameter for com-

Figure 10.8 Variability of consumption rate and lead times

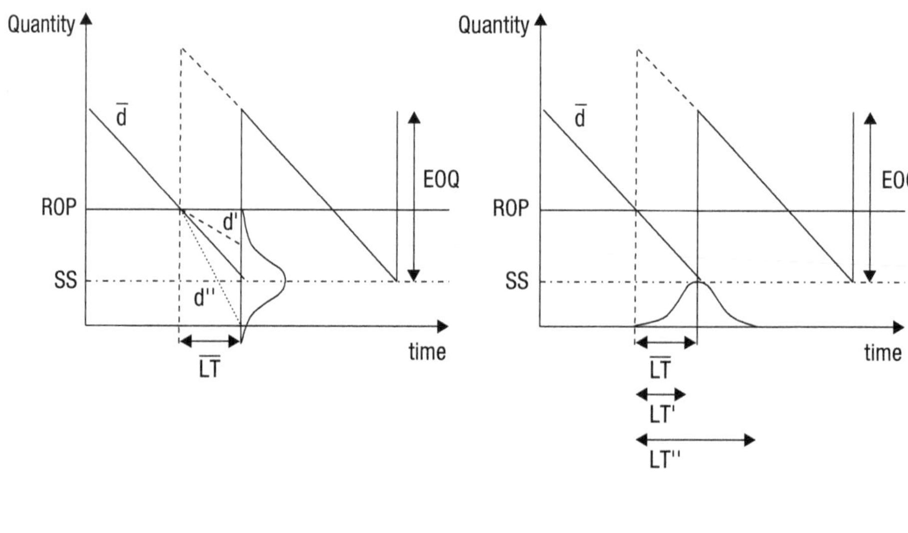

parison: the **Service Level** (SL), i.e. the percentage of stock-out periods considered sustainable with respect to the total of the request periods.

That way, a business that defines a service level of 90% believes it can promptly fulfill incoming orders 9 out of 10 times, considering it acceptable to find itself with insufficient stock 1 out of 10 times. In reality, as we will see below, the crux of the problem lies precisely in the analytic determination of the **desired optimal service level**.

To establish the convenient level of Safety Stock, corresponding to a chosen service level, it is necessary to know the variability of demand (or consumption) and of lead time, i.e. their frequency distribution. The probabilistic distributions most used for the purpose is the Gauss distribution. If, for example, the distribution of consumption can be represented with sufficient approximation by a Gauss (or normal) curve, it is possible to express the amount of the **Safety Stock** (SS) in terms of a coefficient (*K*), to be multiplied by the standard deviation of demand or consumption (σ_d), that ensures the protection from stock-out risk at a certain level of service.

K in fact represents the value of the standardized variable, with mean value = 0 and standard deviation = 1, for which defined cumulative probability levels are obtained: due to the Gauss curve properties we obtain cumulative **probabilities equal to 68.26%, 95.44%** and **99.74%** at values of *K* equal to 1, 2 and 3, respectively, or **service levels equal to 84.13%, 97.72%** and **99.87%,** since the portion of the curve represented by negative σ is already covered by the stock sized on mean demand values (equal to Reorder Point). On this point, see **Figure 10.9**.

Figure 10.9 The Gauss curve in the calculation of safety stock

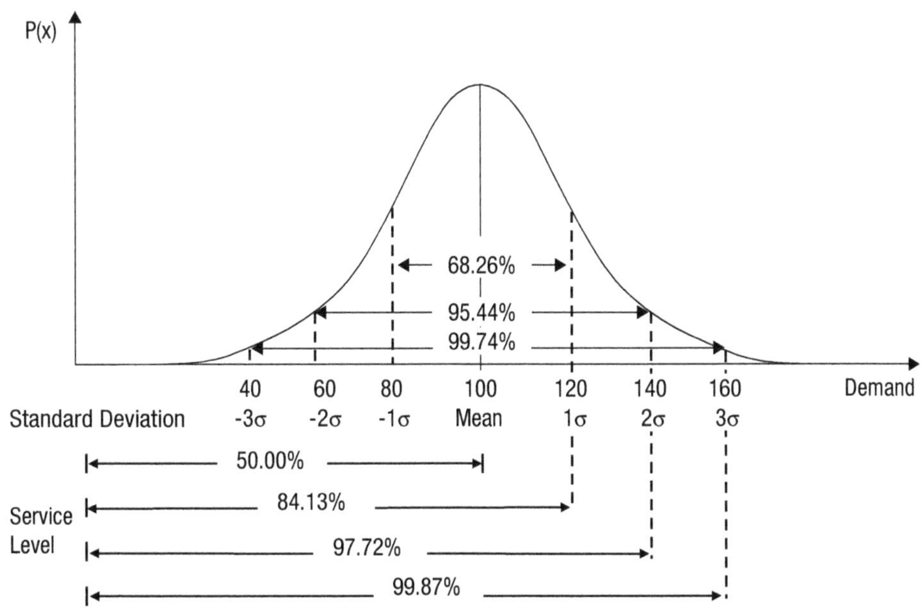

There is thus a probability of 84.13% (50 + 68.26/2), 97.72% (50 + 95.44/2), and 99.87% (50 + 99.74/2), respectively, that actual demand is lower than the sum of the mean value plus the multiplier of the chosen standard deviation, and thus that the demand can be covered by the safety stock held.

$$SL\ 99.87\% = P\,(d < \bar{d} + 3\sigma_d) \qquad [10.14]$$

This way, it is possible to calculate the amount of the Safety Stock linked to a target Service Level, the expression of the company's intention to accept situations of insufficient stock (stock-out). In cases in which only one of the elements at play is affected by uncertainty, be it consumption (d) or lead time (LT), the safety stock is respectively calculated as follows:

$$SS_{LT} = K\sigma_{LT} \times \bar{d} \text{ (units)} \qquad [10.15]$$

and

$$SS_d = K\sigma_d \times \sqrt{LT} \qquad [10.16]$$

While in the first case, the $\mathbf{SS_{LT}}$ is generated by the product of the standard deviation of the lead time for the average consumption (or demand) rate in the unit of time, in the second ($\mathbf{SS_d}$) it is necessary to also consider the possibility that the consumption rate shows deviations from the average (expressed by its standard deviation) in each temporal interval (t), considered uniformly, of which the lead time is made up, that in this case is known and constant. Therefore, for independent variations:

$$SS_d = K \times \sqrt{\sigma^2_{dt1} + \sigma^2_{dt2} + \sigma^2_{dt3} + \ldots + \ldots \sigma^2_{dtn}} = K \times \sqrt{\sigma_d{}^2 LT_n} = K\sigma_d \times \sqrt{LT_n}$$

where n = number of periods (t) making up LT.

In the event that both factors are simultaneously subject to variability, knowing that a random variable resulting from two independent random variables holds a standard deviation equal to the sum of the two variances, due to the law of compound probability, the Safety Stock ($SS_{d,LT}$) is found to be:

$$SS_{d,LT} = K \times \sqrt{\sigma^2_d \overline{LT} + \sigma^2_{LT} \bar{d}^2} \qquad [10.17]$$

where:
σ_d = standard deviation of demand d;
σ_{LT} = standard deviation, expressed in the same unit of measure (periods t) of the lead time LT.

It should be stressed that the quantity of the Safety Stock ultimately depends on three components:

- K, that expresses the **target Service Level** and is thus a parameter that management can control;

- σ_d, that expresses the **uncertainty of demand**, which the company can do little about;
- σ_{LT}, or **variability of lead time**, in which containment or stabilization is the result of interaction and collaboration between the client company (or department) and the supply partner and the reliability of the latter.

The relationship between service level, probability of stock-out periods and probability average stock-out times in a period, see **Box 10.1**.

Box 10.1 Service level definitions

Given a specific period, such as a month, year, etc., service level (SL) is usually defined as a %, such as 85% or 95%. It is possible to calculate SL in different ways:

- **probability of stock-out periods**. Assuming K = 2, the probability that demand is greater than availability in each order cycle is about 2.3% (1 – 97.7), because in the remaining 97.7% of cases it will meet the demand. In this case it means that for a given part number, every 100 orders, a stock-out situation will be generated approximately 2-3 times;
- **average stock-out times in a defined period**. Assuming a period equal to one year, for example, and setting:
  - NSO = number of stock-outs in a period (n)
  - D = demand in the same period (units/period)
  - Q = purchase order quantity (units)
  - N = number of orders in a period = $\frac{D}{Q}$ (times)

$$NSO = \frac{D}{Q}(1 - K\sigma_d) = N\,(1 - K\sigma_d)$$

Imagining 52 weekly orders in a year, NSO = 52 × 2.3% = 1.2 stock-outs in a year

Source: adapted from De Toni and Panizzolo, 2018.

Simplified method to size safety stock

The simplicity of this method hides many **limits of application**. The main limits are the need to estimate the variability distribution of Demand and of the Lead Time (generally represented by a normal curve), and the fact that the Service Level is determined based on past experience, with the conviction that past conditions will remain unchanged in the future. Those limits, despite not affecting the basic validity of the method, often lead the persons responsible for inventory management to prefer **empirical solutions** that tend to determine the ideal service level and the related safety stock. It must in any event be recalled though, that the degree of accuracy and precision of the analyses considered is to be compared with the cost of the inventory management system, according to the general principle of optimization of the cost/benefit ratio. As stated, Safety Stock is related to the target service level, such that with the increase of the latter, the precautionary stock level and the related holding cost also necessarily increases; to the contrary, when it decreases, the probability of experiencing a stock-out increases, with the resulting costs. It can thus be argued from a conceptual standpoint, as already mentioned, that there exists a convenient level of safety stock, equal to the quantity that minimizes the cumulative cost of holding and stock-out.

It is also evident that while for demand and consumption there can be long time series based on which the analyses proposed above can be conducted, this is **more difficult for lead times**. This is due to both the frequency with which the orders are issued and the risk that in the time horizon taken into consideration the supplier or the contractual terms of delivery have changed. Therefore, in general, the estimate of probability that the lead time will be greater than average is conducted empirically by the purchasing managers, that can lead to a "forced" increase of the average lead time. Whatever the method of determination of the two components of safety stock (SS_d e SS_{LT}), it is possible to adopt **simplified approaches** for the determination of the overall safety stock ($SS_{d,LT}$).

In empirical terms, it has in fact been observed that when SS_{LT} is significantly higher than SS_d or vice versa, the value of $SS_{d,LT}$ is equal to approximately the greater SS between the two. Since the variability of demand is generally higher than the variability of LT, it is possible to reasonably size the safety stock on the first between the two. In that case, for the materials with low usage value, for which the approximation is justified, it is possible to use the following **empirical rule** (Corke D. 1969, in Da Villa, 1991):

- Calculate the ratio between SS_d and SS_{LT};
- If $SS_d / SS_{LT} \geq 1.5$, set $SS_{d,LT} = SS_d$;
- If $SS_d / SS_{LT} < 1.5$, set $SS_{d,LT} = 1.5 \times SS_{LT}$.

As shown by **Table 10.4**, the differences calculated according to the approach illustrated in [10.17] and that based on the empirical rule are relatively limited.

Table 10.4 Application of empirical rule for materials with low usage value

SS_d (units)	SS_{LT} (units)	SS_d/SS_{LT}	1 $SS_{d,LT} = \sqrt{SS_d^2 + SS_{LT}^2}$ (units)	2 $SS_{d,LT}$ = Empirical Rule (units)	3 % Difference 2 – 1 / 2 (%)
90	30	3.00	94.9	90	–5.4
80	40	2.00	89.4	80	–11.8
60	40	1.50	72.1	60	–20.2
65	50	1.30	82.0	75	–9.3
50	50	1.00	70.7	75	5.7
30	45	0.67	54.1	67.5	19.9

10.4 The methods of stock control

As introduced in the previous chapter, stock control systems distinguish based on the nature of demand: goods that are independent or dependent of demand. Below, we refer to the former, and to the **look-back logic**, because it is based on the historical performance of the variables considered.

The techniques for reconstituting stock are necessarily used for **independent demand goods**, and as we will illustrate, are the basis of the main lot-sizing logics used also in the planned order receipts phase for demand dependent goods (see Chapter 13). Among the former, the best-known, which are illustrated here, are based on the management choices related to two main variables (Hax and Candea, 1967; Da Villa, 1991; Sianesi, 2011; De Toni and Panizzolo, 2018):

- the **logic (frequency) of stock revision**, that can be **continuous** or **periodic**;
- the **order (purchase or production) quantity**, that can be **fixed** or **variable**.

With reference to the first variable, we distinguish between two approaches:

- the first provides for **continuous monitoring** of the level of stock available at each transaction and is based on the use of a **Reorder Point**, below which a replenishment order is issued;
- in the second, the control of the level of stock takes place at **preset**, **constant intervals**, called **Reorder Cycles**, at the end of which the replenishment order is issued to reach a maximum level appropriately defined in advance.

In regard to the second variable, we distinguish between:

- criteria in which the **quantity ordered is set as constant** for each order – for example calculated according to the formula of EOQ or conditioned by technical or logistical constraints;
- criteria in which the **quantity ordered is variable** and varies from order to order – determined by the difference between a maximum stock desired and the stock available as measured at the time of the control.

By crossing the two variables, it is possible to classify the principal methods of stock management, as illustrated in Figure 10.10.

As described in Figure 10.10, the main stock control methods are:

- **Fixed Reorder Quantity inventory control system**, based on Continuous Control (ROP) and Fixed Order Quantity (EOQ);
- **Fixed Reorder Cycle (or Interval or Review period) inventory control system**, based on a Fixed Reorder Cycle (ROC) and Variable Order Quantity (VOQ);
- **Fixed Reorder Cycle (or Interval or Review period) and Order Quantity inventory control system**, based on Fixed Reorder Cycle (ROC) and Fixed Order Quantity (EOQ);

- **Minimum and Maximum and Fixed Reorder Cycle inventory control system**, based on a Fixed Reorder Cycle (ROC), a Variable Order Quantity (VOQ);
- **Minimum and Maximum and Continuous inventory control system**, based on Continuous Control (ROP) and Variable Order Quantity (VOQ);
- **Simplified Stock Replenishment control system**.

Figure 10.10 A typology of inventory control systems

FREQUENCY OF INVENTORY REVISION	ORDER QUANTITY: FIXED - EOQ	ORDER QUANTITY: VARIABLE - VOQ
CONTINUOUS - ROP	– Fixed Reorder Quantity	– Minimum and Maximum with Continuous Control
PERIODIC - ROC	– Fixed Reorder Cycle with Fixed Quantity	– Fixed Reorder Cycle with Variable Quantity – Minimum and Maximum with Periodic Control – Symplified method

Source: adapted from De Toni and Panizzolo, 2018.

Fixed Reorder Quantity control system (ROP and EOQ)

The functioning of **Fixed Reorder Quantity** (FQ) control system is based on the concepts of Economic Order Quantity (EOQ) of purchase or (EPQ) of production and of Reorder Point (ROP) previously described. On this point, see the example and the **Figure 10.11**. In conditions of variable demand, the procurement (or production) order is issued each time the available inventory reaches the preset Reorder Point, where:

$$ROP = \bar{d} \times \overline{LT} + SS \quad [10.18]$$

with:
ROP = reorder point (units);
$\bar{d}$ = average consumption or demand rate (units/period);
$\overline{LT}$ = average lead time (periods);
SS = safety stock (units), calculated as $K\sigma_{d,LT}$, as previously described.

Upon reaching the reorder level, a procurement (or production) order is issued for a fixed quantity, equal to the calculated EOQ/EPQ. The **average stock** ($\bar{I}$) is equal to:

$$\bar{I} = \frac{Q}{2} + SS = \frac{EOQ}{2} + SS \quad [10.19]$$

In case of EOQ in linear replenishment systems, as already mentioned, the average stock (I) is equal to:

$$\bar{I} = \frac{Q'}{2}' + SS = \left(\frac{\mathrm{T} \times (p-d)}{2}\right) + SS \qquad [10.20]$$

In this model, the **review interval** (Order Cycle) between the issue of the orders is **variable** and is linked to the consumption rate, while the ordered **quantity** tends to be **fixed**; we say it "tends to" because Economic Order Quantities can also vary with the variation of the significant cost elements based on which they are calculated. It is in fact advisable to conduct periodic checks of the same, and of the rates of consumption and lead times to calculate the EOQ, ROP and SS, respectively.

The Fixed Reorder Quantity model only allows for reorder by **independent items**, requires constant control, and is preferable in the presence of inexpensive automatic control mechanisms, and of high order costs and unit values of goods in stock.

Figure 10.11 Fixed Reorder Quantity control systems

INVENTORY SYSTEM PARAMETERS				
ROP (units)	**SS (units)**	**EOQ (units)**	**LT (days)**	
1500	400	4500	4	
INVENTORY RECORDS				
DATE	**Load**	**Withdrawal**	**Balance**	**Orders Release**
.....				
7			2000	
8		200	1800	
9		400	1400	4500
10			1400	
11		600	800	
12			800	
13	4500		5300	
14			5300	
15		500	4800	
16		500	4300	
17			4300	
18		500	3800	
19		450	3350	
20		300	3050	
21		500	2550	
22		600	1950	
23		800	1150	4500
24			1150	
25		300	850	
26		250	600	
27	4500		5100	
28		600	4500	

INVENTORY RECORDS				
DATE	Load	Withdrawal	Balance	Orders Release
29		500	4000	
30		300	3700	
31			3700	

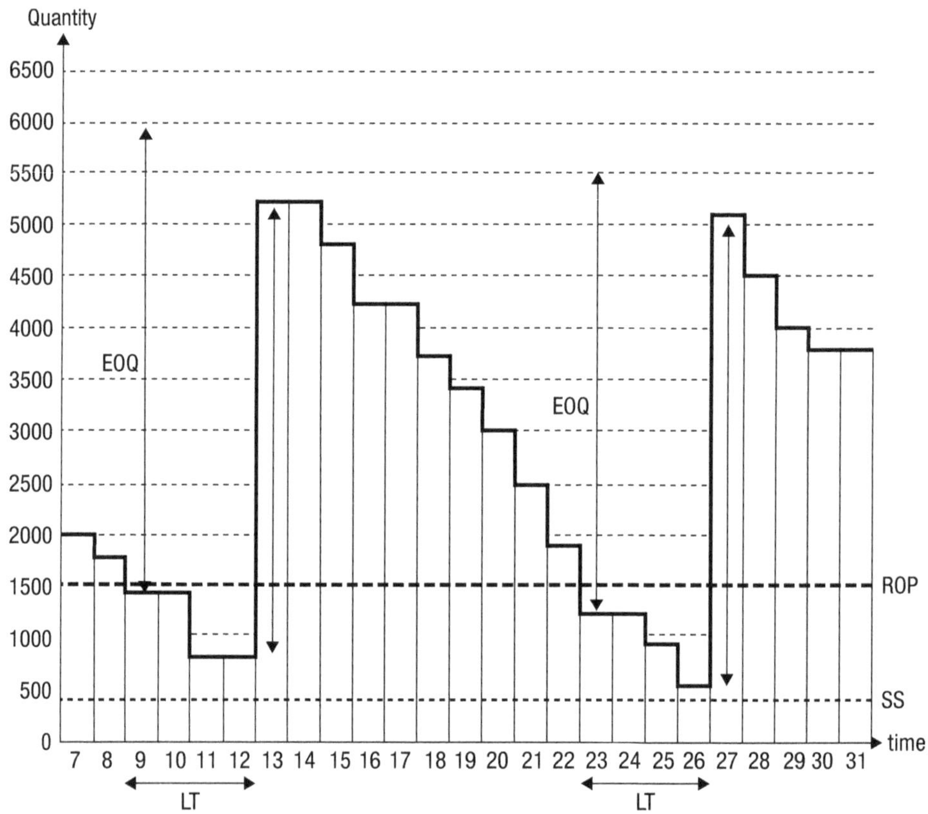

Fixed Reorder Cycle control system (ROC and VOQ)

With this control system, called **Fixed Reorder Cycle** or **Fixed Review Interval (FR)**, the purchase (or production) orders are issued at **regular intervals** of time (1 month, 15 days, etc.), established for **variable quantities** based on variations in demand (or consumption) that occur. In that case, as shown by the example and the **Figure 10.12**, the quantity ordered is equal to the difference between a **Maximum desired (or Target) Stock Level (ML)** (equal to the safety stock added to average consumption expected in a temporal horizon set as equal to the average lead time plus a reorder cycle) and the available inventory at the time of the control.

In the case of variability of both the demand rate and the lead time ($\sigma_{d,LT}$), the model operates due to the following parameters:

$$ML = \bar{d} \times \overline{LT} + \bar{d} \times FR + SS'_{d,LT} = \bar{d} \times (\overline{LT} + FR) + SS'_{d,LT} =$$
$$= \bar{d}\,(\overline{LT} + FR) + K\,\sqrt{(\overline{LT} + FR)\;\sigma_d^2 + \bar{d}^2\,\sigma_{LT}^2} \qquad [10.21]$$

where:
ML = target or maximum stock level (units);
$\bar{d}$ = average consumption rate (units/period);
$\overline{LT}$ = average lead time (periods);
FR = fixed review interval of reorder cycle (periods);
$SS'_{d,LT}$ = safety stock (units), calculated according to [10.17] but considering the standard deviation of the consumption (or demand) rate σ_d in the period composed by FR + LT:

$$SS'_{d,LT} = K\,\sqrt{\sigma_d^2(\overline{LT} + FR) + \sigma_{LT}^2\,d^2} \qquad [10.22]$$

And thus:

$$Q_n = ML - I_n$$

where:
Q_n = quantities ordered at t_n;
I_n = available inventory at time t_n.

Figure 10.12 Fixed Reorder Cycle control systems

		INVENTORY SYSTEM PARAMETERS		
ML (units)	**SS (units)**	**FR (days)**	**LT (days)**	
7500	800	10	4	
		INVENTORY RECORDS		
DATE	**Load**	**Withdrawal**	**Balance**	**Orders Release**
.....				
7			3500	
8		200	3300	
9		400	2900	
10			2900	4600
11		600	2300	
12			2300	
13			2300	
14	4600		6900	
15		500	6400	
16		500	5900	
17			5900	
18		500	5400	
19		450	4950	
20		300	4650	2850
21		500	4150	
22		600	3550	
23		800	2750	

INVENTORY RECORDS

DATE	Load	Withdrawal	Balance	Orders Release
24	2850		5600	
25		300	5300	
26		250	5050	
27			5050	
28		600	4450	
29		500	3950	
30		300	3650	3850
31			3650	

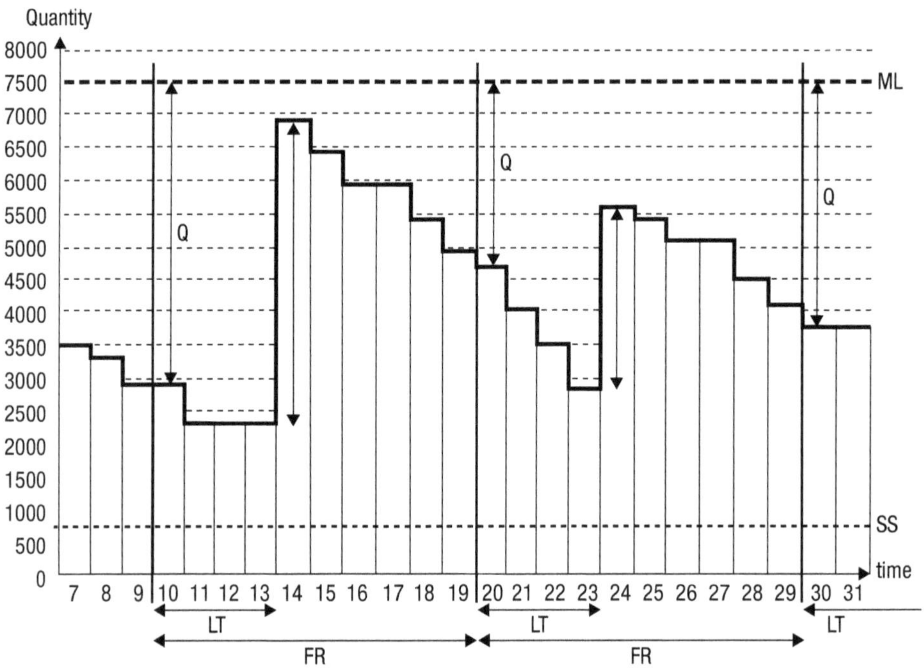

In a simplified model where the demand d is steady, for each FR period, the **average stock** (I) is equal to $Q / 2 + SS'$ where Q represents the average of the Q_i quantities ordered in each generic time n, or the average inventory between its maximum level (I_{max}) and the minimum level (I_{min}). Considering the transit stock coming from the previous order and the demand (or consumption) for the current order:

$$I_{max} = \bar{d} \times (FR + LT - LT)$$

$$I_{min} = 0$$

Therefore:

$$I = \frac{(\bar{d} \times FR + 0)}{2} + SS' = \frac{(\bar{d} \times FR)}{2} + SS' \qquad [10.23]$$

The characteristic that most differentiates this method from the previous one is that the **Safety Stock** must be determined in relation to the variability of expected consumption in the Fixed Review Interval, a constant interval equal to the time between one procurement and the next one ($FR = t_2 - t_1$), summed with the average lead time, and not only based on the latter, as in the Fixed Reorder Quantity model. FR can be defined exogenously, according to logistical convenience, determined for example at a procurement pace linked to the frequency of deliveries established by supplies or by specific production campaigns set at particular time intervals, or optimized from a strictly economic standpoint. In that case, an ***economic FR*** (***FR_e***) is defined, a replenishment interval determined with the aim of ordering on average an Economic Order Quantity:

$$FR_e = \frac{T}{N_e} = \frac{T}{\frac{D}{EOQ}} \; (periods)$$

or, as $D = T \times \bar{d}$:

$$FR_e = \frac{EOQ}{\bar{d}} \; (periods) \qquad [10.24]$$

where:
T = number of days, weeks, etc. composing the period of reference, usually the year (periods);
N_e = optimum number of orders realized in T;
EOQ = Economic Order Quantity (units);
$\bar{d}$ = average demand (or consumption) in the observed period (units/period).

Contrary to the Fixed Reorder Quantity method, in the model under consideration here, the **procurement or production cycle (FR) is fixed**, while the **quantity procured or produced varies**, based on the rate of difference between the Maximum Stock Level (ML) and the available inventory level detected in each reordering instant. This system is much simpler and less costly than the previous one, as there are no cost estimates, unless a FR_e is determined.

This method also allows for **reordering by joint items**, from the same supplier, or production launches for different articles with the same set-up, that for example are linked to the same format. Moreover, this system facilitates the periodic physical control of stock, adapts to non-uniform consumption articles, and often guarantees savings in ordering costs. Indeed, in cases where a level of inventory higher than that of average consumption in the Fixed Reorder Cycle is found, the order is generally postponed until the next period; on the other hand, as illustrated, the **safety stock becomes higher** on average.

An example of calculation of the parameters for stock management is shown in **Box 10.2**.

Box 10.2 Example of calculation of parameters for management of stock of an article

Imagine that for the article taken into consideration we have the following data:

- v = 5 (€/unit)
- i = 20 (%)
- Oc = 20 (€/order)
- Target Servic e Level SL = 97.7%
- LT = 10 (days), known and constant
- T = 230 (days/year)

Considering the historical demand in a month considered significant, expressed by the table shown below, it is possible to determine the following parameters for stock management, rounding to whole values.

Fixed Reorder Quantity Model

$$EOQ = \sqrt{\frac{2DOc}{v \times i}}\ (unit) = 214 \text{ units}$$

$$SS_d = K\sigma_d \times \sqrt{\overline{LT}} = 16 \text{ units}$$

$$ROP = \bar{d} \times \overline{LT} + SS = 66 \text{ units}$$

$$\bar{I} = \frac{EOQ}{2} + SS = 123 \text{ units}$$

Fixed Reorder Cycle Model

$$FR_e = \frac{T}{N_e} = \frac{T}{(D/EOQ)} = 43 \text{ days}$$

$$SS'_d = \sigma_d \times \sqrt{\overline{LT} + FR_e} = 37 \text{ units}$$

$$ML = \bar{d} \times \left(\overline{LT} + FR_e\right) + SS'_d = 301 \text{ units}$$

$$\bar{I} = \frac{\bar{d} \times FR_e}{2} + SS'_d = 145 \text{ units}$$

Day	d_i	$d_i - \bar{d}$	$(d_i - \bar{d})^2$	Day	d_i	$d_i - \bar{d}$	$(d_i - \bar{d})^2$
1	9	4	16	16	7	2	4
2	2	–3	9	17	6	1	1
3	1	–4	16	18	9	4	16
4	3	–2	4	19	stock out	–	–
5	7	2	4	20	stock out	–	–
6	5	0	0	21	5	0	0
7	4	–1	1	22	5	0	0
8	8	3	9	23	4	–1	1
9	5	1	1	24	3	–2	4
10	5	0	0	25	4	–1	1
11	0	–5	25	26	1	–4	16
12	6	1	1	27	2	–3	9
13	5	0	0	28	8	3	9
14	7	2	4	29	3	–2	4
15	10	5	25	30	5	0	0
				mean	5		180

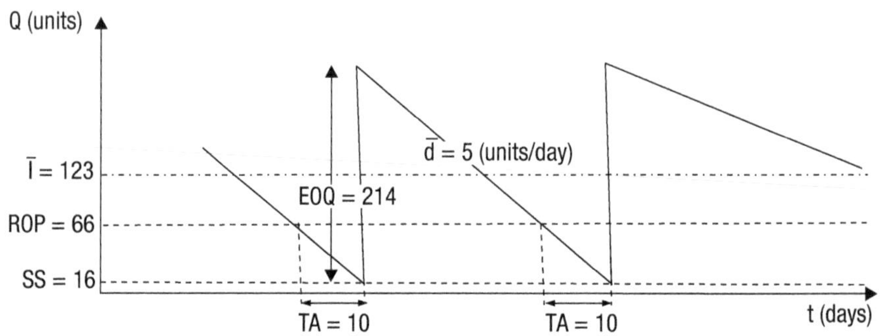

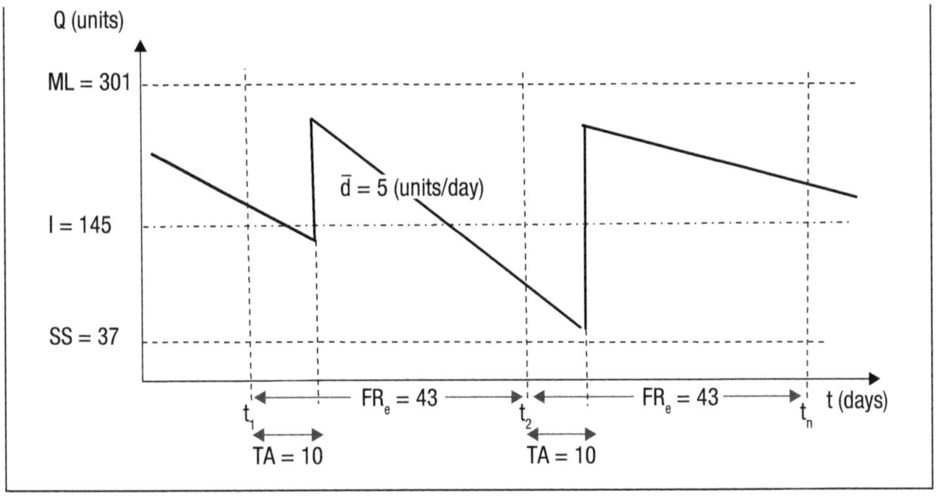

Fixed Reorder Cycle and Fixed Quantity inventory control systems (ROC and EOQ)

With this method, the level of stock is monitored at **Fixed intervals**, as in the previous Fixed Reorder Cycle model. If on the occasion of the periodic review the available stock is lower than a preset level (MinI), a **fixed replenishment order (EOQ)** is issued, otherwise orders are not issued. This is a method similar to the Fixed Reorder Quantity model, with the difference that the control does not take place continuously but at preset frequencies. As illustrated in the example and in **Figure 10.13**, the principal parameters of functioning, mediated by the methods illustrated previously, are:

- $\bar{d}$ = average consumption rate (units/period);
- Q = reorder lot, generally equal to EOQ, as in the Fixed Reorder Quantity model;
- FR = Fixed Reorder Cycle, calculated as in the method illustrated previously, according to logistical constraints or optimized (FR_e);
- $MinI$ = Reorder Point, appropriately modified to take into account the fact that it is a periodic control system (Fixed Reorder Cycles) and not a continuous control system, as in the case of Fixed Reorder Quantity Control.

In that case, the estimate of the Safety Stock (SS′) is defined as in the Fixed Reorder Cycle model, to take into account the average demand in the interval composed by the lead time and the Fixed Interval Review ($FR + \overline{LT}$). In that case, MinI is equal to ML already calculated with the Fixed Reorder model.

$$\text{MinI} = \text{ML} = \bar{d} \times (\overline{LT} + FR) + SS' =$$
$$= \bar{d} \times (\overline{LT} + FR) + K\sqrt{\sigma_{LT}^2\, \bar{d}^2\, (\overline{LT} + FR) + \sigma_{LT}^2\, \bar{d}^2} \qquad [10.25]$$

$$\text{Q} = \text{EOQ}$$

Figure 10.13 Fixed Reorder Cycle and Fixed Quantity inventory control model

INVENTORY SYSTEM PARAMETERS				
EOQ (units)	**SS (units)**	**FR (days)**	**LT (days)**	**MinI (units)**
4500	800	10	4	7500
INVENTORY RECORDS				
DATE	**Load**	**Withdrawal**	**Balance**	**Orders Release**
…..				
7			10000	
8		200	9800	
9		400	9400	
10			9400	0
11		600	8800	
12			8800	
13			8800	
14	0		8800	
15		500	8300	
16		500	7800	
17			7800	
18		500	7300	
19		450	6850	
20		300	6550	4500
21		500	6050	
22		600	5450	
23		800	4650	
24	4500		9150	
25		300	8850	
26		250	8600	
27			8600	
28		600	8000	
29		500	7500	
30		300	7200	4500
31			7200	

Minimum and Maximum with Fixed Reorder Cycle inventory control systems (ROC and VOQ)

This criterion, that is illustrated in the example and the **Figure 10.14**, suggests the issue of **a variable order**, equal to the difference between an inventory level defined as **Maximum stock (upper limit)** and the available stock at time t_n, if availability is lower than an inventory level defined as **Minimum stock (lower limit)**. The lower limit (MinI) is the same as the previous model and the Maximum Stock Level (ML) of the Fixed Reorder Cycle control method. The entity of upper limit (MaxI) is justified by the following considerations. It seems reasonable to assume that the desired size of the average order is represented by an EOQ, due to the implications pre-

Figure 10.14 Min-Max stock - Fixed Reorder Cycle control model

		INVENTORY SYSTEM PARAMETERS		
MaxI	**SS (units)**	**FR (days)**	**LT (days)**	**MinI (units)**
9750	800	10	4	7500
		INVENTORY RECORDS		
DATE	**Load**	**Withdrawal**	**Balance**	**Orders Release**
…..				
7			8000	
8		200	7800	
9		400	7400	
10			7400	2350

INVENTORY RECORDS

DATE	Load	Withdrawal	Balance	Orders Release
11		600	6800	
12			6800	
13			6800	
14	2350		9150	
15		500	8650	
16		500	8150	
17			8150	
18		500	7650	
19		450	7200	
20		300	6900	2850
21		500	6400	
22		600	5800	
23		800	5000	
24	2850		7850	
25		300	7550	
26		250	7300	
27			7300	
28		600	6700	
29		500	6200	
30		300	5900	3850
31			5900	

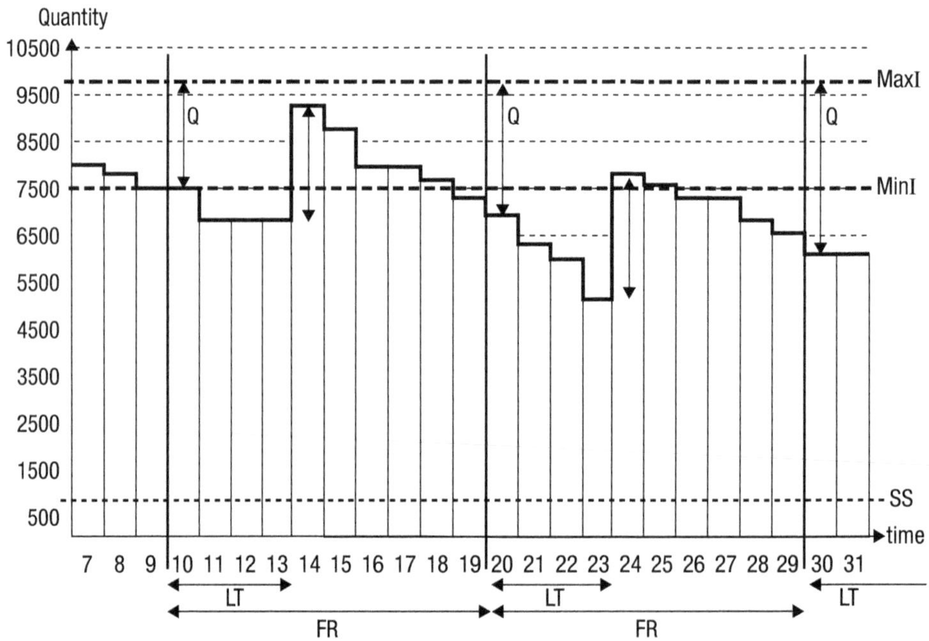

viously discussed with reference to cost optimization; since the size of the real orders depends on the actual demand, the average order will be approximately equal to the difference between the Maximum stock and the Minimum stock (the level below which the order may be issued) plus half of average consumption during a Fixed Reorder Cycle (FR), the time at which the order may be issued.

Therefore:

$$\text{Average order} = \sqrt{\frac{2DOc}{vi}} \cong MaxI - MinI + \frac{\bar{d} \times FR}{2}$$

from which:

$$MaxI = MinI + EOQ - \frac{\bar{d} \times FR}{2}$$

In that case:

$$MinI = \bar{d} \times (\overline{LT} + FR) + SS' \quad [10.26]$$

$$MaxI = MinI + EOQ - \frac{\bar{d} \times FR}{2} \quad [10.27]$$

from which we deduce that the Maximum stock is calculated so as to order Economic Order Quantities (EOQ), on average. Remembering that the economic Fixed Reorder Cycle (FR_e) is calculated as $T / (D / EOQ)$ or $EOQ / \bar{d}$, it is possible to demonstrate that MaxI is equal to MinI + $EOQ/2$:

$$MaxI = MinI + EOQ - \frac{\bar{d} \times FR_e}{2} = MinI + EOQ - \frac{\bar{d} \times \frac{T}{\frac{D}{EOQ}}}{2} =$$

$$= MinI + EOQ - \frac{\bar{d} \times \frac{EOQ}{\bar{d}}}{2} = MinI + EOQ - \frac{EOQ}{2} = MinI + \frac{EOQ}{2} \quad [10.28]$$

Lastly, using the previous notations the ordered quantity in each (n) period is:

$$Q_n = MaxI - I_n \quad [10.29]$$

In this method the Safety Stock ($SS'_{d,LT}$) can be calculated as in the previous one:

$$SS'_{d,LT} = K\sqrt{\sigma_d^2 (\overline{LT} + FR) + \sigma_{LT}^2 \bar{d}^2} \quad [10.22]$$

Minimum and Maximum with Continuous inventory control systems (ROP and VOQ)

This method is the best-known among those derived from the two pure methods explained above (FQ and FR). It is similar to the Min-Max Fixed Reorder Cycle control system, with the difference that the **monitoring of the stock level takes place continuously, not in fixed cycles**. As illustrated in the example and in **Figure 10.15**,

Figure 10.15 **Min-Max Stock - Continuous inventory control system**

INVENTORY SYSTEM PARAMETERS				
MinI (units)	**SS (units)**		**LT (days)**	**MaxI (units)**
1500	400		4	6000

INVENTORY RECORDS				
DATE	**Load**	**Withdrawal**	**Balance**	**Orders Release**
…..				
7			2000	
8		200	1800	
9		400	1400	4600
10			1400	
11		600	800	
12			800	
13	4600		5400	
14			5400	
15		500	4900	
16		500	4400	
17			4400	
18		500	3900	
19		450	3450	
20		300	3150	
21		500	2650	
22		600	2050	
23		800	1250	4750
24			1250	
25		300	950	
26		250	700	
27	4750		5450	
28		600	4850	
29		500	4350	
30		300	4050	
31			4050	

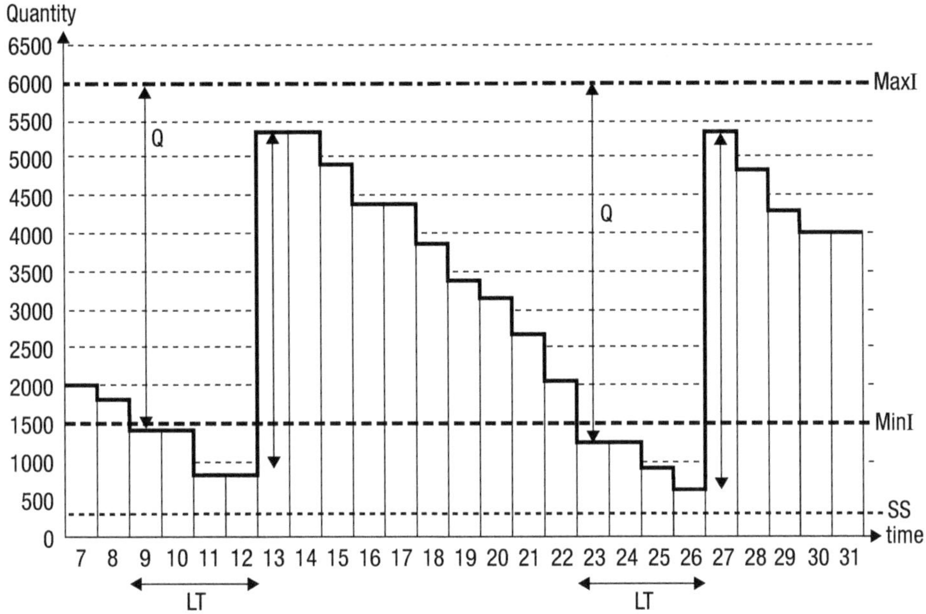

in this case, the inventory Lower Limit (MinI), below which it is necessary to issue a variable order, is equal to the Reorder Point (ROP) previously introduced in the Fixed Reorder Quantity method:

$$MinI = ROP = \bar{d} \times \overline{LT} + SS \qquad [10.17]$$

Any time the available stock reaches the ROP, a variable order to replenish a stock level equal to the MaxI has to be issued, where:

$$MaxI = MinI + EOQ \qquad [10.28]$$

aimed at balancing the stock ordering and holding costs. Max I can also be sized according to logistics constraints, such as the warehouse capacity, or, as already mentioned, trying to balance stock holding and stock-out costs.

Lastly, as in the previous method and using the same notations, the order quantity (Q):

$$Q_n = MaxI - I_n \qquad [10.31]$$

In this method, the amount of the Safety Stock (SS) is equal to that of the Fixed Reorder Quantity inventory control model, i.e.

$$SS'_{d,LT} = K \times \sqrt{\sigma_d^2 \overline{\mathrm{LT}} + \sigma_{LT}^2 \bar{d}^2} \qquad [10.17]$$

Simplified stock replenishment control system (ROC and VOQ)

This is a **simplification of the Fixed Reorder Cycle** method, in which the optimal level is defined as such:

$$ML = K \times dmax \qquad [10.32]$$

where:
$dmax$ = maximum historical (not episodic) consumption (or demand) in the time interval equal to the sum of the average lead time ($\overline{LT}$) and of the Fixed Reorder Cycle (*FR*);
K = safety coefficient determined empirically, for example as a percentage fraction of average coverage.

The method, in analogy with the system from which it derives, suggests ordering at any preset point in time (FR), and replenishing the available stock to the maximum level, for a quantity Q_n equal to the difference between the maximum level and the available inventory I_n at time t_n.

Therefore:

$$Q_n = ML - I_n \qquad [10.33]$$

10.5 Main characteristics of the stock control methods

Returning to the distinctions posed in regard to the characteristics of the methods illustrated, we stress that:

- the methods based on **continuous control:**
  - **guarantee economic advantages** linked to Safety Stock that is on average lower than the periodic control methods and the opportunity to make reorders based on EOQ;
  - on the other hand, they **require constant monitoring** of stock at each transaction and do not allow for joint reorders;
- the methods based on **periodic control:**
  - **offer advantages from a management standpoint**, requiring monitoring of stock at defined cycles, optimized from the standpoint of any administrative or logistics constraints, they allow for efficiencies linked to **joint reorders** and potentially to scheduled deliveries in the same days;
  - on the other hand, they **require higher safety stock** and guarantee the optimization of the quantities ordered only if reordered at intervals based on the economic Fixed Reorder time (FR_e).

The methods illustrated are only some among the many developed in the literature, that at times are more complex because they are able to take additional variables into account, such as the presence of quantity discounts or the launch of production campaigns.[4]

Moreover, in business practice, there are very frequently **mixed criteria**, adapted to the specific characteristics of the relevant situation, and sometimes simplified. On this point, it seems useful to suggest that the choice between the broad number of existing methods must be made by observing multiple aspects, such as the history of the article (frequency of consumption, cyclical and seasonal nature of demand), its critical aspects, linked to both value (actual cost) and its importance in the transformation cycle (opportunity cost), or the existence of constraints imposed by the transport and supply system, in terms of convenient orders. The choice cannot but be made based on prior experience, tests, and simulations, identifying the criteria that guarantee the best results in terms of overall economic convenience and service level, taking into account the managerial skills available as well. Indeed, although the most sophisticated models are often excellent from the standpoint of completeness and methodological rigor, it is worthwhile to remember **that the needs in terms of simplicity and frequency of use often lead to the adoption of less complex solutions**, driven principally by the recognition that only simplicity of application provides sufficient guarantees of actual adoption.

[4] On this point, see Vollmann et al., 1992.

11 Production Planning and Control: The Master Production Schedule

by *Alberto Grando*

11.1 The Master Production Schedule

As we have already mentioned, once the Aggregate Plan is authorized, a medium-term horizon is adopted, and the production planning activities take shape in the preparation of the **Master Production Schedule – MPS**. The MPS draws on inputs from the orders book and the sales forecasts for the medium-term, and considering the constraints on Available Production Capacity determined in the authorized Aggregate Plan; it consists of the disaggregation of the Aggregate Plan into a more detailed plan, both in terms of the objects and the physical quantities considered, product families or single products (codes), and not aggregate product lines or families, and in terms of the time horizon embraced, that is shorter and a sub-set of the previous one, generally measured in months or weeks.

As concerns the frequency of planning and the time horizon, a refreshment of the situation is generally carried out daily or weekly, with horizons linked to the cumulated lead-times for the articles subject to planning, the management practices adopted, and the processing ability of the information systems used. Concerning the subject of the master plan, it is useful to present some considerations regarding the method of response to the market offered by a given production system.

As illustrated in **Figure 11.1**, comparing the response time accepted by the market (**delivery time**) and the time required by the cycle that goes from planning to procurement and production of a product in a broad sense (or **lead time**), we can identify five types of production, already mentioned during the classification of production processes (see Chapter 2) (Wortman, 1983).

1. **Make to Stock (MTS)** or Assemble to Stock (ATS) or also Delivery to Order (DTO): this is the case in which the client is not willing to wait, and therefore it is necessary to produce on forecast, targeting a stock level. In this case, the subject of the forecasts and MPS are families or articles of finished products, since it becomes necessary to plan the entire amount of production to stock, not having any other way to immediately satisfy market requests. This occurs with the production of footwear and clothing, as these are seasonal products, and with many other standard products for which the market requests immediate availability.
2. **Assemble to Order (ATO)**: this is the frequent case in which the client is willing to wait a sufficient amount of time to cover the final phases of assembly, for configurable products in which **postponement policies** are adopted. Therefore,

Figure 11.1 The logics for responding to market demand

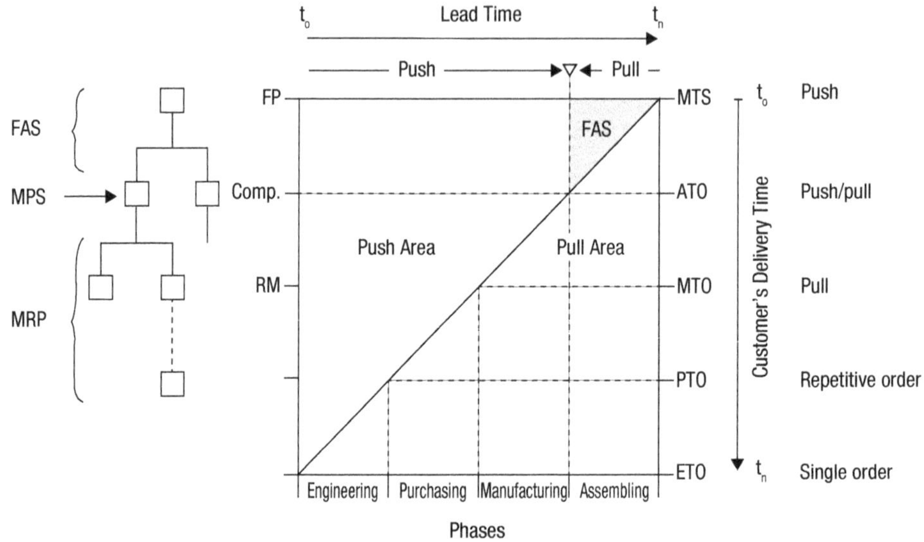

the system is divided into two sections, which are decoupled, as in the example in **Figure 11.1**, from a stock point called the **Decoupling Point** (sometimes defined the Customer Decoupling Point – CDP or Customer Order Decoupling Point – CODP) that represents the point of penetration of the client order into the planning process. With it not being possible to act otherwise due to time restrictions, in the **upstream** section, based on forecasts linked to product families, the assembly components are produced that thus make up the object of the plan, while **downstream**, those components are assembled based on the customer-order specifications using FAS – **Final Assembly Schedule** modules. We see this, for example, in industries such as furniture, and other products that can be configured based on specific orders from the client starting from neutral sets of components.

3. **Make to Order (MTO)**: if the delivery time accepted by the client is sufficient to cover the entire cycle of transformation, all of the phases of production are launched from the orders and scheduled with FAS modules and from the consequent Materials Requirement Planning – MRP (Chapter 13) explosion, while only purchases of materials and components are managed with forecasts and constitute the subject of the master plan; the latter, it consists of a purchasing plan deriving from the estimate of consumption of raw materials and components based on calculations carried out with the Aggregate Plan. The **decoupling stock point**, that will have to have a better assortment in order to allow the launch of production with sufficient time, is placed at the level of **raw materials and components purchased**. This category, for example, includes the producers of industrial machines or very customized items produced based on client specifications, such as some clothes.

4. **Purchase to Order (PTO)**: this is the case in which, due to the uncertainty of demand and/or the long delivery times granted by the market, it is possible to **procure materials to orders**. So this is a production based on repeat orders and there is no MPS since the flow is entirely governed by **FAS logics** that operate on the orders book and extend over the medium-term. Here there is practically no need for forecasts, except in general terms for the sizing of the available production capacity of critical resources and the development of a consequent Aggregate Plan. This is the case, for example, of those who operate in the field of complex systems, personalized machines, tailoring activities, or in cases in which raw materials and components have a high cost, so it is not convenient to manage them in stock or in situations of operating based on technical specifications from the client to personalize general projects filed by the Technical Office.
5. **Engineer to Order (ETO)**: this case is similar to the previous one, in which the customer **delivery time** goes to the point of **embracing the design phase** as well. This happens with single orders, and as in the previous case, planning is carried out through the Aggregate Plan, based on a sales estimate of the probability of obtaining orders, with the goal of sizing production capacity; for example, to determine the size of work teams and plan potential activities carried out by third-party partners, according to the widespread practice of subcontracting. The medium-term plans are triggered by client orders, i.e. by the orders acquired, that are often planned based on state of progress. The potential MPS activates the MRP for the production of the non-customized materials plan, while the Operational Plans lead with FAS modules the MRP explosion for the codes specified by the order once it is acquired. To follow the progress, **Project Management** techniques are often adopted. Examples are given by civil engineering and construction site works.

The ratio between the delivery time accepted by the client and the lead time is commonly defined as the **Planning Index**. If this is lower than 1, it does not allow for complete execution of the production cycle in the residual time between the order date (OD) and the promised delivery date (PDD), but requires the prior availability of materials and/or semi-finished products.

Figure 11.1 shows a summary of the logics described, placing them along the diagonal corresponding to the intersection of the possible delivery intervals granted by the client, compared to the time necessary to execute the different phases that lead from planning to the completion of the product. Using the terminology adopted in the management of physical flows, above the diagonal there is the prevalence of **push systems**, ones that operate on forecasts, and below, that of **pull systems**, based on orders. For each type of production model, between the two flow sections (push and pull) there is the **decoupling stock point**. On the right there is the principal way of responding to demand, while on the left we see the object placed under control of the MPS.

The cases of **PTO** and **ETO**, not requiring forecasts, except for the sizing of production capacity or the preparation of work teams and of the equipment used, avoid control by the MPS in a strict sense, and all of their phases are directly planned by the Final Assembly Schedule (FAS) or, as has already been mentioned, by Project Management techniques. In other types of response to demand – **MTS, ATO**, and **MTO** – the MPS defines the quantities and production times for the objects under its immediate control – finished products, semi-finished products, components, and purchase materials, respectively – delegating to the **MRP – Materials Requirement Planning** explosion the planning of the articles located at the lower levels of the respective bills of materials. In the cases of **ATO** and **MTO**, the tail phases are also governed by **FAS modules**.

11.2 Master Production Planning and Time-Phased Record

With the goal of simplifying the preparation of the different production plans (PP, MPS, and Scheduling) and their management over time, a technique called **Time-Phased Record** is used.

With reference to the MPS, for each planning period, the planned production is calculated starting from the Sales Forecast data (gradually eroded by Actual Orders) and from the Available Inventory for each period, calculated based on the initial On Hand Inventory.

On this point, see the example shown in Table 11.1, which illustrates the case of a planning horizon consisting of 12 weeks.[1] For each week, the Sales Forecast (SF_w) is constant and equal to 200 units, for a total of 2,400 units in the quarter considered. As illustrated in Table 11.1 (a), having calculated an MPS able to produce for each week the amount entered in the Sales Forecast, for a total of 2,400 units, the expected **Available Inventory Balance** will be constant and equal to the initial **On Hand Inventory** of 400 units. At the end of each weekly period, the Available Inventory (AI_w) is in fact equal to the Available Inventory for the previous week (AI_{w-1}) to be summed with what is planned in the current week (MPS_w) deducting the forecast for sales in the same week (SF_w), such that:

$$AI_w = AI_{w-1} + MPS_w - SF_w \qquad [11.1]$$

In the event that AI_w has a negative balance, expected backorders are generated.

As concerns the Available Inventory, in the presence of **Actual Orders different from the Sales Forecast**, normal practice is to consider the greater value among the two, in the event of reliable forecasts that are expected to be confirmed or exceeded by the orders book. Therefore, in that case the Available Inventory at the end of each week (AI_{wi}) is equal to the summation of the Available Inventory of the previous week (AI_{w-1}) and any quantities planned in the MPS (MPS_w), after deducting, for each period, the greater of the Sales Forecast (SF_w) or actual Orders (O_w).

$$AI_w = AI_{w-1} + MPS_w - MAX\,(SF_w, O_w) \qquad [11.2]$$

[1] The discussion will follow is partially adapted from the approaches and the examples described in Vollman et al., 1992.

Table 11.1 **MPS and Time-phased Record**

(a) MPS and Available Inventory

Part Number x	**Weeks**	**1**	**2**	**3**	**4**	**5**	**6**	**7**	**8**	**9**	**10**	**11**	**12**	**Total**
On hand Inventory: **400**	Forecast	200	200	200	200	200	200	200	200	200	200	200	200	2400
	Available	400	400	400	400	400	400	400	400	400	400	400	**400**	
	MPS	200	200	200	200	200	200	200	200	200	200	200	200	2400

(b) MPS and Available Inventory when Actual Orders differ from Sales Forecast

Part Number x	**Weeks**	**1**	**2**	**3**	**4**	**5**	**6**	**7**	**8**	**9**	**10**	**11**	**12**	**Total**
On hand Inventory: **400**	Forecast	200	200	200	200	200	200	200	200	200	200	200	200	2400
	Customer orders	200	200	**400**										
	Available	400	400	200	200	200	200	200	200	200	200	200	**200**	
	MPS	200	200	200	200	200	200	200	200	200	200	200	200	2400

If an Actual Order were greater or lesser than the Sales Forecast in a period, with the MPS equal, the Available Inventory (or Available Inventory Balance) would increase or decrease. Lastly, only in the case of significant forecasting errors (underestimation), would there be a risk of stock-out.

In Table 11.1 (b) it is imagined that, as orders are gradually accumulated, in period 3 the Actual Order is equal to 400 and therefore Available Inventory decreases to 200. Changes with respect to forecasts can obviously also refer to the quantities in MPS, such that the Available Inventory plays the role of buffer in regard to the unreliability of the Sales Forecast and the MPS.

As already mentioned, the progress of the Planning Horizon entails the development of new Sales Forecasts and/or the acquisition of new Actual Orders, and as a consequence an update of the MPS. In the example shown in Table 11.2 (a), it is imagined that despite estimating overall sales of 2,400, the expected sales will be subject to **seasonal variations**.

In this case, the planner opts for a **Level planning strategy** (Chapter 6), which has a constant MPS set at 200 units per week. Stock is accumulated in the first 6 periods, when the quantities planned at the MPS level exceed the forecast sales, to be used in the subsequent 6 periods when demand is greater than the quantities in MPS. Part (b) of the same Table illustrates a **Chase planning strategy**, in which the quantities planned at the MPS level are equal to the sales foreseen, minimizing the Available Inventory. In both cases, the plan starts and finishes with the same Inventory, and the quantities produced in MPS are equal to those of the Sales Forecast, but over

Table 11.2 **Seasonal demand and alternative MPS**

(a) Seasonal sales and Level MPS approach

Part Number x	**Weeks**	**1**	**2**	**3**	**4**	**5**	**6**	**7**	**8**	**9**	**10**	**11**	**12**	**Total**
On hand	Forecast	100	100	100	100	100	100	300	300	300	300	300	300	2400
Inventory: **400**	Available	500	600	700	800	900	1000	900	800	700	600	500	**400**	
	MPS	200	200	200	200	200	200	200	200	200	200	200	200	2400

(b) Seasonal sales and Chase MPS approach

Part Number x	**Weeks**	**1**	**2**	**3**	**4**	**5**	**6**	**7**	**8**	**9**	**10**	**11**	**12**	**Total**
On hand	Forecast	100	100	100	100	100	100	300	300	300	300	300	300	2400
Inventory: **400**	Available	400	400	400	400	400	400	400	400	400	400	400	**400**	
	MPS	100	100	100	100	100	100	300	300	300	300	300	300	2400

(c) Seasonal sales and Lot-sized MPS

Part Number x	**Weeks**	**1**	**2**	**3**	**4**	**5**	**6**	**7**	**8**	**9**	**10**	**11**	**12**	**Total**
On hand	Forecast	100	100	100	100	100	100	300	300	300	300	300	300	2400
Inventory: **400**	Available	300	200	100	600	500	400	100	400	100	400	100	**400**	
	MPS				600				600		600		600	2400

the horizon of the plan the costs associated with it, already presented in Chapter 6, change significantly.

An additional method of drawing up the MPS is shown in **Table 11.2 (c)**. In this case, each time the Projected Available Balance goes below 100 units, **a batch** of 600 units is planned, attempting to balance the stock holding costs with those linked to possible backorders and stock-out. As shown in the table, given seasonal demand, the batches have different coverages over the planning horizon, equal to 4 weeks for the first batch and 2 weeks for the subsequent batches.

11.3 Rolling plans and Order Promising

As mentioned in Chapter 6, the MPS, like other plans (PP and FAS) is constantly updated on a **rolling basis**, maintaining the same horizon of visibility – confirmed horizon equal to the product lead time – moving forward at a weekly rate, in our case.

Therefore, starting from **Table 11.2 (c)** imagine that actual final sales in week 1 amount to 200 units, instead of 100 units, and thus the On Hand Inventory for week 2 goes from 300 to 200 units. Imagine also that the refreshment of the MPS, for a horizon of 12 weeks from week 2 to week 13, incorporates a change in the Sales Forecast, such that from week 2 to week 6, growth is estimated to be from 100 to 200 units per week, and for week 13 the expectation will remain at 300 units, as illustrated in Table 11.3 (a).

It is evident that if the expected increase in sales were confirmed and covered by orders, at the end of the period of 12 weeks (from 2 to 13), the sales would amount to 700 units more than initially foreseen. In the absence of a revision of the MPS, 9 periods of **stock-out** would occur, represented in Table 11.3 (a), with severe implications in terms of cost and reputation. In Table 11.3 (b), we see a possible revision of the MPS, that by maintaining batches of 600 units and a re-order level of 100 units, to satisfy the new Sales Forecast requires anticipating the first batch in week 2, increasing the number of batches produced from 4 to 5, and ending up with an Available Inventory of 100 units. The possibility to increase the batches must however be verified in terms of Production Capacity, available materials, and so on.

The examples presented to this point show the cases in which in order to deal with the client orders anticipated in the Sales Forecast, a plan is drawn up that guarantees immediate availability.

In many cases, clients place planned orders for non-immediate delivery dates, and in the context of the supplier-client negotiation, the **capacity to promise** the quantities requested on a Planned (or Promised) Delivery Date (PDD) becomes critical, which if possible should be close to the Required Delivery Date (RDD), as illustrated previously in section 4.2.

In that case, for the supplier company it is important to perform calculations that

Table 11.3 New forecast and adjusted rolling MPS

(a) Seasonal sales, Lot-sized MPS and revised forecast

Part Number x	**Weeks**	**2**	**3**	**4**	**5**	**6**	**7**	**8**	**9**	**10**	**11**	**12**	**13**	**Total**
On hand Inventory: **200**	Forecast	200	200	200	200	200	300	300	300	300	300	300	300	3100
	Available	0	-200	200	0	-200	-500	-200	-500	-200	-500	-200	**-500**	
	MPS			600				600		600		600		2400

(b) Seasonal sales, Lot-sized MPS adjusted to face the revised forecast

Part Number x	**Weeks**	**2**	**3**	**4**	**5**	**6**	**7**	**8**	**9**	**10**	**11**	**12**	**13**	**Total**
On hand Inventory: **200**	Forecast	200	200	200	200	200	300	300	300	300	300	300	300	3100
	Available	600	400	200	600	400	100	400	100	400	100	400	**100**	
	MPS	600			600			600		600		600		3000

allow it to confirm, in the presence of a backlog of orders (quantities) for future deliveries, the possible **Planned (or Promised) Delivery Dates**. Using the example reported in Table 11.2 (c), Table 11.4 (a) shows a Lot-sized MPS, in which in addition to the original Sales Forecast, understood here as forecasted quantities to be delivered in each single week, we also see the backlog of actual orders received by customers and known at the beginning of the first week, composed respectively of 100, 60, and 40 units, promised for delivery, in the first three weeks. In order to quantify each week the quantities of product actually available to cover new orders, and as a consequence the Promised Delivery Dates that can be confirmed to clients for new orders, we use an MPS function called **Available To Promise (ATP)**, illustrated below in simplified form.

As shown in the example in Table 11.4 (a), the ATP allows for calculating for each week covered by the horizon of the Plan, the quantities of product available for new orders, because they are not yet covered by actual customer orders. Knowing the product availability (for quantities and periods) it is possible to confirm a Promised Delivery Date to the client.

Table 11.4 (a) shows the case of an ATP calculation performed in week 1. Given an On Hand Inventory of 400 units and a cumulative Order Backlog of 200 units (Actu-

Table 11.4 **Available To Promise**

(a) Sales Forecast, Lot-sized MPS, Orders Backlog

Part Number x	**Weeks**	**1**	**2**	**3**	**4**	**5**	**6**	**7**	**8**	**9**	**10**	**11**	**12**	**Total**
	Forecast	100	100	100	100	100	100	300	300	300	300	300	300	2400
	Customer orders	100	60	40										
On hand Inventory: **400**	Available	300	200	100	600	500	400	100	400	100	400	100	**400**	
	ATP	**200**			600				600		600		600	
	MPS				600				600		600		600	2400

ATPw1 = 400 - (100 + 60 + 40) = **200**

(b) ATP at week 1

Part Number x	**Weeks**	**1**	**2**	**3**	**4**	**5**	**6**	**7**	**8**	**9**	**10**	**11**	**12**	**Total**
	Forecast	100	100	100	100	100	100	300	300	300	300	300	300	2400
	Customer orders	100	80	60	40	40								
On hand Inventory: **400**	Available	300	200	100	600	500	400	100	400	100	400	100	400	
	ATP	**160**			**520**				600		600		600	
	MPS				600				600		600		600	2400

ATPw1 = 400 - (100 + 80 + 60) = **160**
ATPw4 = 600 - (40 + 40) = **520**

al Orders for 100, 60, and 40 units) to be delivered in the first three weeks, it is possible to calculate an Available To Promise equal to 200 units. Since the Sales Forecast is equal to 100 units in weeks 2 and 3, it is expected that in these weeks orders will arrive for an additional 100 units.

The calculation of the ATP foresees that:

- in **week 1** the ATP is equal to the sum of the beginning On Hand Inventory (OHI_{w1}) and the quantities potentially planned in MPS (MPS_{w1}), deducting the sum of the Actual Orders ($\Sigma_{wi,n}O_{wi,n}$) booked until the subsequent MPS [11.3];
- in the **following weeks**, the ATP must be calculated only in the weeks in which production is foreseen of a planned quantity in MPS and will be equal to any quantities planned in the MPS (MPS_{wi}), deducting the sum of the actual Orders ($\Sigma_{wi,n}O_{wi,n}$) booked until the subsequent MPS [11.4];
- in the **weeks in which no MPS** is foreseen, the ATP is not calculable, and to define the volumes that can be promised to cover possible orders, the orders received since the last ATP previously calculated are deducted;
- in the **absence of orders**, of course, the ATP quantities coincide with those planned in MPS;
- in the case **of negative ATP values**, it is brought to zero, if possible drawing the necessary quantities from the ATPs of the previous periods.

As a consequence, having defined:

ATP_{wi} = Available To Promise in a generic week i;
OHI_{w1} = On Hand Inventory at the beginning of the planning horizon;
MPS_{wi} = Quantities potentially planned by MPS in each generic week i;
O_{wi} = Actual Booked (Promised) Orders in each generic week i;

$$ATP_{w1} = OHI_{w1} + MPS_{w1} - \sum_{w1,n} O_{wi} \qquad [11.3]$$

$$ATP_{wi} = MPS_{wi} - \sum_{w1,n} O_{wi} \qquad [11.4]$$

This way, considering the data shown in Table 11.4 (a), in the first week, by applying [11.3] it is possible to calculate the following ATP quantities:

$ATP_{w1} = 400 + 0 - (100 + 60 + 40) = 200\ units$

If in the course of week 1 additional orders to be delivered come in, as shown in Table 11.4 (b), such that in the first five weeks the cumulative backlog increases from 200 to 320 units (Actual Orders for 100 + 80 + 60 units in the first three weeks and 40 + 40 units in the fourth and fifth weeks). As in the previous case, since the Sales Forecast in the first five weeks is equal to 500 units, it is expected that between the second and fifth weeks additional orders equal to 180 units will be booked. Applying [11.3] and [11.4], it will be possible to update the ATP for week 1 and calculate the figure for week 4 in which a new MPS batch is produced, respectively equal to:

$ATP_{w1} = 400 + 0 - (100 + 80 + 60) = 160\ units$
$ATP_{w4} = 600 - (40 + 40) = 520\ units$

While for the other weeks for which an MPS is foreseen and orders have not yet been received, the quantities of ATP and MPS coincide:

$ATP_{w7}\ and\ ATP_{w9} = 600 - 0 = 600\ units.$

Continuing with this example, as already shown in **Table 11.3 (b)**, imagine that after having determined the final numbers for week 1, the MPS has been modified to adjust it to a **new Sales Forecast**, increasing from 100 to 200 units the MPS quantities to produce between weeks 2 and 6. In the meantime, further orders to be delivered have been booked: 100, 100, 80, and 40 units, respectively in weeks 2, 3, 4, and 5. Therefore, at the beginning of week 2, we have a cumulative order backlog of 320 units and an ATP equal to 520 units. In week 5, the ATP is 560 units. In **Table 11.5** we show the example of the calculation of the ATP relating to weeks 2 and 5.

To this point we have supposed that the actual orders gradually acquired from week to week covered the Sales Forecasts. However, orders do not necessary always completely confirm the Sales Forecast, and clients can request Required Delivery Dates (RDD), planning their needs far ahead of time. Starting with **Table 11.5**, imagine having obtained additional orders for 100 units in week 2 and having the final figures for that week. In the new planning horizon – weeks 3 to 14 – new orders are booked for an additional 300 units in week 3, 800 units in week 9, and 350 units in week 11, as illustrated in **Table 11.6**.

Based on the On Hand Inventory for week 2, equal to 600 units (200 + 600 – 200), deducting the greater between Actual Order or Sales Forecast, the Available Inventory in week 3 is equal to 200 units and that for week 4 is equal to 0.

As concerns the calculation of the ATP quantities in week 8, having an ATP equal to the MPS quantities (600 units), the quantities ordered in week 9 (800 units) do not seem satisfiable. However, observing the cumulative ATP, it is possible to accept and

Table 11.5 **Seasonal sales, Lot-sized MPS adjusted to face the revised forecast and ATP at week 2 and 5**

Part Number x	**Weeks**	**2**	**3**	**4**	**5**	**6**	**7**	**8**	**9**	**10**	**11**	**12**	**13**	**Total**
	Forecast	200	200	200	200	200	300	300	300	300	300	300	300	3100
	Customer orders	100	100	80	40									
On hand Inventory: **200**	Available	600	400	200	600	400	100	400	100	400	100	400	**100**	
	ATP	**520**			**560**			600		600		600		
	MPS	600			600			600		600		600		3000

ATPw2 = (200 + 600) - (100 + 100 + 80) = **520**
ATPw5 = 600 - 40 = **560**

fulfill the orders for week 9, that will be covered by the ATP for week 8 (600 units), to which we can sum a portion of the ATP for week 5, that will thus be reduced from 560 to 360 units, while the ATP for week 10 will be equal to 250 units.

There exist **various ways of calculating the ATP**. Among the many, we mention the use of **cumulative ATP**. Table 11.7 represents a simplified cumulative ATP, in which starting from the ATP of week 2 of Table 11.5, for each period the ATP_{wi} is calculated, equal to the ATP_{w-1} increased by the MPS_{wi} quantities, where present, and reduced by the Actual Orders O_{wi}.

As can be seen from Tables 11.6 and 11.7, due to the new orders, negative availability is generated in some periods, such that it will be necessary to perform the refreshment of the MPS, that anticipates some production batches, or negotiate potential delays or division of deliveries with some clients. This is where the **dynamic aspect** of the system lies, that allows for performing evaluations on the principal options in the management of the Promised Delivery Dates (PDD) (Secchi, 2000). By way of example, see Box 11.1.

Table 11.6 **Seasonal sales, Lot-sized MPS adjusted to face the revised forecast and ATP at weeks 5, 8 and 10**

Part Number x	**Weeks**	**3**	**4**	**5**	**6**	**7**	**8**	**9**	**10**	**11**	**12**	**13**	**14**	**Total**
	Forecast	200	200	200	200	300	300	300	300	300	300	300	300	3200
	Customer orders	400	80	40				800		350				750*
On hand Inventory: **600**	Available	200	0	400	200	-100	200	-600	-300	-650	-350	-650	-950	
	ATP			**360**			**0**		**250**		600			
	MPS			600			600		600		600			2400

* Delta Actual Orders on Sales Forecast

ATPw5 = 600 - 40 - 200 = **360**
ATPw8 = 600 - 600 = **0**
ATPw10 = 600 - 350 = **250**

Table 11.7 **Calculation of cumulative ATP**

Part Number x	**Weeks**	**3**	**4**	**5**	**6**	**7**	**8**	**9**	**10**	**11**	**12**	**13**	**14**	**Total**
	Forecast	200	200	200	200	300	300	300	300	300	300	300	300	3200
	Customer orders	400	80	40				800		350				750*
On hand Inventory: **600**	Available	200	0	400	200	-100	200	-600	-300	-650	-350	-650	-950	
ATP week 2 = **520**	ATP	**120**	**40**	**600**	**600**	**600**	**1200**	**400**	**1000**	**650**	**1250**	**1250**	**1250**	
	MPS			600			600		600		600			2400

* Delta Actual Orders on Sales Forecast

Box 11.1 Cumulative ATP and decision on allocation of available capacity

Imagine that in the month of May, a company has the following orders book (Chart a). Given a Starting On Hand Inventory of 600 units, it is possible to deduce a **Cumulative ATP**, as shown in Chart (b), based on the MPS Scheduled Receipts policy (batches of 300 units) and the commitments deduced from the Orders Book.

Chart (a) Company's Orders book

Orders book May		
Days	Orderd quantities (units)	Customer I.D.
4	300	001
9	200	009
14	150	002
18	150	001
18	100	008
22	150	014
26	250	003
26	100	002
28	200	012
31	200	003

Chart (b) Cumulative ATP for May

Cumulative ATP May				
Day	On Hand Inventory (units)	MPS Scheduled Receipts (units)	Customer Orders (units)	Cumulative ATP (units)
1	600			600
2				600
3				600
4			300	300
5				300
6				300
7				300
8				300
9		300	200	400
10				400
11				400
12				400
13				400
14			150	250
15				250
16				250
17		300		550
18			250	300
19				300
20				300
21				300
22			150	150
23				150
24				150
25		300		450
26			350	100
27		300		400
28			200	200
29				200
30				200
31		300	200	300

As mentioned above, the profile of the Cumulative ATP allows for verifying each day the product availability and consequently checking whether to accept additional orders, guaranteeing a Promised Delivery Date (PPD).
In the event of unavailability, it is also possible to implement interventions to attempt to satisfy client requests. For example, imagine that during the month of May a new order comes from client 006 for 150 units on day 25. Due to the cumulative ATP, the order could be accepted and satisfied. However, as illustrated in Chart (c), clients 002 or 003 would be penalized, receiving a delivery of less than the request for 50 units.

Chart (c) Cumulative ATP variation due to the introduction of the new order

	Cumulative ATP May			
Day	On Hand Inventory (units)	MPS Scheduled Receipts (units)	Customer Orders (units)	Cumulative ATP (units)
....		....	....	....
23				150
24				150
25		300	150	300
26			350	-50
27		300		250
28			200	50
29				50
30				50
31		300	200	150

Taking into account the importance of the various clients, the planner may thus act to evaluate the **possible options** to accept all of the orders received, and to **dynamically verify, through "what-if" analyses, the impact** of the existing orders on the portfolio. As shown in Chart (d), it is in fact possible:

- to negotiate with client 002 the deferment of the entire delivery until day 27;
- if the client were willing to accept divided delivery, the supplier could sustain the cost of a double shipment, delivering 50 units on day 26 and the balance of 50 units on day 27;
- alternatively, the supplier could ask client 003, who made an order of 250 units for day 26, if he is willing to receive 200 units on that date, to then be completed with the 50 missing units in the shipment already set for May 31.

Chart (d) Planner's options to satisfy all May orders

	Cumulative ATP May			
Days	On Hand Inventory (units)	MPS Scheduled Receipts (units)	Customer Orders (units)	Cumulative ATP (units)
....		....	....	....
23				150
24				150
25		300	150	300
26			250	50
27		300	100	250
28			200	50
29				50
30				50
31		300	200	150

Cumulative ATP May				
Days	On Hand Inventory (units)	MPS Scheduled Receipts (units)	Customer Orders (units)	Cumulative ATP (units)
....		....	....	....
23				150
24				150
25		300	150	300
26			300	0
27		300	50	250
28			200	50
29				50
30				50
31		300	200	150

Cumulative ATP May				
Days	On Hand Inventory (units)	MPS Scheduled Receipts (units)	Customer Orders (units)	Cumulative ATP (units)
....		....	....	....
23				150
24				150
25		300	150	300
26			300	0
27		300		300
28			200	100
29				100
30				100
31		300	250	150

As shown in the example, the ATP allows for assigning a promised date to the client as a function of a **dynamic verification of availability** conducted on the finished products. In the case in which, having analyzed the possible options, that availability is insufficient to cover further orders on the Required Delivery Dates (RDD), to be able to indicate a possible Promised Delivery Date (PDD) to fulfill the order it is possible to use a second function, called **Capable To Promise (CTP)**.
Looking back at the Bill of Materials of the finished product, the **CTP performs a double check**:

- **availability of the part numbers produced at the first level**, i.e. of the components necessary to produce or assemble the finished product, through the launch of an MRP – Material Requirements Planning procedure;
- **availability of the production capacity necessary** to perform those terminal phases of production or assembly, through the launch of a CRP – Capacity Requirements Planning procedure at finite capacity.

With the CTP modules, the identification of the first date that can be promised to the client, thus takes into account:

- the availability of the materials and components necessary to produce the finished product;
- the available capacity to proceed to the end phases of production or assembly of the finished product, starting with that availability;
- the lead times necessary to perform those final operations and make the finished product available as ATP.*

* Further details are provided in Secchi, 2000 and Vollmann et al., 1992.

11.4 MPS and management decisions

The drafting of the MPS, in the presence of a range of products produced by the same production unit, implies a series of **decisions** that refer principally to:

- **lot sizing**, the size of the product lots for each article planned;
- **sequencing**, the variation of the production sequences, taking into account the performance of the level of stock of materials and components, set-up times and costs, capacity and efficiency of production resources, and so on;
- **management of the priorities** assigned to the products, i.e. of potential advances or delays of production orders for finished products, considering the trend of stock of finished products and the dynamic of orders;
- **marginal changes to production capacity** (overtime, subcontracting) in the context of the constraints set by the Aggregate Planning;

Literature and business practice have developed multiple planning algorithms. Merely by way of example, we present the case of an MPS developed to determine the quantities to produce of different articles that share the same production resource (a department or machine), with the goal of guaranteeing the same Coverage, taking into account for each article the different dynamic of demand.

That method is applied differently in the presence of **continuous process**, such as in the case of a paper mill that produces different types of paper or a phase of dyeing that alternates different colors, etc., and production in **repetitive lots**, as in the case of mechanical production.

Imagine having to produce in a generic Time period (T_i) 6 different articles using the same resource, and that for each article the information shown in **Table 11.8** is available, i.e.:

T_i = Period covered by production plan;
St_j = Unit Standard Time for each j product (hours/unit);
BI_j = Beginning Available Inventory for the j products (units);
D_{ij} = Demand for the j products in the T_i (units);
APC_i = Given Available Production Capacity in the planned period T_i (hours);
LS_j = Lot size (or batch) for each j product (units);
P_{ij} = Production volumes for the j products in the T_i (units).

In the case of **continuous production**, as illustrated in **Table 11.9**, the process goes forward:

- assigning to each article a portion of the hours relating to a shared resource (machine or labor) available or scheduled, obtained through the calculation of the **Aggregate Coverage Time (ACT)**. This is equal to the ratio between overall Beginning Inventory, expressed in standard times, summed with the Available

Table 11.8 Production plan based on coverage times (Data)

		Mix of part numbers to be produced						
		A	B	C	D	E	F	
St_j	Unit Standard Time (hours/unit)	4	1.6	10	1.8	3	7	
BI_j	Beginning Inventory (units)	62	125	38	150	120	98	
D_j	Monthly Demand (units)	30	42	15	48	39	42	
LS_j	Production Lot size (units)	60	52	32	38	45	50	
APC_i	Available Production Capacity (given hours)							**965.0**

Table 11.9 Production plan based on coverage time for continuous production

		Mix of part numbers to be produced						
		A	B	C	D	E	F	Total
BI_j	Beginning Inventory (hours)	248	200	380	270	360	686	**2144.0**
D_j	Monthly Demand (hours)	120	67.2	150	86.4	117	294	**834.6**
$D_j \times ACT$	Required Coverage (hours)	447	250	559	319	436	1095	
$D_j \times ACT$	Required Coverage (units)	112	156	56	179	145	156	
P_{ij}	Production to be scheduled (units)	50	31	18	29	25	58	
P_{ij}	Production to be scheduled (hours)	199	50	179	52	76	409	**965.0**

Production Capacity of the period, and the Overall Demand for the period, also expressed in standard hours:

$$ACT = \sum_j (BI_j \times St_j) + APC_i / \sum_j (D_j \times St_j) \quad [11.5]$$

$$ACT = (2{,}144 + 965) / 834.6 = 3.725$$

- dividing Available Production Capacity such that the Coverage Times of the articles at the end of the production period (T_i) are the same, that is, using the Aggregate Coverage Time (ACT) to schedule each article J in period T_i, so as to guarantee for each the coverage calculated based on the Coverage Time.

$$P_{iJ} = (D_j \times ACT) - BI_j \quad [11.6]$$

On the other hand, in the case of **batch process**, as illustrated in **Table 11.10**, the process goes forward by:

- assigning to each article, produced in sequence by the same production resource, a portion of the hours relating to the shared resource (machine or labor) available

Table 11.10 Production plan based on coverage times for batch production

		Mix of part numbers to be produced						
		A	B	C	D	E	F	Total
LS_j	Production Lot size (units)	60	52	32	38	45	50	
LS_j	Production Lot size (hours)	240	83.2	320	68.4	135.0	350	**1196.6**
$ct_j = BI_j / D_j$	Coverage time	2.1	3.0	2.5	3.1	3.1	2.3	

		Mix of part numbers to be produced				
		A	F	C	B	Delta
LS_j	Production Lot size (units)	60	50	32	52	
LS_j	Production Lot size (hours)	240	350	320	83,2	
RPC_i	Required production capacity (Σ hours)	240.0	590.0	910.0	**993.2**	28.2

or scheduled, obtained through the calculation of the **Coverage Time for each article** (ct_j). The latter is equal to the ratio between the Beginning Inventory of each article J, expressed in standard times, and the period Demand for each article, also expressed in standard time:

$$ct_j = BI_j / D_j \qquad [11.7]$$

- using the coverage time for each article (ct_j) to define the sequence according to which production in standard lots is to be launched, expressed in standard times, starting with the article with the lowest ct_j, until the exhaustion of the production capacity for the period (APC).

In this case, given the lot constraints, the right sequence is A, F, C and B. The 28 hours missing to complete batch B have to be recovered in the next T_i scheduling period, together with D and E, after repeating the same procedure.

12 Bill of Materials and Master Data

by *Raffaele Secchi*

12.1 Introduction

Every company has a specific set of **master data** – regarding Bill of Materials, production cycles, and production centers – that is absolutely fundamental for the integrated management of all the manufacturing activities. The elaboration of this set of master data does not take place under the responsibility of a single function, but rather is a very complex process, that involves multiple actors at different moments: during the development of a new product, as a consequence of a redesign that substantially modifies the structural characteristics of an existing product, after a change of the existing production structure, and so on.

Considering the relevance, but also the complexity related to the generation and maintenance of the master data, all the most recent information systems include specific modules dedicated to their management. The most developed functions are those related to the Bill of Materials. They ensure that the data is correct and valid, avoiding duplications of the data within the information system, and they allow the operations managers to carry out fast queries to get useful information. In addition to the analysis of these functionalities, this chapter will also examine the management of Bills of Material for products offered with numerous configurations characterized by variants and options. In these situations, the development of a Bill of Materials for each product – a solution generally adopted in companies with standardized production – is not practicable and the problem has to be addressed differently.

The chapter ends with the analysis of the other technical data related to production cycles and production centers. This data is also fundamental, especially in relation to production planning and management control activities. In the first case, it is used to verify the availability of production capacity after the explosion of the requirements carried out through the MRP system (Chapter 13) and to check the allocation of the workloads to production resources when issuing orders to the departments; in the second case, the data is used to monitor the resources absorbed during the transformation process.

12.2 The Bill of Materials (BOM)

The data contained in the BOM constitutes the primary source for the procedures relating to production planning and materials procurement. Although this is the most traditional area of use, the richness of information contained in the BOM can also be used for many other purposes (Mather, 1987):

- product cost: the BOM provides the data to determine the product cost, at least as regards the variable part relating to the materials used;
- inventory accounting: in manufacturing companies characterized by flow production, the consumption of raw materials and components is not usually recorded when the materials are picked, but a posteriori, through the BOM, when the finished products are put in stock (backflushing);
- substitution of missing components: in case of stock-out of a component (due to a supply delay or problems occurred during the manufacturing steps), through appropriate calculations on the BOM it is possible to determine which finished products cannot be produced or decide which product will be assigned the maximum priority as a function of the components available;
- technical support: the BOM represents an indispensable tool, if appropriately updated, to keep track of any changes in the manufacturing process, to ensure effective product maintenance or replacement of defective parts;
- support for assembly activity: when the product structure is defined consistent with the sequence of the assembly steps, the BOM can usefully complement, and possibly replace, the drawings specifically issued to indicate the correct ways of assembling the product;
- management of product changes: when it becomes necessary to modify a product, the BOM becomes essential to identify all of the items that are involved in the revision process, based on their links.

The central role of the BOM, demonstrated by the multiple uses that have been highlighted here, has led IT vendors to develop modules with specific functions to generate and manage it. Before analyzing those functions in detail, we should first provide a summary description of the information that can generally be found in a BOM.

12.2.1 Technical and management data

The BOM can be considered as the computerized integration of a set of technical and managerial data that is stored in two specific archives (parts master data and product structure records) of the information system.

Within the **article master data**, a distinction is usually made between:

- article master data in a strict sense:
  - the part number, necessary for the unique recognition of the article in the information system;
  - the full description of the article;
  - the origin, depending on whether the article is produced internally or acquired externally;

  - the unit of measure;
  - the number of the drawing in which the article is present;
  - any indicators expressing the criticalities of the code (for example, the class within an ABC classification on consumption);
- technical data:
  - the dimensional characteristics of the article (overall dimensions, weight, etc.);
  - the material of which the article is made;
  - the product reject coefficient, that specifies how much the article's requirements are to be increased due to known production rejects or non-conformities;
- logistics data:
  - the batch policy, i.e. the choice of the criteria for replenishing the article;
  - the size of the production or purchase batch;
  - the safety stock, which indicates the quantity of the item to be kept in stock as a precautionary measure, in order to ensure business continuity even in the presence of unplannable events (variability in demand or consumption, variability in suppliers' lead time, plant downtime, etc.);
  - the production or purchase lead times, which indicate the time between the release of the production or purchase order and the time when the item is available for use;
  - the production departments or suppliers that are involved in the requirements explosion for the item;
- accounting data:
  - the warehouse where the item is stocked;
  - the standard price or cost of the item, for the computation of specific cost accounting configurations;
  - the cost center where the item is manufactured or loaded for accounting purposes.

The **product structures** records, on the other hand, contains all the information regarding the architecture of the product according to the links between the single items that make it up. In detail, the main data concern:

- the usage, that expresses the quantities of the item of lower level necessary to realize a unit of the item of immediately higher level;
- the level, that expresses the position of the item within the structure. Conventionally, the level takes on value zero in the case of a finished product, up to the n-th value for the last component. In order to properly perform the material requirement planning procedure, it is of primary importance to set the Low Level Code. It represents the item's lowest position in one or more Bills of Materials subject to planning as each component can appear at multiple levels of the BOM and in multiple BOMs. It is used to determine the sequence in which the materials are planned during the MRP run. The MRP system plans the items with the low-level code of 0 first and then those with a low-level code of 1 and so on. The lower

the code is the higher the item is in the product structures, 0 being a top level item;

- the process reject coefficient, that accounts for the extra-consumption of the lower level component (child) due to process rejects during the production of the higher level code (parent); in the case of processing of raw materials, the coefficient is used to take into consideration any technological waste that cannot be eliminated. From the standpoint of the logical organization of the data, it is advisable for the quantity anticipated in the usage coefficient to be considered net of process waste/rejects. On the one hand, this favors the accounting for extra-consumption; on the other, it facilitates the change of the data relating to waste in the case of technological innovations of machines or management interventions aimed at improving the yield coefficient of the production process;
- the conditions (or dates) of validity, that allow to define the assumptions that make a link active between two codes within one or more BOMs. It can be specified that a BOM link is valid for a date range, a revision level (revision A to revision C), a batch number, or a serial code range (100100 to 100900);
- the lead time correction term, that allows to modify the planned lead time for the item in order to determine, depending on the actual assembly process, the date at the latest when the production or purchase order can be released. An example can clarify what has been stated here. In the case of construction of electromechnical devices made up of cabinets that contain numerous cables and circuit boards, the latter are placed structurally at the lower levels of the Bill of Materials, while the assembly sequence calls for them to be assembled at the end of the cycle. During planning, following the structure of the product, the boards would be procured well before their actual use. By using the lead time correction term, it is possible to postpone the order issue date so that the production or procurement of the boards takes place when they are actually needed to feed the assembly of the product.

12.2.2 The management functions of the Bill of Materials

As already pointed out, all the main software packages offer specific modules for the generation and updating of the article master data and the product structure archive.

In more analytical terms, there are some functions that support the user in the **creation of the BOM** for a new article. They facilitate the assignment of the part number to the new entity upon checking for involuntary duplications, they help the user to enter the article master data and the product structure in the respective archives, allowing real-time checking of any anomalies during the development of the BOM links.

A second set of functions is aimed at managing all the **changes affecting the BOM** as a whole or single items that make it up. The most frequent case concerns a massive change related to a specific item. With a single update operation, the user can make the change on all the links where that item is present. In the case of cancellation/replacement of items, the system must verify that it is possible, i.e. that

Figure 12.1 BOM of a generic product

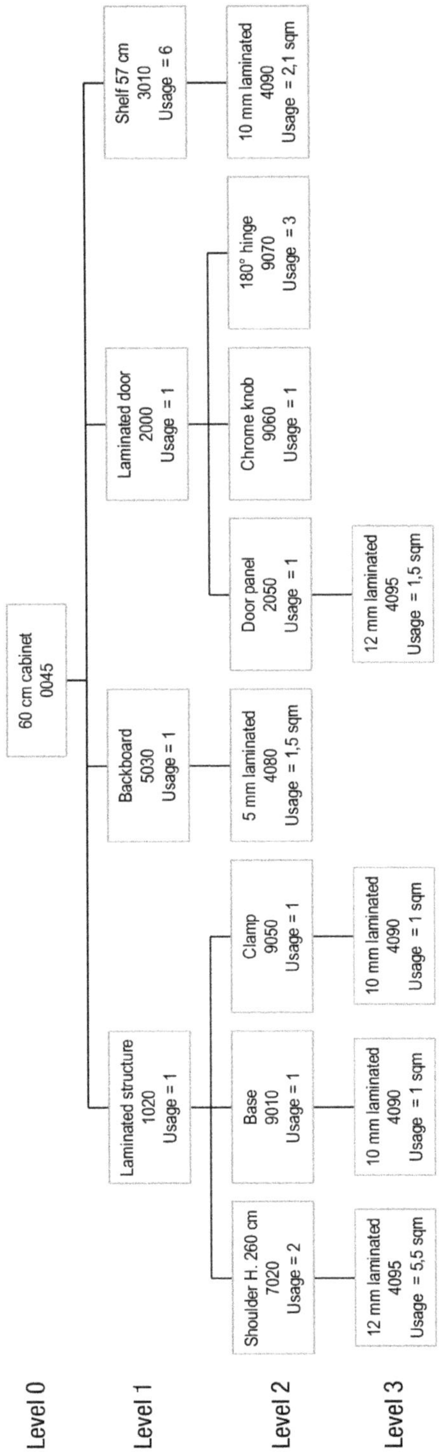

the BOM to be changed does not have open positions (it does not appear in some production or procurement order still existing or there is stock of that product in the warehouse).

Lastly, the most recurring functions regard the possibility of **interrogating the BOM** in order to obtain precise information on the article master data and the product structure links. In order to exemplify the various forms of query, consider the generic BOM illustrated in Figure 12.1, that provides the classic tree-form representation of the product.

The most common possibilities of querying the product structure archive regard:

- **explosion**: scalar search of all the components that have a link of dependence to a specific article object of the explosion. As shown in Table 12.1, the search can identify only the parts placed at the next lower level of the BOM (single-level BOM), or as shown in Table 12.2, it can be cascaded to all the levels of the BOM (multi-level BOM). The explosion procedure is used during the MRP calculation to transfer the requirements from a "parent" item to the corresponding "children" items. Moreover, it can be used to make "local" corrections to production plans following the occurrence of stock-out situations of some codes or variations of consumption. In this case, without re-launching the MRP calculation, it is possible to carry out a scalar explosion of the codes that are to be produced and to check the feasibility based on the availability of the lower-level items;
- **where-used report** (implosion): search of all finished or semi-finished products where a given item is used. Unlike the explosion that shows the parent-child links, the implosion identifies the child-parent links. Again, a single-level (Table 12.3) or multi-level (Table 12.4) implosion is possible. The implosion facilitates the management of the BOM changes, since it is possible to automatically analyze "in which parents a specific child is present" and update the BOM of all the finished products affected by the change accordingly;
- **summarization**: computation of the quantity of all the components needed to produce a finished product or assembly at a specific level of the Bill of Materials. As shown in Table 12.5, the structural links between parts are not shown, but only the total requirements for each item present in a finished product or assembly. Such report can be useful to highlight the gross requirements at all levels during planning, and potentially, to organize the operations for picking articles from the inventory;

Table 12.1 **Example of single-level explosion**

Item: 1020	Description: Laminated structure		
Code	**Description**	**Quantity**	**Unit of measure**
7020	Shoulder H. 260 cm	2	unit
9010	Base	1	unit
9050	Clamp	1	unit

Table 12.2 **Example of multi-level explosion**

Level	Code	Description	Quantity	Unit of measure
0	0045	60 cm cabinet	1	unit
. 1	1020	Laminated structure	1	unit
. . 2	7020	Shoulder H. 260 cm	2	Unit
. . . 3	4095	12 mm laminated	5.5	sqm
. . 2	9010	Base	1	unit
. . . 3	4090	10 mm laminated	1	sqm
. . 2	9050	Clamp	1	unit
. . . 3	4090	10 mm laminated	1	sqm
. 1	5030	Backboard	1	unit
. . 2	4080	5 mm laminated	1.5	sqm
. 1	2000	Laminated door	1	unit
. . 2	2050	Door panel	1	unit
. . . 3	4095	12 mm laminated	1.5	sqm
. . 2	9060	Chrome knob	1	unit
. . 2	9070	180° hinge	3	unit
. 1	3010	Shelf 57 cm	6	unit
. . 2	4090	10 mm laminated	2.1	sqm

Table 12.3 **Example of single-level implosion**

Item: 4095 Description: 12 mm laminated				
Level	**Code**	**Description**	**Quantity**	**Unit of measure**
1	7020	Shoulder H. 260 cm	5.5	sqm
1	2050	Door panel	1.5	sqm

Table 12.4 **Example of multi-level implosion**

Item: 4095 Description: 12 mm laminated				
Level	**Code**	**Description**	**Quantity**	**Unit of measure**
2	7020	Shoulder H. 260 cm	5.5	sqm
. 1	1020	Laminated structure	–	–
. . 0	0045	60 cm cabinet	–	–
2	2050	Door panel	1.5	sqm
. 1	2000	Laminated door	–	–
. . 0	0045	60 cm cabinet	–	–

Table 12.5 **Example of summarized issue of Bill of Materials**

Item: 0045	**Description: 60 cm cabinet**		
Code	**Description**	**Quantity**	**Unit of measure**
1020	Laminated structure	1	unit
2000	Laminated door	1	unit
2050	Door panel	1	unit
3010	Shelf 57 cm	6	unit
4080	5 mm laminated	1.5	sqm
4090	10 mm laminated	4.1	sqm
4095	12 mm laminated	7	sqm
5030	Backboard	1	unit
7020	Shoulder H. 260 cm	2	unit
9010	Base	1	unit
9050	Clasp	1	unit
9060	Chrome knob	1	unit
9070	180° hinge	3	unit

- **validity check of BOM links**: indicate whether a link between multiple items is active and, consequently, which item should be correctly used among the different versions that have undergone subsequent changes. The validity check is generally based on the date, but can also be carried out on the basis of on the batch number or serial number of a given item. For example, it is really helpful when you need to replace a component that is no longer functional on a product belonging to a particular batch. Based on the active link at the time the product was produced, it is possible to exactly verify which specific version of the component must be used for the maintenance work.

12.3 Planning Bill

The term **Planning Bill** generally refers to an artificial group of items that is divided into the classic structure of a BOM (thus the term "bill") but is used to facilitate and improve the planning process (thus the term "planning") (De Toni et al., 2013). These are fictitious BOM specifically developed to mitigate the problem arising from the limited reliability of the forecasts. Through their use, the forecast reliability can improve by making the object of the forecasts more aggregated. This is based on the entirely intuitive, but also analytically demonstrable fact that forecasts on aggregated quantities are better than those for disaggregated ones. Therefore, the

Planning Bills represent a tool aimed at improving planning processes on a forecasting basis in order to maximize the fulfillment of customer orders and thus make up a significant support for adopting advanced operations practices.

12.3.1 Super Bill

The best-known and most widely used Planning Bills are the so-called **Super Bills**, that do not refer to any physically identifiable objects since they consider all the different variants in which it can be produced. We use the expression "variant" to indicate a specific characteristic or function of the product (such as the color of the body of a motor vehicle, the engine type, etc.), while we use the expression "option" when we refer to the sum of alternatives that distinguish the variant (for example, blue, white, or black for the color of the body, or gasoline, diesel, or turbodiesel for the type of engine).

The process of creating a Super Bill is based on an articulated set of phases that entails:

- the analysis of the commonalities;
- the creation of the Modular Bills;
- the creation of the Super Bill.

The first phase consists of the identification of the parts that are common to all the products analyzed and the parts that characterize the specific version of the finished product. The most used approach to the analysis of the similarities is known as the **Matrix Bill of Material**. It requires the setup of a matrix in which the rows represent subassemblies, components, and raw materials, and the columns represent finished products. The analysis of the matrix allows for verifying how many times an item is used in the finished product, thus facilitating the separation of the common parts from the parts that characterize single versions.

To provide an example, imagine carrying out the analysis of the commonalities through the use of a Matrix Bill of Materials with reference to four versions of cranes characterized by different lifting capacities (20 and 40 tons) and different arm extensions (30 and 40 meters). Figure 12.2 shows the BOM for the single versions of the finished products exploded at a hypothetical first level.

Table 12.6 shows the analysis of the commonalities conducted with the Matrix Bill of Material method. The cabin and the fifth wheel, that appear in all the finished products and are characterized by the same usage coefficient for each finished product, can be grouped as common parts. The mini platform, the light counterweights, and the medium tires, on the other hand, represent items that characterize both of the versions with 20 ton lifting capacity. The maxi platform, the strong counterweights, and the high capacity tires, conversely, characterize the 40 ton versions. The short arm assembly is specific for the models having an extension of 30 meters, and the long arm assembly for the other two models of the finished product.

Figure 12.2 Bills of Materials of four different crane configurations

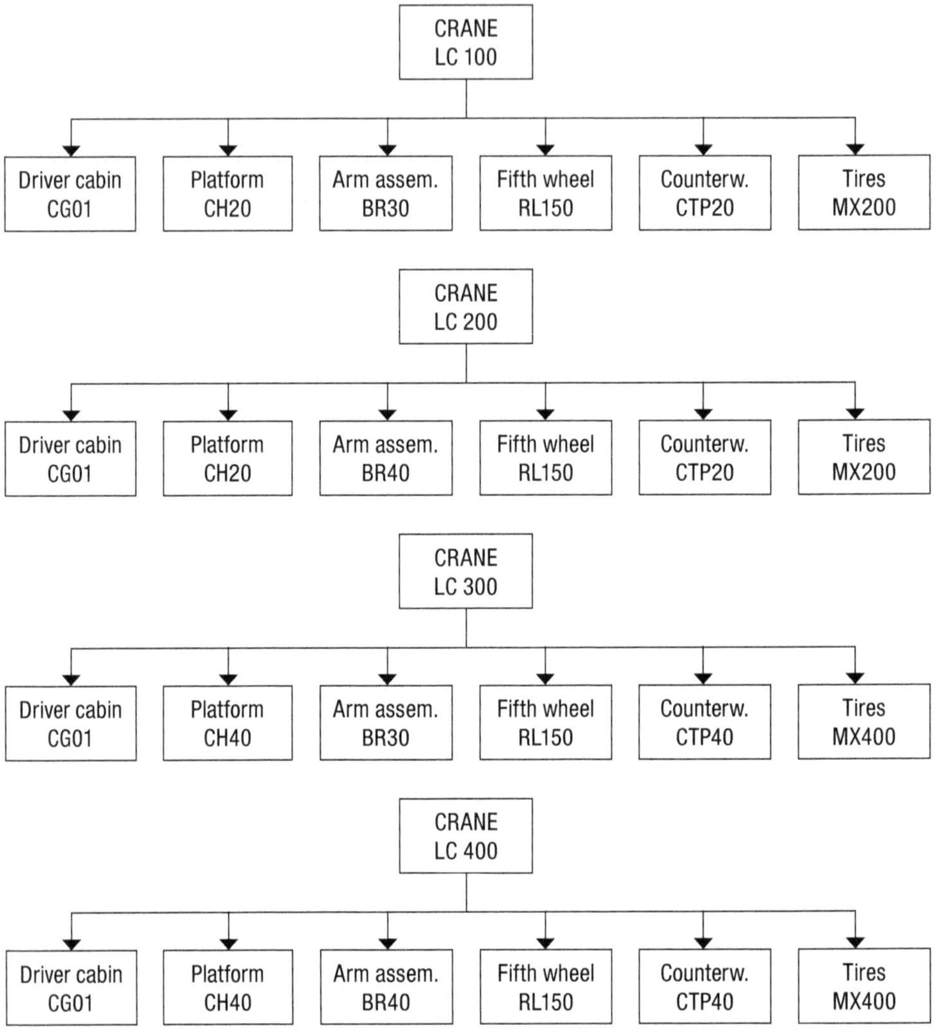

When the complexity is rather limited, the analysis of the commonalities can be carried out through spreadsheets in which the data contained in the product Bills of Materials can be entered. Beyond certain levels of complexity it becomes necessary to use specific software that, through the functions generally used in the management of relational databases, are able to rapidly and accurately identify the existing commonalities between the finished products.

The results of the analysis constitute the output for the generation of the Modular BOMs, i.e. for the second phase of the process that leads to the creation of the Super Bills. As a function of the degree of commonality and specificity that emerges from the matrix, it is possible to identify the modules constituted by sets of components

Table 12.6 **The analysis of the similarities according to the Matrix Bill of Materials method**

Code	Description	Models			
		LC100	LC200	LC300	LC400
CG01	Driver cabin	*	*	*	*
CH20	Mini platform (20 tons)	*	*		
CH40	Maxi platform (40 tons)			*	*
BR30	Short arm assembly (30 meters)	*		*	
BR40	Long arm assembly (40 meters)		*		*
RL150	Fifth wheel diameter 150 cm	*	*	*	*
CTP20	Light counterweight (20 tons)	*	*		
CTP40	Strong counterweight (40 tons)			*	*
MX200	Medium capacity tires	*	*		
MX400	High capacity tires			*	*

that can be planned as single groups. The module that groups together the common components, regardless of the final configuration of the finished product, is called **Common Parts Bill**; the other modules related to the different variants on which the customer's choices are expressed are called Modular Bill. The **Modular Bill** thus constitutes a fictitious Bill of Materials that contains all the items of a same variant. Figure 12.3 illustrates the Bill of Materials for common parts and the modular Bills of Materials as they emerge from the analysis of the commonalities conducted previously.

In the example illustrated above, the modularization process can be considered complete since each first level component has found a specific position, either within the Common Parts Bill or in a Modular Bill. In the event that the four cranes have a specific first level component for the single configuration, then it would be necessary to proceed to a subsequent explosion of the components at the second level to verify the existence of further commonalities between the components and potentially to allocate them in a Modular Bill (and less likely, in the Common Parts Bill). Therefore, each time a specific component is present for a single configuration, the process can be reiterated at a lower level of the Bill of Materials to determine if some commonalities exist with respect to the modules identified previously.

The modules represented in Figure 12.3 can be used for the construction of the Super Bill relating to the cranes analyzed. Level zero of the Super Bill, represented in Figure 12.4, has the medium product in its fictitious configuration, that is not representative of any physically producible crane. At the lower levels are the modules identified previously through the modularization process.

Figure 12.3 Representation of Bill of Materials by common parts and Modular Bills

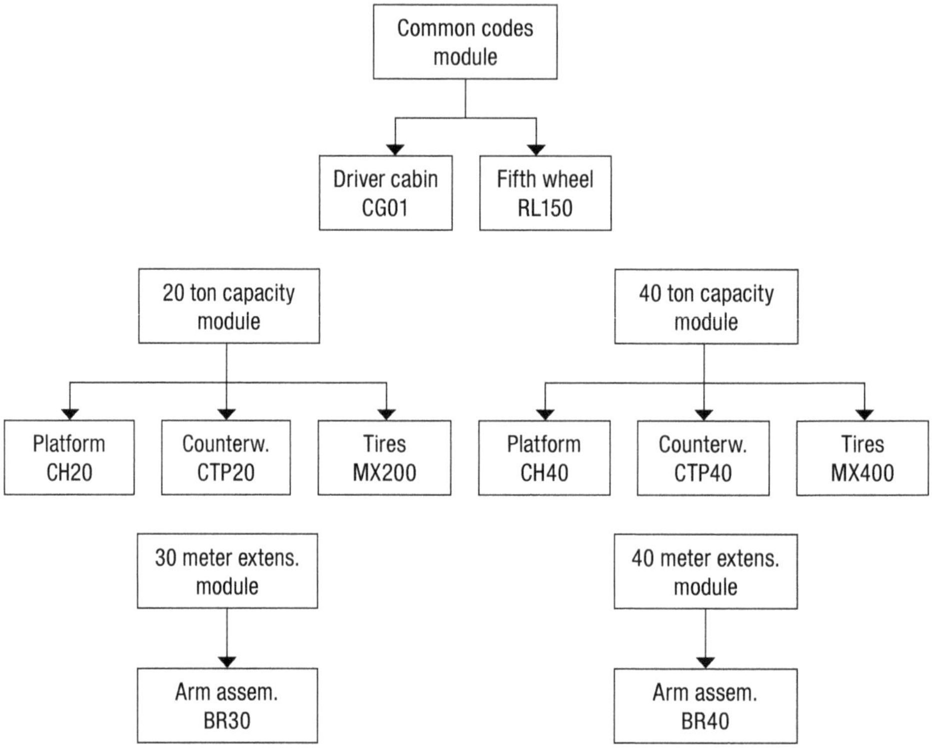

Figure 12.4 The Super Bill

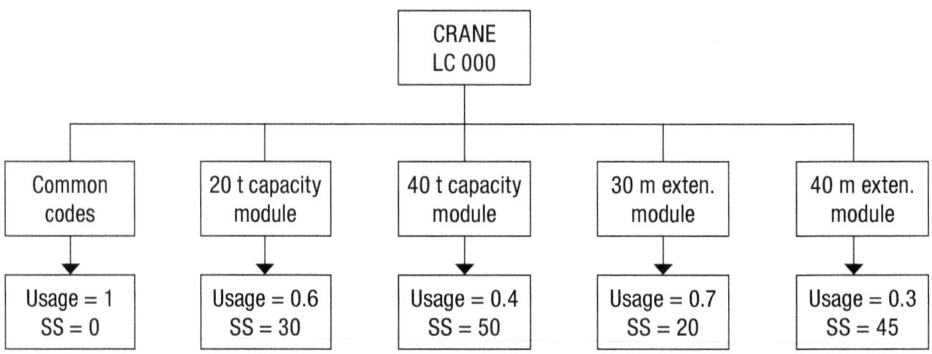

For each module, an average decimal usage is defined, that represents the popularity forecast for the single option, obtained on the basis of the historical sales series of the finished products, if necessary adjusted according to particular market trends, and the relative safety stock that allows for dealing with any changes in the mix generated by an incorrect forecast of the usage coefficients. The common codes module appears in the Super Bill with a usage equal to one, since the use of the items grouped in this module is independent of the specific configuration of the finished product chosen by the customer. The safety stock is considered zero since it is assumed that the quantities actually launched in production in the short term will never be greater than the amounts planned in the medium term within the master production schedule. The calculation of usage coefficients and safety stocks, given the significant impact it has on the level of service, production costs, and financial assets, must be carried out by the top management and periodically reviewed when formulating the production budget for the new financial year.

This simple example makes it easy to recognize the complexity of the process that leads to the creation of Super Bills. However, it is also necessary to emphasize the significant advantages that can be achieved. In addition to the improvement in the technical data management process, there are also significant effects on production planning processes. Super Bills allow for mitigating the planning problems in those companies where production or procurement lead times are longer than the customers' delivery time and a wide range of finished products is obtained from the combination of numerous variants. In these contexts, the formulation of a Master Production Schedule related to finished products can lead to significant inefficiencies: the sales forecast for a single finished product is highly unreliable and what has already been purchased or launched in production is usually just partially coherent with the order portfolio that in the meantime has been consolidated. The use of Super Bills allows to contrast the limits now underlined since the risks associated with the launch of "blind" procurement or production orders, i.e. before the actual orders have been received, are mitigated by the fact that the medium-term production plan is no longer formulated on the basis of the forecasts of finished products, but on those of the modules, that are characterized by a degree of reliability that is certainly greater. Moreover, the separation of the common items from the specific item reduces the risk of an uncontrolled increase of the stock of common components. Indeed, they also become the subject of aggregate forecasts rather than specific forecasts linked to the demand for each variant, that is certainly less reliable.

In some cases, it is more convenient to create Super Bills through **Kit Bills** (Mather, 1987), rather than through a preliminary modularization process as illustrated above. This opportunity arises if the following conditions occur (Da Villa and De Toni, 1990):

- the sum of the finished products can be divided into a certain number of product groups that do not have particular optional features;
- each of these groups is representative of a set of finished product items for which common group items and specific product items can be identified;
- it is possible to make forecasts on the "popularity" of each finished product within the group to which it belongs.

Under these conditions, it is possible to identify a kit representing the common codes (common parts list) and as many kits as the number of products considered within the group, each representing specific product codes. The code assigned to the kit thus becomes subject to planning as are the codes assigned to the modules obtained through the modularization process. The process of creation of the Super Bill is analogous to what we have seen previously, with the only difference that in the place of the modules expressing the different options, we find the kits of the specific product codes.

Although the kitting process is certainly simpler than the modularization process, it still requires a careful analysis when determining commonalities within a group of finished products. By way of example, consider the four BOMs shown in **Figure 12.5**. For the purpose of constructing a Super Bill through a kitting process of the four products, two different alternatives can be pursued. In the first, shown in **Figure 12.6**,

Figure 12.5 Bills of materials for four versions of cranes

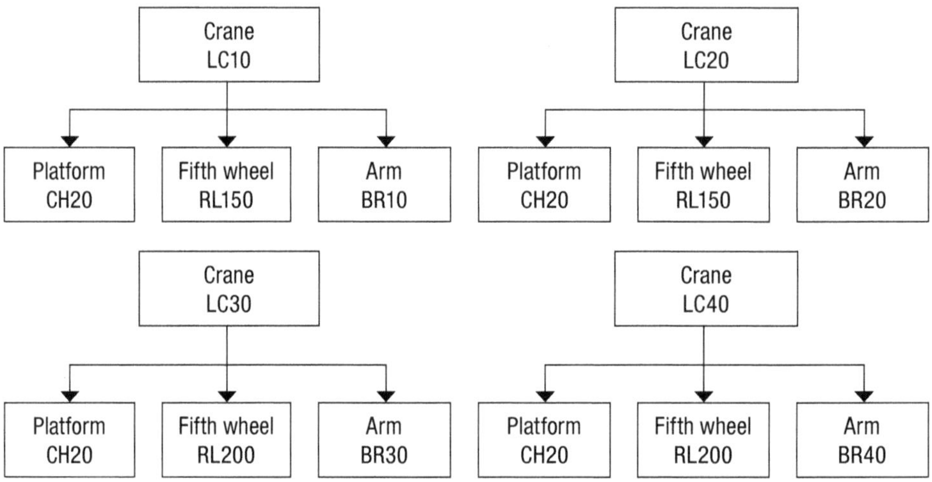

Figure 12.6 Creation of a Super Bill based on minimum commonalities

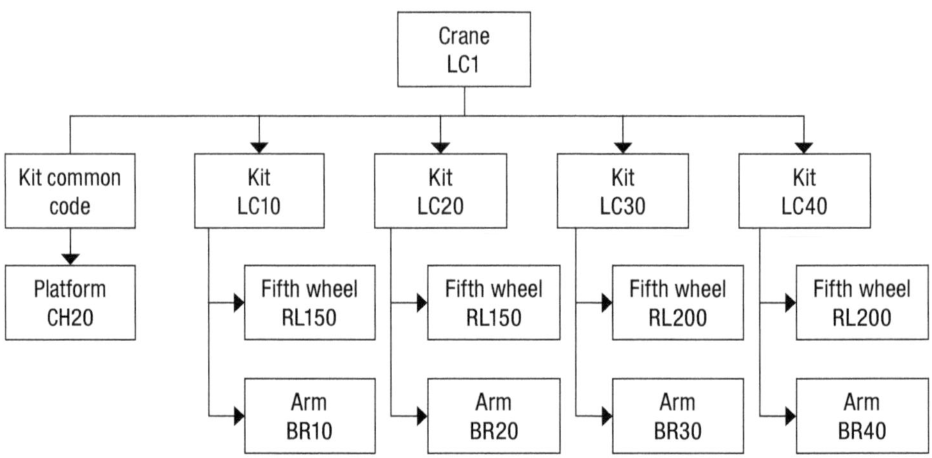

an approach based on **minimal commonalities** is chosen, whereby only a single code (the platform) is considered in the kit of common items, and four other kits are identified, each containing product-specific items.

A second option, shown in Figure 12.7, is based instead on the determination of **maximum commonalities**: this involves identifying two distinct reference groups, each characterized by a kit with two common items and two kits of product-specific items.

The choice of the minimum commonalities must be made when the aim is to face the problem of the reliability of the sales forecasts - in the example, the forecasts in fact regard a single aggregate, LC1 – even though this penalizes the company from a financial perspective, since it is necessary to make a significant investment in safety stock relating to all the specific items. On the other hand,

Figure 12.7 Creation of a Super Bill based on maximum commonalities

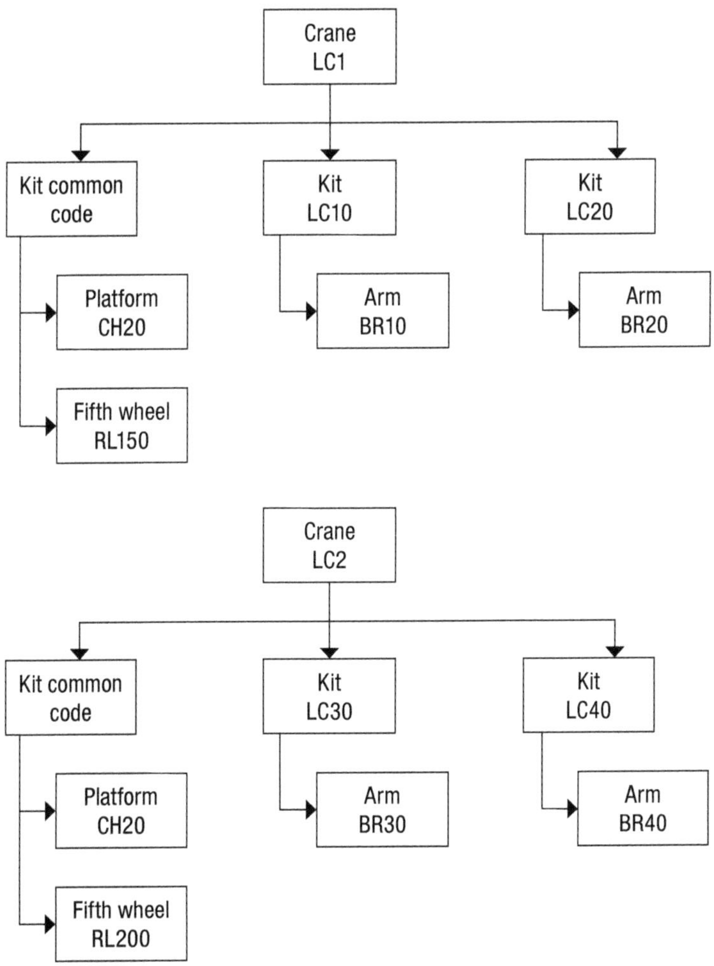

the choice of the maximum commonalities must be preferred when the possible common items have a certain financial relevance; in this case, these same items become the subject of planning without safety stock, and with a popularity coefficient equal to one.

The Kit Bill can be effectively adopted even when products at the first level of the BOM have a large number of low value items (such as fittings, small parts, etc.) that are used for the production of higher-level assemblies. In this case, it is advisable to plan the kit items as a group, even if in the subsequent production process the items are used according to specific sequences for each product.

12.3.2 Family Bill

The **Family Bill** can be defined as a sort of Super Bill for finished products belonging to the same product family or line. By family we generally mean a group of finished products whose similarities in terms of design and production facilitate sales forecasting, aggregate planning and sometimes cost control. Product families can in turn be grouped, through other Family Bills, into higher groups such as sales markets or distribution channels, until they insist on a single parent code that normally coincides with the company's turnover.

The structure of the Family Bill is made up of a zero level code that represents the family itself and some branches of the zero level code that are the finished products. Through the structure of the Family Bill it is possible to develop a plan, for example a MPS, related to the different elements of the family. This plan represents the starting point for the Rough-Cut Capacity Panning (RCCP), which takes into consideration the forecast of the overall production volume at the aggregate level to attain a rough calculation of capacity requirements. This is the figure to be compared with the actual production capacity. In case of significant differences between these two dimensions (both in annual global terms and in nearer time buckets) any actions to close the gap must be planned.

The principal advantages associated with the use of Family Bills are related to the following aspects:

- the formulation of aggregated sales forecasts, characterized by greater reliability, from which sales forecasts of individual finished products can be subsequently derived through the use of popularity coefficients;
- the consistency check of the sales forecasts generated at different levels (overall turnover, families, and finished products);
- the possibility of comparing the actual aggregate sales data with forecasts in order to highlight possible volume and mix variance.

The adoption of the planning BOM requires a preliminary assessment on how the different finished products can be aggregated into families. This entails defining the object on which the salesmen are called to provide their forecasts. The higher you set the level on which the salesmen define their forecasts – which concern the zero-level codes of the Family Bills – the more correct these are, since fewer and more aggregated figures to be forecasted are required. To the contrary, when the forecasts concern items at a lower level of the BOM, that is, when the zero-level codes are greater in number, it may be easier to adjust the popularity percentages of the various Family Bill child codes obtained on a historical basis. This is useful especially when the future market trend differs in general from the pattern followed in the past. This leads to the need for subsequent manual interventions over time to correct the percentages calculated automatically.

12.3.3 Pseudo Bill

Companies operating according to an engineering to order logic and producing complex products with long lead times do not usually know all the details of the product structure when the customer places the order. For the purpose of planning, a **pseudo BOM** can be created with a structure similar to analogous products constructed in the past, that temporarily substitutes the BOM that will result at the end of the design process. The items in the pseudo BOM are in part already defined and known and in part estimated. The latter are marked as transitory and are substituted by the actual items as the engineering department gradually defines them. The pseudo BOM used in the normal MRP run allows the planner to identify the critical path synchronizing purchasing as well as production activities and meeting the customer's delivery date. The lead times of the transitory codes usually include not only the production lead times but also the engineering lead times.

12.3.4 Inverted Bill

The Inverted Bill constitutes a particular application of the Planning BOM that takes place when most or even all of the finished products require a particular raw material or component. The aggregate demand forecasts in terms of consumption of the raw material or component are extremely more reliable than the forecasts on the individual end items. The creation of the **Inverted Bill** involves assigning a pseudo code to the common raw material or component and creating parent-child links between the common raw material or component and the final products.

12.4 Product configurators

Products characterized by a large number of variants, that in turn can be split into numerous options selected by the customers to customize their final choice, can be effectively managed through a **product configurator**. It allows for the automatic creation of a BOM, and in some cases of the relative production cycle, according to the configuration of the finished product chosen by the customer. The use of a product configurator presupposes the definition of a **product archetype** whose BOM, as illustrated in **Figure 12.8**, is made up of fixed, unmodifiable links and variable links. The latter are used to file all the possible options associated to a definite variant; during the customer order entry, only one of the variable links is activated for the generation of the BOM.

Designing a product configurator implies a detailed activity of analysis of the optional features that can give rise to variable links.

First, among the different options, it is necessary to identify the only combinations that lead to configurations consistent with a given product range. These combinations need to be made explicit through feasibility matrices whose articulation allows for accepting or rejecting the choices made by the customer.

Second, it is necessary to define for each optional feature a **domain of validity**, meaning that the client's choice can only occur within a given range defined a priori by the firm. The domain can be established in such a way as to require a choice between possible values, a positive or negative choice, or the definition of a numerical value within a precise range. Validity domains are usually entered into the configurator using simple Boolean formulas (for example: and, or, not). It is important to underline that the validity domain of an option can be influenced by the choices previously made on other optional characteristics. As an example, consider the production of a door for which the choice of the handle color is influenced by the decision regarding the color of the lacquer of the panel; with a cream lacquer the choice is between a black and a brass handle, while with a blue lacquer between a brass or a chrome handle. To manage these situations, it is necessary to build matrices which

Figure 12.8 Example of Configuration Bill

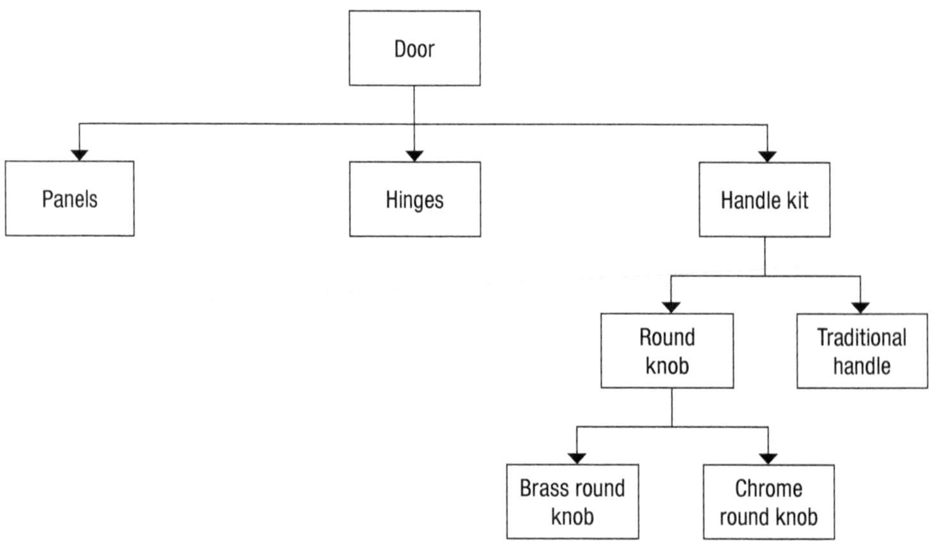

include the validity domain resulting from the combination of the validity domains of the options previously selected.

There is also the case where a choice made previously restricts the validity domain of an optional characteristic to a single compatible value. For example, the selection of a 60 Hz transformer creates a constraint also for the choice regarding the electrical motor, whose validity domain is necessarily reduced to the only frequency compatible with that of the transformer. In these cases, it is appropriate to define rules that determine the inheritance of the chosen value for another option: the link between transformer and motor remains variable, but is automatically activated when the higher level characteristic is chosen.

The choice of the different options is usually managed through a questionnaire in which all possible alternatives leading to different versions of the finished product are reported. In some cases, the questionnaire is directly available within the application and therefore it is possible to configure an order in real-time. The questions appear according to the choices made by the customer, obviously restricting the validity domain or even precluding those alternatives that are incompatible with decisions already made.

The definition of the various answers leads to the instantaneous creation of the part number, through a series of suffixes that are gradually added to the "root" that identifies the undifferentiated article, and of the BOM for the calculation of requirements. Similarly, the work cycle is also configured and can be different according to the options chosen by the customer. All this, taking place according to automatisms, drastically reduces the time dedicated to these activities and the risks of error. It is clear, however, that the potential time reduction must be evaluated in the light of the preliminary investment required for the identification of options, the definition of validity domains and feasibility matrices, the structuring of questionnaires to support the process of order configuration.

The most suitable production contexts for the adoption of product configurators are the **ATO** ones (Chapter 11): the company produces a product that is configured by the customer through the selection of a wide range of options that are the result of possible combination of common components (typically called modules) stocked at the decoupling point. Since the variety of finished products offered to the customer is as wide as the combination of all possible variants, it would be extremely difficult to manually generate all of the configurations requested by the customer. Therefore, the configurator makes the management of BOMs within the information system more efficient and reduces potential errors resulting from manual product configuration activities.

In **MTS** contexts, on the other hand, configurators do not find valuable application since the finished product is produced based on forecasts and the variety is usually limited compared to ATO contexts. In **MTO** contexts, manual configuration of the finished product is very difficult since even the lowest BOM levels are produced on customer order. It is difficult to define standard configuration rules to develop a software able to perform the automatic configuration of the product.

12.5 Production cycles

The production cycles (**routing**) describe the manufacturing processes of the products through the definition of the steps (method), the necessary resources (work centers, tools, equipment, etc.) and the times (cycle times) required for each operation. The data contained in the cycles essentially respond to two main purposes:

- to provide detailed instructions to the operators performing manufacturing and assembly activities;
- to provide other company functions (primarily planning and industrial accounting) with data that specifying which resources are used and in what quantities (for the purposes, respectively, of ex-ante planning of the resources to be allocated to the production process and ex-post evaluation of the resources actually used).

These two goals, specifically the first one, vary significantly depending on the production context. For example, if this is organized according to the principles of MTO, the absolute priority becomes the need to provide operators with precise and detailed information concerning the methods of execution. On the contrary, in the case of repetitive production (MTS), that need is significantly reduced.

With these differences in mind, we now proceed to illustrate the information set of the production cycle under the assumption of the greater complexity associated with non-repetitive production. The data contained in a production cycle can be subdivided into header data, which refer to the cycle as a whole, and operation-level data, which pertain to the individual phases involved. In relation to the first set generally are considered:

- cycle description and identification data;
- data on the nature of the cycle (for example, standard cycle, alternative cycle, cycle related to an engineering change);
- data on the part or product to which the cycle relates;
- data relating to the reference drawing/scheme/part program.

At the level of the single production operations, the following are managed:

- descriptive and identification data of the operation (sequence number, code, indication of whether it is a basic or alternative operation);
- data on nature of operation (for example, direct, auxiliary, setup);
- operating instructions for each operation, plus control plans for operations subject to quality checks;
- identification data for the main center or workstation;
- list of alternative centers/workstations;
- identification data for materials, tools, and equipment;
- time data specifying the use of production equipment;
- planning data (scheduled start and finished dates, if applicable).

Time data is normally divided between **process time** (machine time and/or labor time required to produce one unit) and **preparation time** of the machine or system to execute the specific operation (changeover or setup time). A problematic aspect regards the variability of the setup time if it depends not only on the operation to be executed, but also on the one just completed. The impossibility of defining a standard setup time a priori implies the inclusion of an average time in the cycle, adequately evaluating, if possible, the frequency with which the most anomalous conditions occur. Sometimes the cycle also includes the time associated with the queues that are normally created at the entrance to the work centers. This follows from the fact that a WIP buffer is tolerated at the entrance to each work center in order to reduce any machine downtime associated with the lack of input.

Other important parameters for planning regard time constraints affecting two or more successive operations. As already illustrated, cycles normally propose rather simple structures where a progressive number indicates the temporal sequence of the various operations within the cycle. More sophisticated applications provide an overlapping coefficient that indicates the amount of activity that must be completed in the previous phase before the next phase can begin. This solution satisfies the specific needs of parallelizable processes, typically as in the case of job shops.

Similarly to what has already been illustrated for BOMs, information systems have functions for querying the cycles archive. For example, it is possible to obtain an overview of the total working time of a finished product or a subassembly through the recomposition of the cycle, that is split for each Bill of Materials level. Such an interrogation allows to monitor the transformation costs or labor costs. Another query regards, for example, the extraction of all the items that pass through a given work center. Such information can be extremely useful to rapidly verify the workload allocated to a work center, or to simulate the impact of make or buy choices that involve a set of items.

12.6 Production centers archive

Production centers are represented by a group of homogeneous resources that perform a specific operation or activity. The expression "homogeneous resources" is intentionally broad since the center can be represented, depending on the case, by a single machine tool (numerical control lathe), by a group of similar machines (lathe department), by an assembly line, by a single operator or by a work team.

The elementary data commonly stored in **production center archives** is the following:

- descriptive and identification data of the work center;
- data on the availability of the center, according to the work calendar that provides for a certain number of days, shifts and hours per period;
- average data on utilization and efficiency of the center, useful during the scheduling process to determine the actual production capacity available in a given period;

- data on hourly cost, usually split between machine cost, labor cost, and cost of auxiliary setup activities;
- data on the machines that make up the center, if the latter consists of several production units that are each subject to scheduling activities. In this case, for each unit an availability calendar is prepared that can be managed by exceptions with respect to that of the center;
- data on the individual operators or technical personnel working in the center, to provide information for industrial accounting procedures or administrative personnel management.

13 Stock Control Methods for Dependent Demand Goods and the Choice of the Appropriate Method

by *Alberto Grando*

13.1 Look Ahead approach: Material Requirements Planning logic

The methods of inventory management illustrated in Chapter 10, called look back, are based on the principle of replenishment of stock running out. The characteristic trait that unites those methods lies in the need to constantly maintain a suitable level of inventory. Therefore, on the one hand, those methods entail significant investments in working capital, while on the other, they are based on simplifying hypotheses (estimable demand or consumption based on past track records, linear withdrawal, normal distributions) necessary for the statistical solution of the problem, but that cannot always be verified in reality.

Imagine having to use look-back techniques to manage components and raw materials whose demand depends on the demand for the finished product, as illustrated in **Figure 13.1**. Based on frequent consumption and small quantities for the latter, the need for the component and the raw material will show more intense and rare spikes over time, generating long periods of useless overinvestment in stock for the part-numbers placed at the lower levels of the bill of materials. However, if the requirements of the dependent demand components are planned according to a **look-ahead technique**, issuing their orders sufficiently in advance to cover the related lead time, the level of inventory will be significantly lower.

The methods based on stock replenishment (look back) can thus be conveniently used where it is not possible to do otherwise or the hypotheses on which they are based are credible, such as in the case of inventory management of finished products subject to fragmented and constant demand, or of low value, frequent consumption materials and components. In manufacturing companies, the very characteristics of the process often lead to phenomena that diverge considerably from the cited hypotheses.

- A first element of divergence can be found in the phenomenon of the **bullwhip effect,** whereby, despite the presence of level demand for finished products, due to the use of lot sizing practices, the (dependent) requirement for sub-assemblies and components placed at increasingly lower levels in the **Bill of Materials (BOM)** shows a discontinuous trend, with consequent effects on the inventory system, that can be significant, as illustrated in **Table 13.1**. On the other hand, in the case of a component that, due to a design inspired by criteria of modularity, is common to multiple finished products (and thus appears in the BOM of various products) its

Figure 13.1 Look back and look ahead approaches for dependent demand goods

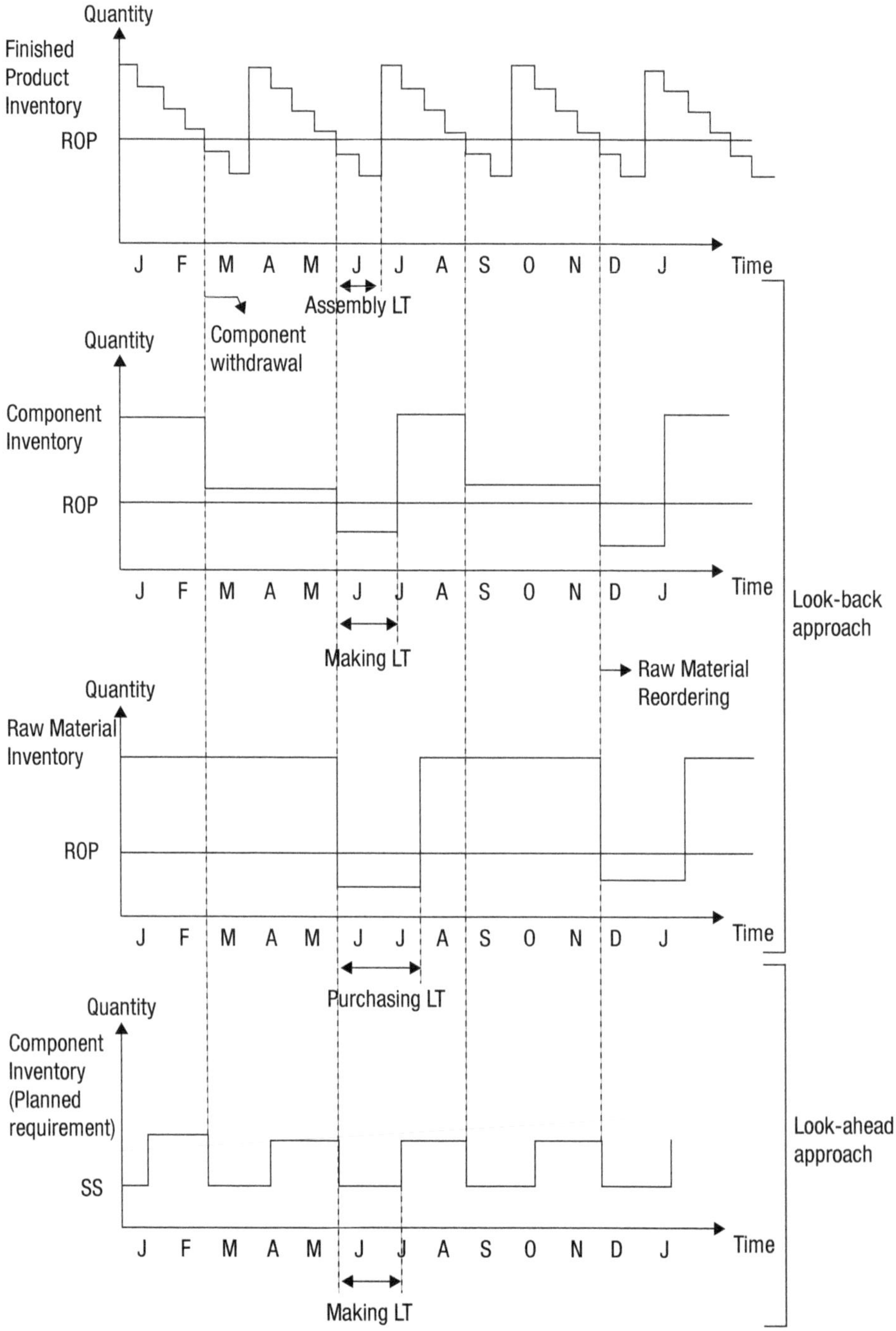

Source: adapted from Orlicky, 1975.

Table 13.1 Level demand and upstream lot sizing practices

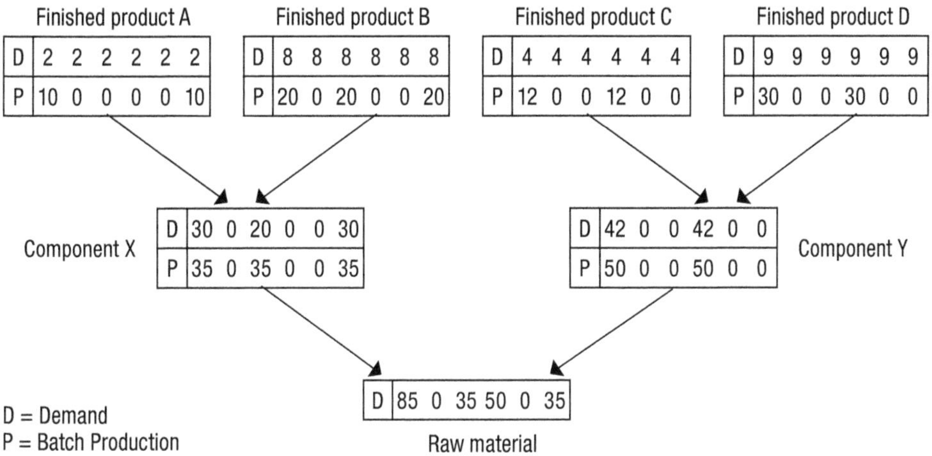

Source: adapted from Brandolese et al., 1991.

consumption will show a gradually damped discontinuity, until tending towards sufficient stationariness and eventually suggesting the adoption of criteria of replenishment management.

- Secondly, we must observe the effect of the phenomenon of **decreasing availability**; it can be found that, as a result of the law of compound probability, with the increase of the number of components that flow into a finished product, the likelihood of its availability decreases. It follows that there will be either an expensive level of safety stock, or frequent cases of missing pieces and risk of stock out.

The second approach to stock management, called look ahead or by requirements, aims to exceed the described limits, and consists of the technique of **Material Requirements Planning – MRP** (Orlicky, 1975; Vollmann et al., 1984; Plossl, 1994).

The **requirements logic** assumes knowledge of the **Master Production Plan (MPS)** of finished products, or of the items subject to planning,[1] and bases its validity on the possibility of calculating the need for semi-finished products, components, and materials used, in a cascading manner for dependent demand articles.

The adoption of an MRP approach requires an information management system able to provide data and information that are complete, timely, and reliable. The **system inputs** are generated by the **multiple actors** involved in the overall manufacturing operations.

[1] As shown in Chapter 11, in case of the adoption of ATO, MTO strategies, the items under MPS control are not the Finished Product, but components positioned at different levels of the BOM.

- Engineering Departments for the development of the product's BOM, the internal Lead Times and the Work Cycles.
- Production Planning for the MPS/FAS and the analysis of Available Production Capacity.
- Procurement for the data relating to purchase Lead Times and specifications.
- Logistics for the Inventory level control and so forth.

In addition to some configuration parameters, the functioning of an MRP system entails the availability of information reported in **Figure 13.2** split among inputs and outputs.

The **Inputs** are:

- the **Master Production Schedule** (**MPS**) and **Final Assemble Schedule** (**FAS**) of the codes subject to the planning;
- the **Orders Issued** to internal Departments or to external Suppliers, not yet closed, whose (known) quantities have not yet been entered into stock (but for which the future date of entry into stock is known);
- the **Bill of Materials** (or product structure file or product tree) – also called recipe or formula in the process industries like Chemical, Pharma or Food – describing the explosion of a product into all of its components, based on an increasingly de-

Figure 13.2 **Inputs and outputs of the MRP logic**

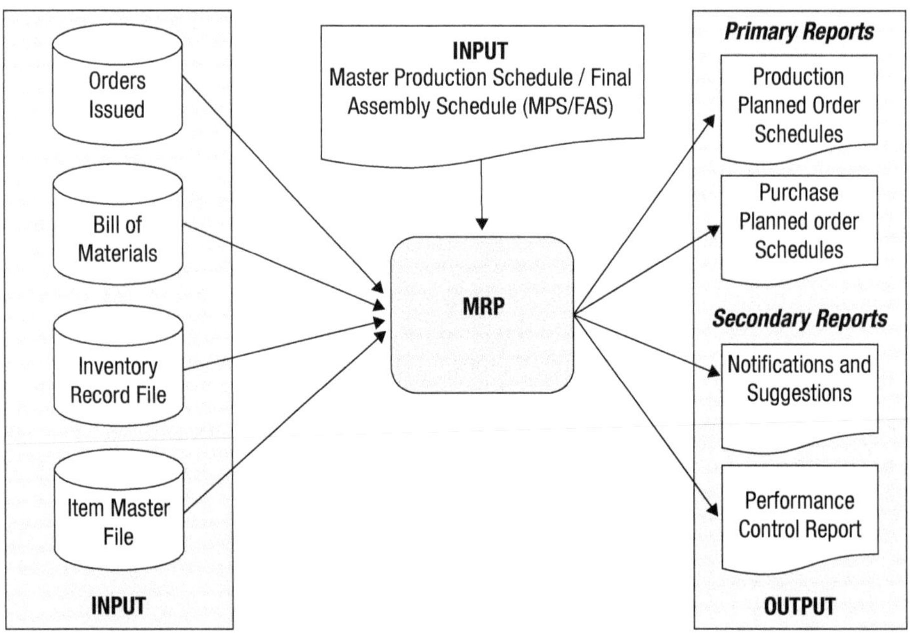

Source: adapted from De Toni and Panizzolo, 2018; Chase et al., 2004.

tailed scheme (groups, sub-groups, assemblies, sub-assemblies, etc.), with the goal of determining the ultimate requirements of each material or component;
- the **Inventory Record File**, that registers the level of on-hand stock for each part-number (component and material) updated at any transaction (Inventory status data);
- the **Item Master File** containing, as shown in Chapter 12, a series of technical and managerial information such as the part number, part name, low-level code, unit of measure, design charge number, drawing reference, standard cost, lead times etc.

The **Outputs** are split into Primary Reports and Secondary Reports.

Primary Reports are:

- **Production Planned Order Schedules**, also called Manufacture or Assemble Order Proposals, relating to needs of materials and components necessary to cover what is planned by the MPS/FAS, calculated from the MRP explosion based on the available stock. Being Planned Orders, until they are transformed into **Production Actual Orders** issued to the manufacturing departments, they can be changed or adjusted to take into consideration any variation in the inputs;
- **Purchase Planned Order Schedules**, also called Purchase Order Proposals, relating to the needs of materials and components purchased from suppliers. Those planned orders are managed by the Procurement Department, that will transform them into **Purchase Actual Orders**, modifying their quantity, grouping them, etc., as a function of the goals of convenience and optimization in the relationship with suppliers.

The **Secondary Reports** are:

- **Notifications and suggestions**, that is, feedback information sent to the Planner on a broad number of cases such as the need for rescheduling to guarantee planned dates, indications on the need to issue orders, modifying the status of planned orders to confirmed orders, etc.;
- Information to structure **Reports for the performance control**, relating to delays, frequency of rescheduling, missing parts, etc.

13.2 The MRP record processing

The MRP processing consists of the **time-phased calculation** of components and materials requirements at each level of assembly or manufacture, starting with the MPS of finished products (or other planned item) to be manufactured and assembled until reaching the maximum level of explosion, through the use of the data included in the BOM. This way, the MRP, operating by periods (or Time Buckets) and levels (from 0 to n), allows for rigorous control of the timing of each planned order (thanks to pre-calculated lead times) and the amounts in stock (net balances of the parts in Inventory Record File), that are arriving or need to be ordered. At times, the Gross Requirements are appropriately increased in order to take into account the reject coefficients (of products or processes) indicated in the part-numbers and structures archives, calculated based on normal defectiveness detected in the inspection upon acceptance or in the shop-floor, for the items procured or those produced internally, respectfully.

In each Time bucket of a MPS and for each part-number belonging to a BOM, the MRP operates through the elaboration of the following (Plossl, 1995).

- **Gross Requirements**, i.e. the planned usage during a period, equals to the Planned Order Release – Order date of the part-number "parent" (level n in the BOM) multiplied by the Usage Quantity that links the parent to a part-number "child" (level n+1) in the BOM.
- **Scheduled Receipts**, i.e. the replenishment orders (open or issued) due at the beginning of each period, coming from Planned Orders Released in the past.
- **On-hand Inventory** (or Projected Available Balance), i.e. the current and projected available inventory level at the end of each period, equals to On-hand Inventory at the beginning of the period (or Projected Available Balance at the end of the previous period) – the Gross Requirements during the period + the Scheduled Receipts expected at the beginning of the same period.
- **Net Requirements**, i.e. the Gross Requirements net of the Projected Available Balance; Net Requirements equals the Projected Available balance when negative.
- **Planned Order Receipts – Due date**, i.e. the replenished orders planned at the beginning of each period, equal to the Net Requirements in the absence of Lot-sizing policies.
- **Planned Order Releases – Order date**, i.e. the Planned Order Receipts offset by the related Lead Times.

To better clarify the functioning of the model, consider the following example. Let us suppose a company producing air filters plans to produce a product for tractors, codified by the part number (A).

To define the MPS of the Air Filter (A), planners take into consideration the larger between Sales Forecast and Customer Orders already acquired, as well as the On Hand Inventory, split between beginning and ending inventories, as reported in Table 13.2.

The other elements necessary for the functioning of an MRP system are:

- the BOM per product (A), with the indication Usage Quantity per level, represented in Figure 13.3. At level 1, the air filter is made through the assembly of 2 components (part numbers): end caps (B) and the filter element (C). At level 2, the filter element (C) is in turn made by a cylinder of wire mesh (D) and an internal pleated paper cartridge (E);

Table 13.2 MPS of the air filter (A)

Level 0	Part Number: A	Weeks	1	2	3	4	5	6	7	8	9	10
		Forecast	100	100	150	150	200	200			100	100
Lead Time: 1 week		Customer orders	150	80	150		100		100			
On hand stock: 280		Projected available balance	280/ 130	130/ 30	30/ 0							
		Net Requirements			120	150	200	200	100		100	100
		Planned order receipts			120	150	200	200	100		100	100
		MPS		120	150	200	200	100		100	100	

Figure 13.3 Air filter A – Bill of Materials and related Information

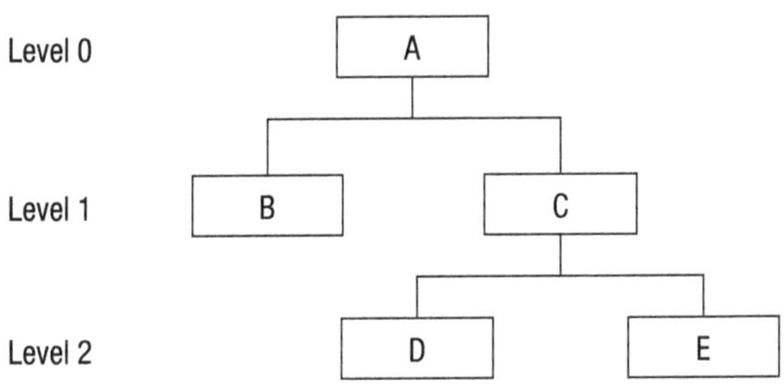

BOM Level	Part Number	Description	Usage Quantity	Unit of measure	Lead Time	On hand Inventory
0	A	Air filter	–	pcs	1	280
. 1	B	End cap	2	pcs	1	300
. 1	C	Filter element	1	pcs	2	400
. . 2	D	Wire mesh	1	pcs	1	200
. . 2	E	Pleated paper cartridge	1	pcs	2	300

- the On Hand Inventory (Projected available balance) for each part number belonging to the BOM of product (A), as well as the related production and procurement lead times.

It is then possible to proceed with the explosion of the requirements starting from level 0 (finished product – Air Filter) to reach the n_{th} level, thus launching the purchase and/or production orders for the various components needed. **Table 13.3** illustrates that:

- The **Planned Order Receipts** entails the assembly of 120 units of finished product (A) in the 3rd week, 150 in the 4th, 200 in the 5th and 6th, and 100 in the 7th, 9th and 10th. Since the assemble lead time of the product at level 0 is equal to one week, it is thus necessary to launch the MPS (**Planned Order Releases**) one week in advance, requesting 120 units in the 2nd week, 150 in the 3rd, etc.
- In each n period, the **Planned Order Releases** (MPS) of the finished product Air Filter (A), consisting of the assembly of 2 units of end cap (B) and 1 unit of filter element (C), instantly generates the **Gross Requirements** of the two components (B) and (C) in the same n period. This is equal to the planned units of (A), reported in the Planned Order Releases, multiplied by the Usage Quantity of the single part-number, i.e. 2 and 1 for components (B) and (C), respectively.
- To obtain the **Projected Available Balance**, the quantities of materials already ordered but not yet received (Scheduled Receipts), and the On-hand Inventory at the end of the n-1 period are to be added and subtracted, respectively, from the Gross Requirements thus calculated (netting phase), being careful to respect the timing of the arrivals. The **Net Requirements** occur when the Projected Available Balance become negative.
- Once the **Net Requirements** is calculated, equal to the Planned Order Receipts in the absence of lot-sizing policies, it is necessary to launch the **Planned Order Releases**, being careful to issue them advance based on the relevant lead times. The same procedure applies to components (D) – wire mesh – and (E) – pleated paper cartridge – of the second level, and so on down to the part-numbers of the n_{th} level.

13.3 Parameterizing MRP

In **parameterizing an MRP system**, it is also necessary to consider the **main elements** that influence the dynamic of requirements over time (Orlicky, 1975; Vollmann et al., 1984; Plossl, 1994).

a. The calculation of Gross Requirements in the presence of external (Independent Demand Goods) requirements.
b. The determination of Projected Available Balance and thus the Net Requirements in the presence of Safety Stock.
c. The use of Adjusted Lead Times in the Planned Order Releases.
d. The more used Lot-sizing logics in the Planned Order Receipts.
e. The calculation of Net Requirements in the presence of part-numbers belonging to different Bills of Materials.
f. The Bottom-up Pegging procedure.
g. The processing refreshment frequency: Regeneration and Incremental systems.

a) The calculation of Gross Requirements in the presence of external requirements

The Gross requirement of a component can be generated both by internal and external requirements. The former is represented by the demand for the higher-level item in a BOM (parent) on which it depends, according to the deterministic links expressed by the Usage Quantity, taking the name of internal requirement. The latter, by an exogenous incremental requirement, independent of the links in the BOM, that may be induced by the need to have spare parts, consumables, or components destined to after sales operations (Independent demand goods). This takes the name of **external requirements**. In that case, the quantities of external requirements have to be added to internal Gross Requirements, the result of the explosion of the BOM, allowing for the purchase or production of quantities able to also cover requirements linked to spare parts or other uses not expressed by the BOM links.

By way of example, see the case shown in Table 13.4, which shows the case of production of the Air Filter from Table 13.3 (same BOM and MPS), for which, however, the distributors of tractor spare parts show orders for replacement caps (B) and cartridges (E) (external requirement) on average equal to 10% and 20% of their respective internal Gross Requirements.

b) The determination of Net Requirements in the presence of Safety Stock

In the most common applications, the Projected Available Balance incorporates the Safety Stock, to face, for instance, possible delays in the Scheduled Receipts quantities. In this case, the manifestation of the first Net Requirements does not emerge in the presence of a Projected Available Balance reaching zero, but upon reaching a preset level of **Safety Stock**. On this point, see the example shown in Table 13.5, that again uses the example of Table 13.3 (same BOM and MPS), assuming that for the part numbers C and E a Safety Stock is required of 150 and 180 units, respectively.

c) The use of Adjusted Lead Time

Concerning this point, we recall that planned lead time has the aim of guaranteeing the availability of materials as of the date of the first net requirements, i.e. at the Planned Order Receipts – Due date. There are **three factors** that can lead to a lead time adjustment.

- First, in case of **time buckets** MRP systems: the planned lead time generally does not coincide with actual lead time, since MRP systems operate on plan horizons marked by periods (buckets), that are generally weekly, to which the actual lead times, generally measured in days, are rounded off.
- Secondly, in the case of different components needed to assemble a "parent" part number, each of them with different lead times, two options to manage the lead times are usually adopted. The first one, called **Forward Schedule Logic** (Cutting approach), aims at scheduling as early as possible all part-numbers adjusting the respective lead times to the one required as first. On this point, it should be stressed that a prudential attitude, aimed at overestimating lead times, can translate into increases of work in process stock (desirable if seen as a precaution, but

Table 13.3 An example of materials requirements planning

Level 0	Part Number: A	Weeks	1	2	3	4	5	6	7	8	9	10
		Planned order receipts			120	150	200	200	100		100	100
		MPS		120	150	200	200	100		100	100	

Level 1	Part Number: B	Weeks	1	2	3	4	5	6	7	8	9	10
Usage quantity: 2		Gross requirements		240	300	400	400	200		200	200	
		Scheduled receipts										
On hand stock: 300		Projected available balance	300 300	300 60	60 0							
		Net requirements			240	400	400	200		200	200	
		Planned order receipts			240	400	400	200		200	200	
Lead Time: 1 week		Planned order release		240	400	400	200		200	200		

Level 1	Part Number: C	Weeks	1	2	3	4	5	6	7	8	9	10
Usage quantity: 1		Gross requirements		120	150	200	200	100		100	100	
		Scheduled receipts			100							
On hand stock: 400		Projected available balance	400 400	400 280	280 230	230 30	30 0					
		Net requirements					170	100		100	100	
		Planned order receipts					170	100		100	100	
Lead Time: 2 weeks		Planned order release			170	100		100	100			

Level 2	Part Number: D	Weeks	1	2	3	4	5	6	7	8	9	10
Usage quantity: 1		Gross requirements			170	100		100	100			
		Scheduled receipts										
On hand stock: 200		Projected available balance	200 200	200 200	200 30	30 0						
		Net requirements				70		100	100			
		Planned order receipts				70		100	100			
Lead Time: 1 week		Planned order release			70		100	100				

Level 2	Part Number: E	Weeks	1	2	3	4	5	6	7	8	9	10
Usage quantity: 1		Gross requirements			170	100		100	100			
		Scheduled receipts										
On hand stock: 300		Projected available balance	300 300	300 300	300 130	130 30	30 30	30 0				
		Net requirements						70	100			
		Planned order receipts						70	100			
Lead Time: 2 weeks		Planned order release				70	100					

Table 13.4 **The calculation of gross requirements in the presence of external requirements**

Level 1 **Part Number: B**	**Weeks**	**1**	**2**	**3**	**4**	**5**	**6**	**7**	**8**	**9**	**10**
Usage quantity: 2 **External requirement: 10%**	Gross requirements		264	330	440	440	220		220	220	
	Scheduled receipts										
On hand stock: 300	Projected available balance	300 300	300 36	36 0							
	Net requirements			294	440	440	220		220	220	
	Planned order receipts			294	440	440	220		220	220	
Lead Time: 1 week	Planned order release		294	440	440	220		220	220		

Level 1 **Part Number: C**	**Weeks**	**1**	**2**	**3**	**4**	**5**	**6**	**7**	**8**	**9**	**10**
Usage quantity: 1	Gross requirements		120	150	200	200	100		100	100	
	Scheduled receipts			100							
On hand stock: 400	Projected available balance	400 400	400 280	280 230	230 30	30 0					
	Net requirements					170	100		100	100	
	Planned order receipts					170	100		100	100	
Lead Time: 2 weeks	Planned order release			170	100		100	100			

Level 2 **Part Number: D**	**Weeks**	**1**	**2**	**3**	**4**	**5**	**6**	**7**	**8**	**9**	**10**
Usage quantity: 1	Gross requirements			170	100		100	100			
	Scheduled receipts										
On hand stock: 200	Projected available balance	200 200	200 200	200 30	30 0						
	Net requirements				70		100	100			
	Planned order receipts				70		100	100			
Lead Time: 1 week	Planned order release			70		100	100				

Level 2 **Part Number: E**	**Weeks**	**1**	**2**	**3**	**4**	**5**	**6**	**7**	**8**	**9**	**10**
Usage quantity: 1 **External requirement: 20%**	Gross requirements			204	120		120	120			
	Scheduled receipts										
On hand stock: 300	Projected available balance	300 300	300 300	300 96	96 0						
	Net requirements				24		120	120			
	Planned order receipts				24		120	120			
Lead Time: 2 weeks	Planned order release		24		120	120					

Table 13.5 Determination of Net Requirements in the presence of Safety Stock

Level 1 **Part Number: B**	**Weeks**	**1**	**2**	**3**	**4**	**5**	**6**	**7**	**8**	**9**	**10**
Usage quantity: 2	Gross requirements		240	300	400	400	200		200	200	
	Scheduled receipts										
On hand stock: 300	Projected available balance	300 300	300 60	60 0							
	Net requirements			240	400	400	200		200	200	
	Planned order receipts			240	400	400	200		200	200	
Lead Time: 1 week	Planned order release		240	400	400	200		200	200		

Level 1 **Part Number: C**	**Weeks**	**1**	**2**	**3**	**4**	**5**	**6**	**7**	**8**	**9**	**10**
Usage quantity: 1	Gross requirements		120	150	200	200	100		100	100	
	Scheduled receipts			100							
On hand stock: 400 **Safety stock: 150**	Projected available balance	400 400	400 280	280 230	230 150	150 150	150 150	150 150	150 150	150 150	
	Net requirements				120	200	100		100	100	
	Planned order receipts				120	200	100		100	100	
Lead Time: 2 weeks	Planned order release		120	200	100		100	100			

Level 2 **Part Number: D**	**Weeks**	**1**	**2**	**3**	**4**	**5**	**6**	**7**	**8**	**9**	**10**
Usage quantity: 1	Gross requirements		120	200	100		100	100			
	Scheduled receipts										
On hand stock: 200	Projected available balance	200 200	200 80	80 0							
	Net requirements			120	100		100	100			
	Planned order receipts			120	100		100	100			
Lead Time: 1 week	Planned order release		120	100		100	100				

Level 2 **Part Number: E**	**Weeks**	**1**	**2**	**3**	**4**	**5**	**6**	**7**	**8**	**9**	**10**
Usage quantity: 1	Gross requirements		120	200	100		100	100			
	Scheduled receipts										
On hand stock: 300 **Safety stock: 180**	Projected available balance	300 300	300 180	180 180	180 180	180 180	180 180	180 180			
	Net requirements			200	100		100	100			
	Planned order receipts			200	100		100	100			
Lead Time: 2 weeks	Planned order release	200	100		100	100					

otherwise uselessly costly). To the contrary, more updated systems allow to start the order of each part number according to the **Backward Schedule Logic**, i.e. scheduling as late as possible, by postponing the commitment of raw materials and components, with the aim of guaranteeing a perfect synchronization of the arrivals, aimed at reducing the work in process.
- Third, in the case of adoption of **Safety Lead Times** instead (or together) a Safety Stock. In this case, the Planned Order Releases – Order date take into account the Safety Lead Times, therefore orders are planned earlier in order to receive them, as Scheduled Receipts, into the Projected Available Balance (Inventory) before the actual date that the MRP would otherwise suggest.

d) The main Lot-sizing Policies

A further aspect deserving examination are the different policies can be adopted in the **Planned Order Receipts phase**. In the previous examples, we assumed the issue of assembly, manufacture, and purchase orders equal to the Net Requirements calculated for each part number placed at each level, made suitably in advance as a function of their respective lead times. In this case, the quantities emerging at the Net Requirements row are exactly the same of the ones at the Planned Order Receipts. In reality, there are **different criteria** for choosing the appropriate sizing of batches to be ordered (grouping phase, or lot sizing), depending on the characteristics of the operations system in question and the specific management policies based on which the system needs to be parameterized. The most common criteria for reordering are the following.

- The **Lot-for-lot criterion (L4L)** authorizes the issue of orders of a size equal to the Net Requirements for the period, guaranteeing the minimization of holding costs; it is useful in the presence of special articles, with strong variable demand, sometimes based on client orders, or articles that are particularly costly. In addition, this policy is suggested for intermediate levels of the BOM to avoid, in case of larger batches, a cascade effect, that increases the Gross Requirements at the lower levels of the BOM, similar to the bullwhip effect.
- The **Fixed Order Quantity criterion (FOQ)**, based on non-economic fixed quantities, meets the need of satisfying maximum lot constraints, for example dictated by available warehouse or container space, or to the contrary a minimum lot, with the aim of respecting the constraints dictated by the supplier or the transport system in terms of standard package units, transportation units, or in case of high set up times.
- The **Economic Order Quantity criterion (EOQ)** is based on the need to minimize the costs associated with the ordering and holding stock, following the EOQ formula already illustrated (Chapter 10); it is commonly used in the presence of articles with uniform consumption and medium-to-high value.
- Finally, the **Fixed Cycle Requirements (FCR)** or *Fixed Period Requirements* (FPR) or *Fixed Coverage* (FC) *criterion*, each time the need for a reordering ap-

pears, suggests the issue of an order for a quantity equal to the requirements accumulated in a defined time horizon (n periods or time buckets). This is what happens with replenishment policies made according to monthly procedures, such that the requirements for four weekly periods are accumulated, or to optimize the demand management. Contrary to criteria based on Fixed or Economic Quantity, in which the quantities ordered are constant and the reordering cycle differs, this method authorizes orders at preset frequencies (Fixed Cycles) for variable quantities.

Table 13.6 shows an example of the functioning of the four lot-sizing methods described, starting from the same **Net Requirements**, while Figure 13.4 and Table 13.7 report a full example of MRP explosion, considering different Lot Sizing policies and Safety Stock impact.

Other lot sizing criteria have been developed such as the Last Unit Cost or based on more complex algorithms as, for example, the Wagner-Whitin one (Brandolese et al., 1991; Sianesi, 2011).

e) The calculation of Net Requirements in the presence of a part number belonging to different bills of materials

As Figure 13.5 shows, one or more components can belong to different levels of the **same BOM or multiple BOMs**, as in the case of components that are normalized (screws, gaskets, lamps, etc.) or the fruit of design practices oriented to modularity.

In that case, the Gross Requirement of the common part number (A) is computed only once, as a summation of the different Gross Requirements, coming from the product of the respective Planned Order Releases – Order date by their Usage Quantity indices, as illustrated by the "path" of the horizontal lines of Figure 13.5. The calculation of the **Net Requirement of the common part number** (A), performed deducting the quantities contained in the single Projected Available Balance are carried out as a function of the needs generated at the lower level, consistent with the logic of the Low Level Code already mentioned (Chapter 12).

Table 13.6 **Examples of Planned Order Releases bound by different Lot sizing policies**

LEVEL N Part Number: X	WEEKS	1	2	3	4	5	6	7	8	9	10
Lead Time: 2 weeks	Net requirements	0	0	200	200	250	250	200	200	150	100
Lot Sizing Policies											
Lot sizing: L4L	Planned order receipts	0	0	200	200	250	250	200	200	150	100
Lot sizing: EOQ = 600	Planned order receipts	0	0	600	0	600	0	0	600	0	0
Lot sizing: FQ = 400	Planned order receipts	0	0	400	0	400	400	0	400	0	0
Lot sizing: FC = 4 weeks	Planned order receipts	0	0	900	0	0	0	650	0	0	0

Figure 13.4 Bill of Materials and related Information

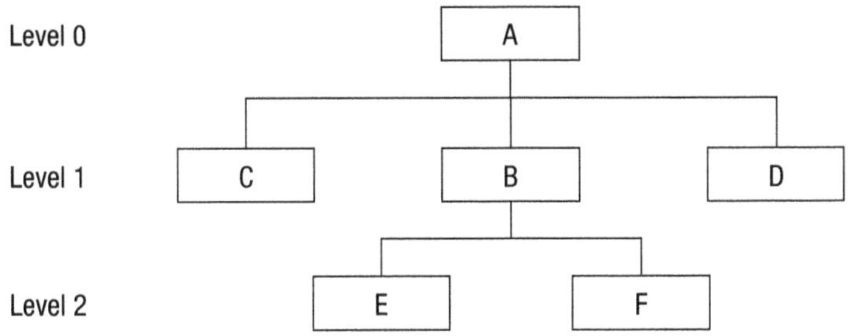

BOM Level	Part Number	Usage Quantity	Lead Time (weeks)	On hand Stock (units)	Safety Stock (units)	Lot Sizing Policy
0	A	–	1	280	0	L4L
. 1	B	2	1	300	50	L4L
. 1	C	1	2	400	100	EOQ = 400
. 1	D	1	2	600	200	EOQ = 200
. . 2	E	2	1	1000	0	FQ = 800
. . 2	F	4	1	1200	300	FC = 4 weeks

Table 13.7 Example of MRP explosion with different Lot Sizing Policies and Safety Stock

Level 0 Part Number: A	**Weeks**	**1**	**2**	**3**	**4**	**5**	**6**	**7**	**8**	**9**	**10**
	Forecast	100	100	150	150	200	200	200	150	100	100
	Customer orders	150	80	150		100		100			
On hand stock: 280 Safety stock: 0	Projected available balance	280 130	130 30	30 0							
	Net requirements			120	150	200	200	200	150	100	100
Lot sizing: L4L	Planned order receipts			**120**	**150**	**200**	**200**	**200**	**150**	**100**	**100**
Lead Time: 1 week	Planned order release		120	150	200	200	200	150	100	100	

Level 1 Part Number: B	**Weeks**	**1**	**2**	**3**	**4**	**5**	**6**	**7**	**8**	**9**	**10**
Usage quantity: 2 B/A	Gross requirements		240	300	400	400	400	300	200	200	
	Scheduled receipts										
On hand stock: 300 **Safety stock: 50**	Projected available balance	300 300	300 60	60 50	50 50	50 50	50 50	50 50	50 50	50 50	
	Net requirements			290	400	400	400	300	200	200	
Lot sizing: L4L	Planned order receipts			**290**	**400**	**400**	**400**	**300**	**200**	**200**	
Lead Time: 1 week	Planned order release		290	400	400	400	300	200	200		

Level 1	**Part Number: C**	**Weeks**	**1**	**2**	**3**	**4**	**5**	**6**	**7**	**8**	**9**	**10**
Usage quantity: 1 C/A		Gross requirements		120	150	200	200	200	150	100	100	
		Scheduled receipts			100							
On hand stock: 400		Projected available	400	400	280	230	430	230	430	280	180	
Safety stock: 100		balance	400	280	230	430	230	430	280	180	480	
		Net requirements				70		70			20	
Lot sizing: EOQ = 400		Planned order receipts				**400**		**400**			**400**	
Lead Time: 2 weeks		Planned order release		400		400			400			

Level 1	**Part Number: D**	**Weeks**	**1**	**2**	**3**	**4**	**5**	**6**	**7**	**8**	**9**	**10**
Usage quantity: 1 D/A		Gross requirements		120	150	200	200	200	150	100	100	
		Scheduled receipts			200	100						
On hand stock: 600		Projected available	600	600	480	530	430	230	230	280	380	
Safety stock: 200		balance	600	480	530	430	230	230	280	380	280	
		Net requirements						170	120	20		
Lot sizing: EOQ = 200		Planned order receipts						**200**	**200**	**200**		
Lead Time: 2 weeks		Planned order release				200	200	200				

Level 2	**Part Number: E**	**Weeks**	**1**	**2**	**3**	**4**	**5**	**6**	**7**	**8**	**9**	**10**
Usage quantity: 2 E/B		Gross requirements		580	800	800	800	600	400	400		
		Scheduled receipts	800									
On hand stock: 1000		Projected available	1000	1800	1220	420	420	420	620	220		
Safety stock: 0		balance	1800	1220	420	420	420	620	220	620		
		Net requirements				380	380	180		180		
Lot sizing: FQ = 800		Planned order receipts				**800**	**800**	**800**		**800**		
Lead Time: 1 week		Planned order release			800	800	800		800			

Level 2	**Part Number: F**	**Weeks**	**1**	**2**	**3**	**4**	**5**	**6**	**7**	**8**	**9**	**10**
Usage quantity: 4 F/B		Gross requirements		1160	1600	1600	1600	1200	800	800		
		Scheduled receipts										
On hand stock: 1200		Projected available	1200	1200	5100	3500	1900	300	1900	1100		
Safety stock: 300		balance	1200	5100	3500	1900	300	1900	1100	300		
		Net requirements		260				1200				
Lot sizing: FC = 4 weeks		Planned order receipts		5060				2800				
Lead Time: 1 week		Planned order release	**5060**				**2800**					

Figure 13.5 Requirements and low level code

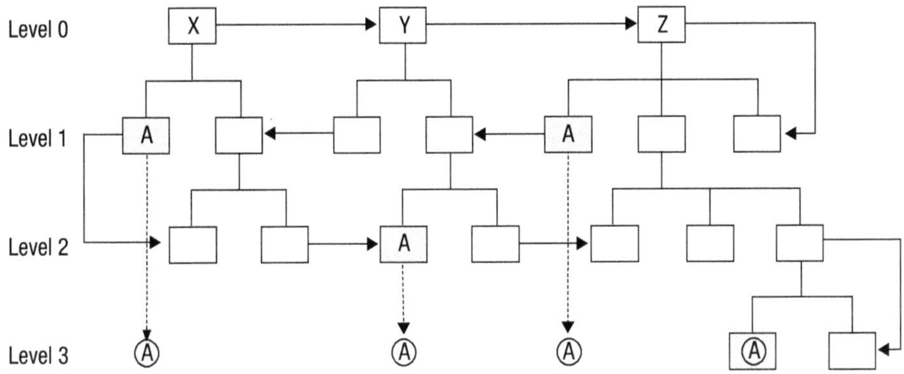

Source: De Toni and Panizzolo, 2018.

f) The Bottom-up Pegging procedure

The Pegging procedure consists of following a bottom-up logic to retrace the MRP record chains, based on the links between needs of parent-child articles memorized during the MRP explosion. In that case, the **Pegging** function allows for reconstructing the needs-orders relationships from the bottom up, according to the inverse sequence **"child's Gross Requirements – parent's Planned Order Releases – parent's Gross Requirements"**; as a consequence, it is possible to analyze the potential impact of an anomaly emerging on a part number at a lower level (delay in delivery, excess discards, machine downtime, etc.) on higher level articles (potential deviation in terms of dates or quantities planned), and through bottom-up rescheduling, decide which must be replanned.

g) The Processing Refreshment Frequency: Regeneration and Incremental systems

Lastly, in terms of frequency and method of rescheduling, the MRP applications can operate according to two different logics: *Regeneration* systems and *Incremental* systems. With **Regeneration systems** the full explosion of requirements is periodically recalculated based on the most recent MPS elaboration. That approach is used in the event of infrequent rescheduling, at the most one per week (even if today MRP systems can operate even daily, given their elaboration speed). It entails the cancellation of the previous records which is substituted in full by the new one, pushing the plan's visibility horizon ahead by one period (batch rescheduling); the main benefit of this method is the possibility to realign scheduling to reflect any changes made to the databases records in the meantime, such as changes in the archives, changes in parameters, unanticipated scraps etc. On the other hand, the system is particularly burdensome due to the significant mass of data to be processed. **Incremental systems**, or **Net Change** systems, entail interventions on limited portions of the explosion, due to partial variations made to the MPS. In that case, requirements are recalculated with reference to only the records affected by the variation (partial explosion), while the remaining part of the calculation and the time horizon (periods or time buckets covered by the scheduling) remain unchanged with respect to the last scheduling cycle. Incremental methods are more suited to frequent calculations (even more than one per day), sometimes induced by the need for simulation, they reduce waiting times for the results, although beyond a certain level of use they however require a regeneration refreshment to realign the databases to avoid the risk (growing over time) of inaccuracy. In that sense, the two logics are often adopted jointly, using the first for periodic realignments and the second for intra-period small variations.

13.4 The evolution of MRP systems

The adoption of an MRP system entails an unquestioned commitment of organizational resources. The complexity of the method and the need to have automatic systems to process information are at times costly, but they translate into significant benefits in terms of accuracy and the timeliness of the entire production information system. The real-time availability of updated records also allows for developing careful decisions based on objective evidence. The adoption of MRP systems allows for greater visibility of what happens in the entire planning horizon and a growing understanding of the problems linked to the priorities and interdependence between orders, requirements, production capacity, available stock, and lead times. Over time, all of this has translated into a marked orientation to planning, that by extending to other areas of the business' activity, has allowed for adopting more structured systems aimed at planning processes regarding all resources and production factors. In that sense, the **MRP systems have gradually evolved**, involving in the planning process first the production capacity resources, and then all of the production resources.

Over time, MRP systems have spread to almost all industrial companies, making up the backbone of production information systems, subsequently evolving into **ERP – Enterprise Resources Planning** class systems (see Chapter 19). The first generation of MRP systems operated at **infinite capacity,** that is, they developed the explosion of requirements virtually in the absence of constraints on capacity, consequently generating workloads for departments or machines and orders for purchase materials that at times were not compatible with existing capacity. That limit led to the need to conduct scheduling interventions downstream of the MRP calculations, through the use of appropriate leveling rules or "**finite capacity scheduler**" packages, that have the goal of generating department load profiles compatible with the related constraints on capacity. These are algorithms aimed at identifying the optimal lot-sizing and sequencing in respect for the delivery dates and the limits on available production capacity. The problem is actually made complex by the need to take into due account the variability of set-up times and of the internal and external lead times. As we have already mentioned, the former are generally related to the sequence of lots to place into production and the latter are related to the batch size and to the available capacity of internal and external resources. This issue is critical in the short and medium-term, and with reference to single production resource (departments, work centers, etc.). With reference to longer time horizons and more aggregated production resources, the compatibility between available capacity and load generated by orders is assumed to be comparable, since the MPS is developed in accordance with the limits on available production capacity checked during the Aggregate Planning (Chapters 6 and 8). Today, many manufacturing planning and control systems operate at finite capacity, they are able to manage the elaboration of requirements through synchronization with the available production capacity (through CRP – Capacity Requirements Planning modules) of the departments or work centers involved.

The first step in the process of developing MRP systems can be traced to the establishment of **Closed-loop MRPs**, in which the system, by exceeding the limits linked to infinite capacity planning, is able to verify the existence of available capacity in the department. Concretely, those systems produce a close dialogue between MRP and CRP, such that, in addition to the orders confirmed and issued, the orders planned by the MRP can also be translated into capacity requirements and into the respective production resource loads, appropriately leveled and compatible with the limits of existing capacity. That method has the merit of lengthening the visibility time horizon of the analyses relating to the resource loads, that were previously limited only to the firm period and resulting from only the orders issued, improving the possibility of leveling the plan itself, optimizing the capacity/load balance of the resources in the medium-term, and guaranteeing simulations concerning the rescheduling (for example, postponing the delivery dates of specific orders) when particular unforeseen events or bottlenecks arise.

As we have already stated, the validity of the computations in question extends over short and medium time horizons, beyond which the levels of accuracy decrease, making it convenient to use less costly and more approximate applications such as RCCP (Rough Cut Capacity Planning). Therefore, the Closed-loop MRP, acting on the links between MRP and CRP, is able to plan the availability of three fundamental resources – materials, resources production capacity, and time – coordinating the MPS with the actual availability of the internal resources.

The process of informatics integration of the areas involved in the planning, management and control of physical flows subsequently led to the spread of the planning method called **MRP II or Manufacturing Resources Planning**. As we can see from the name and the acronym, despite maintaining the same logic underlying the MRP systems, these represent an evolution of the same, extending the influence of the planning process to all production resources, production capacity, investments, personnel, and so on.

13.5 The sizing of the investment in stock and the selection of the most suitable management criterion

As we have seen, there are multiple elements that can influence the size of the investment in stock, that are strictly related. To summarize, the variables able to exercise greater influence can be considered the following (Grando, 1995).

1. Characteristics of materials.
2. Nature of demand.
3. Predictability of demand.
4. Variability of demand.
5. Characteristics of the production process.
6. Duration of transformation cycle (lead time).
7. Service level offered.

Below we will examine each single factor separately.

1) Characteristics of the materials

From this standpoint, it is necessary to take into account both the **characteristics and the usage value of the materials**. In fact, the more specific a material or component is, the more the level of stock will be influenced by the dynamic of demand for the final product to which it belongs. On the other hand, the more generic the materials are, the more the variability of demand for them will be lessened with respect to fluctuations of demand for finished products. The complexity of the product (understood as the breadth and depth of the BOM, i.e. the number of raw materials and components and its division over multiple levels, respectively) influences the level of stock. The more complex a product is, the more likely it is that the risk of stock out situations will arise (in terms of finished product), entailing growing levels of Safety Stock. If, in fact, a finished product is made up of 3 materials with a 90% independent probability of being present, the probability that they will be present simultaneously will be equal to 72,9% as a result of the compound probability rule. As we have seen, this also influences the choice of stock management criteria.

As far as the **Usage Value (UV)**, it is evident that for the few materials with high usage value a rigorous control of inventory will be necessary in order to limit the related investment (that is possible with look ahead methods or implemented with very frequent cyclical inventory controls). For the large number of articles with low usage value, it will be possible to adopt the less costly look back criteria, or empirical criteria that are very convenient and simpler from a management standpoint.

In the assembly of an electric engine, for example, wiring, nuts and bolts, washers, o-rings and miscellaneous small parts can easily be managed following the logic at the reordering level without jeopardizing the goal of limiting the relevant investment, due to their low usage value; on the contrary, for more expensive components, such as reels and wire coils, it will be advisable to adopt stock management criteria such as the MRP.

2) Nature of demand

On this front, as already mentioned, it is necessary to distinguish between **dependent demand goods** (raw materials, components, sub-assemblies, etc.) and **independent demand goods** (finished products, spare parts, and consumables). Ultimately, the characteristic to be determined is the presence or absence of a mathematical relationship with the demand for finished products (or parent items), deducible from the BOM and expressed by its links and the Usage Quantities. For dependent demand articles it is possible to adopt both stock control criteria, look-back or look-ahead, such that the choice will be determined by an analysis of lead times and the expected delivery time, or by other evaluations of convenience such as the usage value and the critical nature of the component. For independent demand products as spare parts and consumables, on the other hand, since it is not possible to determine rigorous relationships with the level of production of finished products, the use of look back techniques is preferred or, for finished products, inevitable.

3) Predictability of demand

The predictability of demand is generally a function of frequency of use. In the presence of **high consumption frequency**, it is possible to make use of the analysis of historical consumption trends to extrapolate the prospective values. In the case of dependent demand goods and high consumption frequency, it will be convenient to adopt look ahead approaches (MRP techniques accompanied by lot sizing constraints), or look back (ROP, ROC, Min-Max) systems, that are simpler and less costly to manage, and in this specific case, sufficiently efficient. In the case of dependent demand goods and low consumption frequency, the choice of criterion should take into consideration the usage value (U.V.): higher values correspond to MRP methodologies (with lot sizing policies like Lot for lot), lower values correspond to look back (or MRP with lot sizing policies like Fixed Quantity, EOQ, Fixed Cycles) systems. In the case of independent demand goods, the frequency of consumption will merely improve the management of look back techniques, the only ones implementable. The general picture of the choice is illustrated in Figure 13.6.

4) Variability of demand

The **variable trends of demand** (or internal consumption) is able to influence the level of stock for two reasons:

- in quantitative terms, because by causing the dispersion of possible demand values with respect to the average computed on the historical records, it influences the sizing of the Safety Stock;

Figure 13.6 **Choosing the inventory management system**

		Demand	
		Dependent	**Independent**
Consumption Frequency	**High**	Look-ahead (look-back)	Look-back
	Low	Look-back if low U.V. Look-ahead if high U.V.	Look-back

- in qualitative terms, because by varying the mix of products requested, it translates into inventory assortment requirements that are at times very broad and complex to be managed.

5) Characteristics of production processes

Regardless of the classifications illustrated, for an examination of the aspects of stock management and control, the production processes can be classified based on the continuity/discontinuity that characterizes the incoming physical flow (supply) and the outgoing physical flow (consumption). We will speak of **continuous supplies and consumption** in the presence of sufficiently high frequencies (daily, weekly, etc.), and of **discontinuous supplies and consumption** when there are sporadic events or in any event not observed with regularity. This way, it is possible to represent four typical cases, characterized by different combinations of supplies and consumption, as appears in **Figure 13.7**.

The production processes positioned in [*A*] require large investments in components, semi-finished and finished products. Those in [*B*] entail costly stock of raw materials, components and semi-finished products, while those in [*C*], mainly raw materials and components. The case represented in [*D*] is typical of some chemical processes that operate by batches requiring mainly stock of raw materials.

Independent of the classification elements proposed, the characteristics that play a significant role in the sizing of stock are the already-cited attributes of versatility, elasticity, and availability. The more rigid, inelastic, or unreliable a system is, the higher the level of inventory will have to be – all other conditions being equal – in order to respond to the qualitative and quantitative variability of demand. The advent and spread of new forms of automation (see Chapter 20), for example associated with sophisticated predictive maintenance models, are considerably changing the levels of

Figure 13.7 Classification of production processes

Consumption \ Supply	Continuous	Discontinuous
Continuous	Continuous Flow processes Assembly line (A)	Manufactured Batch Flow Repetitive Orders (B)
Discontinuous	(D) Process Production in batches	(C) Small batches Single orders

convenience of many production choices. In highly automated production contexts, whose set up costs are lower than stock holding costs, the choices regarding managing stock based on convenient sizing of batches lose most of their relevance and are replaced by other flow management approaches. The most recent industrial experiences also search for flexibility performance not so much in the hardware component, but in the software. This is the case of applications of **Artificial Intelligence** to scheduling processes. Flexibility is guaranteed by the capacity of AI algorithms, which dynamically propose the allocation of jobs to machines that are relatively dedicated, and thus less costly, and in production overcapacity.

6) Duration of transformation cycle

The duration of the transformation cycle (or lead time or throughput time) reflects directly on the amount of the **work in process and transit stock**. We speak of a duration of the work cycle, not the length, because it is the time dimension that more influences the sizing of the stock of semi-finished products. Interwoven transformation cycles, for example, at equal lengths, meaning an equal number of phases involved, entail longer times (for waits or set ups and retooling), generating greater inventory requirements.

7) Service level offered

As already observed, a close correlation exists between investment in stock and service level offered to the market or to the downstream production departments. Logistics systems oriented towards service need distribution structures developed on multiple levels, with a widespread peripheral warehouse, and inevitably face the duplication of inventory. The correlation between service level and stock is fully manifest in the determination of the **Safety Stock** size: each production or distribution facility determines its own inventory profiles taking into account suitable protections for events that are hard to foresee in a deterministic manner, such as delays in the delivery times from suppliers, market requests, machines downtimes, and so on. The problem resides in the unreliability of information, not the service level dictated by the needs of competition. By this, we mean to stress that today, thanks to the sophisticated techniques for managing the information flows, the new advanced systems such as APS – Advanced Planning and Scheduling, WMS – Warehouse Management Systems, MES Manufacturing Execution Systems,[2] implementation of Big Data Analytics and the most innovative forecast tools, information takes on growing importance (Chapter 19). Reliable information and knowledge thus reduce stock requirements, and at equal service offered, generate new management opportunities, made more interesting by the staggering drop in the cost of information with respect to the growing costs and risks of obsolescence linked to physical inventories. Looking forward, the management of operations systems will thus be increasingly aimed at controlling mainly intangible components, favoring information flows over physical flows.

13.6 Empirical approaches

Given the number of variables in place and the complexity to define an optimized inventory control system for different articles subjected to different lead-times, values, usage quantities, etc., in the management practice, hybrid and empirical systems have been developed. Two cases are reported as follows.

Case 1 refers to an electro-mechanical company producing elevators. The need of keeping under control different variables affecting the inventory level has been faced by developing an effective empirical system to control the materials and components stocks exemplified in **Table 13.8**.

Case 2 refers to a company producing components of heavy-trucks. The following tables shows an empirical approach to define the main inventory management parameters.

First, for each part number the choice between look-back and look-ahead logic has been defined, according to the following parameters:

- commonality index in the company's BOMs;
- frequency of the demand, measured by an index expressing the number of requests per week;
- customer delivery time versus lead time;
- need of protection for critical processes.

[2] Secchi, 2000; Secchi e Rossi, 2018.

Table 13.8 Sizing Safety Stock, Reorder Point and Lot

Leadtime (days)	Frequency	Variability	Stock Management Logic	Safety Stock (days of coverage)	Reorder Point (units)	Reorder Quantity (days of coverage)
<= 15	HIGH	LOW	ROP	5	SS + d × LT	A CLASS: MAX 20
	HIGH	HIGH	ROP	10	SS + d × LT	B CLASS: MAX 30
	LOW		ROP	10	SS + d × LT	C CLASS: MAX 40
	EPISODIC		LOOK-AHEAD	–	–	LOT 4 LOT
From 16 to 25	HIGH	LOW	ROP	10	SS + d × LT	A CLASS: MAX 20
	HIGH	HIGH	ROP	15	SS + d × LT	B CLASS: MAX 30
	LOW		ROP	15	SS + d × LT	C CLASS: MAX 40
	EPISODIC		LOOK-AHEAD	–	–	LOT 4 LOT
From 26 to 50	HIGH	LOW	ROP	15	SS + d × LT	A CLASS: MAX 20
	HIGH	HIGH	ROP	20	SS + d × LT	B CLASS: MAX 30
	LOW		ROP	20	SS + d × LT	C CLASS: MAX 40
	EPISODIC		LOOK-AHEAD	–	–	LOT 4 LOT
From 51 to 80	HIGH	LOW	ROP	20	SS + d × LT	A CLASS: MAX 20
	HIGH	HIGH	ROP	25	SS + d × LT	B CLASS: MAX 30
	LOW		ROP	25	SS + d × LT	C CLASS: MAX 40
	EPISODIC		LOOK-AHEAD	–	–	LOT 4 LOT

Table 13.9 reports a synthetic picture of the relevant choices.

As far as the purchased components, the lot size has been defined, as shown in Table 13.10, by creating clusters of items according to the following parameters:

- purchasing lead time;
- frequency of consumption measured by an index expressing the number of requests per week;
- average consumption;
- ABC in value.

For each cluster of purchased items, a batch based on coverage weeks of average consumption has been empirically sized.

In both cases, it is worthy to notice that, while the reported approaches are based on field experience, the rationale behind the adopted "rules" are fully consistent with the presented theories and algorithms, showing once again, that "there is nothing more practical than a good theory."

Table 13.9 Look back and look ahead logics

		DT >= LT		DT < LT	
Common. Index	**Frequency of the Demand**	**Critical Process**	**Non-critical Process**	**Critical Process**	**Non-critical Process**
Common	High (>0.5)	L. back	L. back	L. back	L. back
	Mid (0.2-0.5)	L. back	L. ahead	L. back	L. back
	Low (<0.2)	L. ahead	L. ahead	L. back	L. ahead
Specific	High (>0.5)	L. ahead	L. ahead	L. back*	L. back*
	Mid-low (<0.5)	L. ahead	L. ahead	L. ahead	L. ahead

* Only on specific request by Sales or based on a customer's agreement.

Table 13.10 Purchased component and Lot sizing

Lot size as weeks coverage		ABC Class		
Lead time	**Frequency of Consumption**	**A**	**B**	**C**
<= 2 weeks	High (> 0.5)	1	2	3
	Low (< 0.5)	2	3	4
From 3 to 4 weeks	High (> 0.5)	2	3	4
	Low (< 0.5)	3	4	5
From 5 to 8 weeks	High (> 0.5)	3	4	5
	Low (< 0.5)	4	5	6
>= 9 weeks	High (> 0.5)	4	5	6
	Low (< 0.5)	6	7	8

14 Shop Floor Planning and Control

by *Valeria Belvedere*

14.1 Introduction

As illustrated in Chapter 6, shop floor planning and control regards the phase of the production planning process that takes place once the MRP (and where present the CRP) has been completed, through which the production orders of all of the components and semi-finished products present in the bill of materials are identified in order to implement the MPS (Hax and Candea, 1984; Vollmann et al., 1984; Grando,1995; Sianesi, 2011; De Toni and Panizzolo, 2018). The main phases of shop floor planning and control can be identified as the following:

- **Order release:** regards the verification of the resources necessary to complete a production order and the issue of the same;
- **Order scheduling:** consists of scheduling the order, defining the time when its processing must be started, the workstations where it will be processed, and the materials and respective quantities to be used; the output of this activity is a production plan compatible with the constraints on production capacity of the departments that is able to exploit the available resources in the best manner possible;
- **Order progress:** this is a control activity, that consists of the monitoring of production lines in order to know, among other things, the state of progress of orders, the amount of the work-in-progress and the finished products already obtained, and the degree of resource saturation.

In this phase of production planning, known as **scheduling**, the order is called a **job**. This term indicates a batch of products (or sometimes a single piece) to be processed, of which the work cycle, i.e. the sequence of operations which the product must undergo to be completed, is known.

Scheduling consists of two phases:

- **allocation:** consists of assigning a job to one of the resources available in a given department. This problem obviously does not arise when there is a single resource, as can happen in the case of a single machine that within a machining factory can perform turning operations. If, however, multiple resources are available, it is necessary to identify a criterion according to which the job is assigned to the machine. In this regard, the process can involve verifying the technological suitability of the resources with respect to the product characteristics or, at equal

levels of suitability, observing saturation, with the aim of achieving a better balance in the use of resources;
- **sequencing**: after having assigned the jobs to specific resources, it is necessary to establish the sequence with which they will be processed, taking into account the priorities of the company in relation to the broad range of elements of performance that can depend on this choice.

The scheduling activity is characterized by considerable complexity due to the high number of input data to be considered, such as the type of materials to be used, the number of productive resources, the operating conditions for functioning of the machines, and the delivery dates of the orders to the clients. An additional element of complexity is represented by the multiple elements of performance influenced by the scheduling, which are the following:

- **Lateness**, the time interval between the actual date of completion of a job and the planned date of completion (also called "due date");
- **Tardiness**, to be calculated like Lateness, but considering only the case of delays, i.e. jobs with actual completion date after the planned date;
- **Flowtime**, understood as the overall time necessary to process a job, from the start of the work to actual completion;
- **Makespan**, the time necessary to complete all the jobs in the system, from the start of the first work process to the completion of the last.

In order to quantify these elements of performance, it is assumed we have the following data:

- N: number of jobs to schedule;
- j: identifier of the single job;
- r_j: planned release date of the j-th job;
- c_j: planned completion date (due date) of the j-th job;
- R_j: actual date of release of the j-th job;
- C_j: actual date of completion of the j-th job;
- L_j: equal to (C_j-c_j), i.e. the Lateness of the j-th job;
- T_j: equal to $max(0, L_j)$, i.e. the Tardiness of the j-th job;
- F_j: actual flowtime of the j-th job, equal to (C_j-R_j).

Based on this data, once the production of the scheduled jobs is completed, it is possible to calculate the following indicators, that correspond to the objectives of the scheduling activity:

$$Mean\ Lateness = \frac{\sum_{j=1}^{N} L_j}{N} \quad [14.1]$$

$$Mean\ Tardiness = \frac{\sum_{j=1}^{N} T_j}{N} \quad [14.2]$$

$$Mean\ Flowtime = \frac{\sum_{j=1}^{N} F_j}{N} \quad [14.3]$$

$$Makespan = \sum_{j=1}^{N} F_j \quad [14.4]$$

Box 14.1 shows an example of the calculation of these indicators.

Scheduling techniques can be classified in relation to the different ways of managing materials flows, that are: push; pull; and push/pull. In the first mode, the semi-finished products present in one phase of the process advance to the subsequent phase based on a production plan. In pull systems, on the other hand, the

Box 14.1 Scheduling indicators: an example

Consider four jobs completed in a week, whose data is reported in **Table 14.1**.

Table 14.1 Data concerning 4 jobs

	Job 1	Job 2	Job 3	Job 4
r_j	1 june	1 june	2 june	3 june
c_j	3 june	4 june	5 june	5 june
R_j	1 june	2 june	3 june	3 june
C_j	3 june	5 june	4 june	6 june
L_j $(C_j$-$c_j)$	0 days	1 day	-1 day	1 day
T_j $[max(0, L_j)]$	0 days	1 day	0 days	1 day
F_j $(C_j$-$R_j)$	2 days	3 days	1 days	3 days

- r_j: planned release date of job *j*
- c_j: planned completetation date of job *j*
- R_j: actual release date of job *j*
- C_j: actual completetation date of job *j*
- L_j: equal to $(C_j$-$c_j)$, i.e.the Lateness of job *j*
- T_j: equal to $max(0, L_j)$, i.e. the Tardiness of job *j*
- F_j: actual flowtime of job *j*, equal to$(C_j$-$R_j)$

On the basis of the available data, the following indicators can be computed:

$$Mean\ Lateness = \frac{\sum_{j=1}^{N} L_j}{N} = \frac{0+1-1+1}{4} = 0.25\ days$$

$$Mean\ Tardiness = \frac{\sum_{j=1}^{N} T_j}{N} = \frac{0+1+0+1}{4} = 0.5\ days$$

$$Mean\ Flowtime = \frac{\sum_{j=1}^{N} F_j}{N} = \frac{2+3+1+3}{4} = 2.25\ days$$

$$Makespan = \sum_{j=1}^{N} F_j = (2+3+1+3) = 9\ days$$

advancement to a downstream station only takes place if that station signals upstream the need to be replenished with a certain quantity of input materials. Lastly, push/pull systems constitute a combination of the previous two, in which some processes are carried out based on the actual demand expressed by the downstream departments, while others (for example those that represent bottlenecks) are managed in a push logic (Chapter 11). In the subsequent sections, we will illustrate the main scheduling techniques associated with each of the three logics just presented, i.e. (De Toni and Panizzolo, 2018): optimization methods and heuristic methods (push logic); Kanban system (pull logic); and Optimized Production Technology and Synchro-MRP (push/pull logic).

14.2 Push scheduling systems

14.2.1 Optimization methods

Optimization methods allow for identifying the best possible scheduling with respect to a specific goal, consistent with the constraints on the production process (number of machines, production capacity, availability of materials, job characteristics, production process characteristics, etc.). More specifically, it is possible to distinguish between **analytic and algorithmic techniques**. The former imply the development of functions that optimize the dependent variable. The latter, on the other hand, describe a procedure to follow (the algorithm) in order to reach the best possible solution through a series of predefined steps.

The **objective functions** most frequently used regard the minimization of Makespan, Lateness, Tardiness, Number of Late Jobs, Flowtime, and Work-in-Progress. Other types of objective functions are aimed at maximizing plant saturation, that generally entails the minimization of the overall setup time.

Initially, the scheduling problems faced in business contexts regarded the development of optimization systems focused on the *Makespan*. Subsequently, the desire to develop multiple objective functions, i.e. functions able to take into account multiple objectives that often were conflicting, generated an increase of the degree of complexity of the scheduling problems. In particular, this led to excessive complexity in computation, such as to induce researchers to develop heuristic scheduling systems, i.e. systems able to offer a solution as close as possible to the optimal one, but without the complexity of the optimization systems. An additional reason that led to the development of heuristic systems is that the final outcome of methods based on optimization is strictly connected to numerous factors, including the **characteristics of the available resources, production process, and product** (Chapter 6), whose variability can lead to very different outcomes, such as to make the modeling and calculation efforts not justified.

With reference to the **available resources**, it appears evident that the labor-intensive vs. capital-intensive nature can affect the flexibility of the production system, in relation to both the volumes and the mix of products to be realized in a given interval of time. From this standpoint, being endowed with processes characterized by a low degree of automation can represent an advantage. However, to fully appreciate this aspect, it is necessary to consider the employment contracts and labor union agreements, that define the maximum amount of overtime and the scope of the jobs of the workforce. Likewise, if the process is capital-intensive, the constraints on the scheduling activity depend on factors such as the versatility of the machines used and the degree of balancing of the production capacity of the work centers.

As regards the **characteristics of the production process**, it is easy to recognize that in a flow shop the job is subject to a fixed and predetermined sequence of work processes. To the contrary, a job shop tends to produce customized products, that can imply very different sequences of work processes and very variable processing times depending on the product to be manufactured. The second case clearly entails greater scheduling complexity.

Observing the **product characteristics**, on the other hand, the complexity of the scheduling process varies depending on whether, to complete a product, a single job or a set of jobs must be managed. The first can be the case of a company that produces simple mechanical components through a molding operation. The second can be represented by a company that produces more complex sub-systems, such as the engine of an automobile, whose final assembly implies the production of numerous jobs upstream. In this latter case, the jobs must be planned in a coordinated manner in order to meet the due dates of the engine. Also with reference to the product characteristics, it is necessary to consider additional elements of complexity. First of all, sometimes it is possible to make use of alternative **routing** for the production of an item. This is particularly useful when, for example, one workstation is overloaded, while another, that is also able to process the item, has a low saturation rate. While from a practical standpoint this represents an opportunity, from a modeling standpoint it is a source of complexity that is not easy to manage. A similar consideration applies for jobs that can begin to be processed at the subsequent workstation before the activities of the previous workstation are completed (this is called "preemption").

The approaches to scheduling are generally divided into two categories: static and dynamic. **Static problems** are characterized by less complexity because they regard cases with a defined number of jobs that enter the process simultaneously, having full availability of all of the work centers. As already mentioned, problems of this type can be addressed with optimization techniques (analytic or algorithmic), that generally minimize the Makespan. For these problems, research has dealt with two main cases: single machine and multiple machines.

When the problems managed are more complex and imply overcoming one or more of the simplifying hypotheses presented above, heuristic rules are used, also called **dispatching** or **sequencing rules**. This takes place in **dynamic problems**, in which it is assumed that new jobs can continuously enter the system.

14.2.2 One-machine case

One-machine problems are generally dealt with following a static approach, and thus assuming a defined set of jobs that enter the production process simultaneously (Vollmann et al., 1984). It is also assumed that the overall setup time is not influenced by the sequence of the jobs. In this case, the Makespan is independent of the sequence of the jobs, and thus the optimization criterion used is the average time that each job spends in the work center characterized by the presence of a single machine. Note that the time spent by the single job includes not only the processing in a strict sense, but also the setup time and the waiting time. Consider the case in which the following jobs must be scheduled, for which we know the processing times (that include setup):

- job_1: processing time 2 hours;
- job_2: processing time 10 hours;
- job_3: processing time 6 hours.

The overall processing time is clearly 18 hours. However, the average time spent by the single job in the work center also depends on the waiting time, that in turn is influenced by the sequence. For example, if the sequence is job_1- job_2- job_3, the sum of the waiting time and of the processing one for each job will be the following:

- job_1: processing time 2 hours + waiting time 0 hours;
- job_2: processing time 10 hours + waiting time 2 hours;
- job_3: processing time 6 hours + waiting time 12 hours.

In this case, the mean time spent by a job in the system is equal to:

$$\textit{Mean time in the system} = \frac{2 + 12 + 18}{3} = \frac{32}{3} = 10.66 \textit{ hours}$$

If, however, the sequence is job_3- job_2- job_1, the time spent by each job in the work center becomes the following:

- job_3: processing time 6 hours + waiting time 0 hours;
- job_2: processing time 10 hours + waiting time 6 hours;
- job_1: processing time 2 hours + waiting time 16 hours.

In this case, the mean time in the system is equal to:

$$\textit{Mean time in the system} = \frac{6 + 16 + 18}{3} = \frac{40}{3} = 13.33 \textit{ hours}$$

These examples make it clear that to minimize this amount of time the jobs must be sequenced in increasing order of processing time. In the example shown above, the optimizing sequence is job_1- job_3- job_2. As a consequence, the sequencing rule that allows for finding the optimal solution is the **Shortest Processing Time (SPT)**. The application of this rule allows for achieving additional benefits, in particular the **minimization of the mean number of jobs in the system and of the work-in-progress**. These two measures are in fact directly linked to the waiting times of the jobs in the work center, that represent a significant portion of the total time.

14.2.3 Two-machine case

In this case, the scheduling process must simultaneously take into account two machines (Vollmann et al., 1984). To that end, the following hypotheses are formulated: all of the jobs are processed by both of the machines in the same sequence; all of the jobs enter the process simultaneously and no others are added over time; and, the overall setup time is independent of the job sequence.

In this case, the Makespan is not independent of the sequence of the jobs in the two machines, and in order to optimize it a procedure developed by Johnson is used, that implies the following steps:

- select the job with the lowest processing time on machine 1 or 2. If this time is associated with machine 1, schedule the job as the first. If, however, the job has the lowest processing time on machine 2, schedule it as the last;
- select the job with the second-lowest processing time on machine 1 or 2 and schedule it as illustrated in the previous phase;
- continue following this method until all of the jobs are scheduled.

An example of the application of this procedure is shown in **Box 14.2**.

Since in reality the number of machines can be greater, over time algorithmic approaches have been developed that are adequate for these contexts and that represent an evolution of Johnson's procedure. However, the methods based on optimization imply very strict hypotheses, making them usable in a very limited number of situations. Moreover, the optimization of the Makespan is not always the main goal to pursue. Therefore, in reality heuristic methods are quite widespread, as we will see in the next section.

14.2.4 Heuristic methods with sequencing rules

Suppose that in a job shop with *m* machines, *n* different jobs must be scheduled. The overall number of different possible combinations is equal to $(n!)^m$. It is evident that the degree of computational complexity of this problem justifies the adoption of heuristic methods, based on the use of **dispatching rules**, able to guarantee satisfactory scheduling in relation to a specific goal that the management deems to take priority. The most common dispatching rules are the following (Hax and Candea, 1984; Vollmann et al., 1984; Grando, 1995; Sianesi, 2011):

Box 14.2 Example of scheduling in the two-machine problem

Assume five jobs must be scheduled, which have to be processed by two machines, M1 and M2. The following data on the processing times of the two machines is available in Table 14.2.

Table 14.2 Data concerning five jobs

Jobs	Processing Time M1	Processing Time M2
J1	5 hours	4 hours
J2	3 hours	6 hours
J3	1 hour	10 hours
J4	7 hours	8 hour
J5	9 hours	2 hours

Implementing the procedure of the two-machine problem, the sequence of the jobs to be processed by M1 and M2 can be defined, as described in Table 14.3.

Table 14.3 Proposed scheduling

Steps	Available jobs	Shortest processing time (job-machine)	Scheduled job	Schedule
Step 1	J1, J2, J3, J4, J5	J3-M1	J3	J3-x-x-x-x
Step 2	J1, J2, J4, J5	J5-M2	J5	J1-x-x-x-J5
Step 3	J1, J2, J4	J2-M1	J2	J1-J2-x-x-J5
Step 4	J1, J4	J1-M2	J1	J1-J2-x-J1-J5
Step 5	J4	J4	J4	J1-J2-J4-J1-J5

- **First Come/First Serve (FCFS)**: the jobs are processed in the order they enter the system;
- **Shortest Processing Time (SPT)**: the job scheduled first is the one with the shortest processing time in the phase considered;
- **Least Work Remaining (LWR)**: the job scheduled first is the one with the shortest overall processing time until the end of the production process;
- **Earliest Due Date (EDD)**: priority is given to the job with the earliest due date;
- **Critical Ratio (CR)**: the job scheduled first is the one with the lowest CR, calculated as follows:

$$Critical\ Ratio\left(CR\right) = \frac{Due\ date - Current\ Date}{Lead\ Time\ Remaining}$$

- **Fewest Operations Remaining (FOR)**: the first job to process is the one with the fewest number of subsequent processing operations;
- **Smallest Slack Time (SST)**: priority is given to the job with the smallest slack time, calculated as follows:

$$\textit{Slack Time (ST)} = \textit{Due Date} - \textit{Setup Times} - \textit{Processing Times}$$

- **Smallest Slack Time per Operation (SST/O)**: priority is given to the job with the smallest mean slack time per operation. Assuming that *k* operations are necessary to complete the job, this indicator is calculated as follows:

$$\textit{Slack Time per Operation} = \frac{\textit{Due Date} - \textit{Setup Times} - \textit{Processing Times}}{k}$$

- **Least Setup Time (LSU)**: priority is given to the job with the least setup time;
- **Next Queue (NQ)**: the job scheduled first is the one that creates the smallest queue in the subsequent work centers. The queue can be calculated either as the number of jobs waiting or the hours of processing of the same.

It is evident that the performance achieved varies as a function of the rule adopted. For example, the Least Setup Time can be appropriate in production contexts focused on the reduction of production costs. The Earliest Due Date, on the other hand, allows for reducing delays, and thus can be recommended to companies that compete on time. The Shortest Processing Time and its variants allow for reducing the overall work times and work-in-progress, but can have adverse effects on the dependability of delivery, at least for some jobs.

14.3 Pull scheduling systems

The **Kanban system** is a peculiar element of just-in-time, that postulates the need to produce only the quantities requested by the (internal or external) client and only in response to an actual order. Kanban represents the operational mode in which companies plan and control production activity in departments from a just-in-time standpoint. According to this system, each department can start a production activity only if the one downstream requests the production of a certain component, in defined quantities. This approach is implemented with the use of *kanbans* (that literally means "cards" in Japanese), that are divided into two categories:

- **Withdrawal kanbans**: these authorize the transfer of a set of pieces, that can be placed inside a standard container, from the outgoing warehouse of a department to the incoming warehouse of another;
- **Production Kanban**: these authorize the processing of a container of semi-finished pieces, placed in the incoming warehouse of a work center, to replenish the material withdrawn.

Observe **Figure 14.1** to understand how the Kanban system allows for activating the information flows between different departments, and thus to regulate production activities (Grando, 2018).

Imagine a department, as illustrated in **Figure 14.1**, that produces mechanical pieces that must undergo two work processes (A and B), in a simple Kanban cycle that links two workstations, respectively dedicated to the two work phases. All of the standard containers (or pallets) that circulate in the departments contain 200 semi-finished pieces and have a card attached. The sequence of the Kanban cycle can be summarized as follows:

1. The finished products warehouse must fulfill a client order of 200 type 1 mechanical pieces, subject to A and B processing. The operator thus will transmit the withdrawal order of a standard container of pieces already processed in work center A to work center B, that will perform the second processing operation on the pieces.
2. According to the rules of the Kanban system, upon withdrawal of the container with the 200 finished pieces, the card attached to it (Kanban L1-B), will be detached and entered into a board placed at the work center (B), that collects all of the cards relating to the pallets to be processed, by product type and chronological sequence.
3. Through an indication of priority (in the most frequent case highlighted by different colors on the board), by taking the production card (Kanban L1-B), the work center B operator receives the order to process 200 pieces; to do so, he/she must begin to withdraw the semi-finished products coming from A, placed in the storage point upstream of his/her station. Imagining a time of 20 seconds per piece, after a little more than an hour the operator executes the production order of 200 pieces. He/she will thus have a container full of finished products, downstream of his/her work center, and an empty container – previously full of semi-finished products – upstream of that same work center; all of the containers upstream of workstation B have a withdrawal card (Kanban M1-BA), because they were previously transferred from work center A.
4. Once the processing has been carried out, the operator at work center B will detach the withdrawal card (Kanban M1-BA) from the container, that is now empty, placed upstream of his/her station, thus indicating the request to work center A to substitute the empty container with a full one; the full container is placed in the storage point downstream of the previous work station (A), and therefore will be transferred from work center A to work center B. To carry out this operation, the Kanban M1-BA will be transferred from work center B to work center A, accompanying the empty container that must be refilled, to then return to work center B attached to the full container.
5. At work center A, the production Kanban associated (Kanban L1-A) with the container to be moved downstream, is substituted with the withdrawal card Kanban M1-BA, transferring it to the buffer point placed upstream of the station further downstream (B). Once these operations have been carried out, work

Figure 14.1 Scheme of the Kanban system

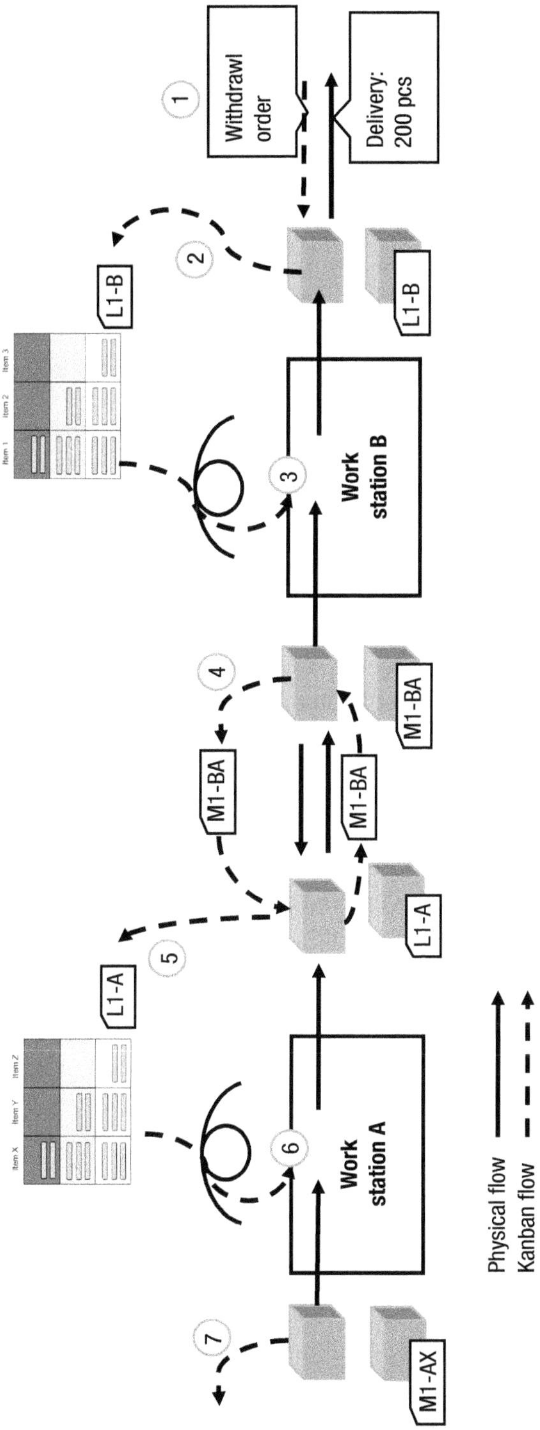

center B will find itself in the same situation as a little more than an hour ago, since both containers, upstream and downstream, will be complete and ready for a new order. At the same time, the production card Kanban L1-A, that has been freed up thanks to the substitution with the withdrawal card Kanban M1-BA, is placed on the board at work center A, as happened previously for work center B.
6. As a consequence, the workers at center A remove the production Kanban L1-A from the board, based on priority, they complete a batch of semi-finished pieces identical to those sent to center B, they attach the card and position it at the outgoing storage point of their work center A, replenishing the stock withdrawn through the previous transfer of the container to the subsequent station. To carry out the work process, the worker at work center A will use the pieces to be processed placed at his/her upstream storage point.
7. Upon using all of the pieces, the work center A worker will in turn request input materials to be processed from the department even further upstream, thus reiterating the Kanban chain that governs the pull system.

As can be seen from the flow diagram, the production Kanbans (Kanban L1-B and Kanban L1-A) always remain at their work station (respectively B and A), they are detached from a full container delivered to the downstream station, they pass through the board that expresses the order of priority between different products and are re-attached following the completion of a new container. On the other hand, the withdrawal Kanbans (Kanban M1-BA), representing a transfer order, move from downstream (work center B) to upstream (work center A) each time that an empty container is brought back, and conversely, from upstream to downstream, each time a container just produced is delivered.

It is evident that, by using the Kanban system, the production activities are managed according to a **pull logic**, since each department is authorized to produce an item only if a downstream work center requests its production. It can also be seen that the number of Kanbans – whether for production or withdrawal – de facto determines the quantity of semi-finished products (or work-in-process) present in the production system. It is therefore necessary to establish what the **correct number of Kanbans** is, in order to avoid, on the one hand, excess accumulation of work-in-process, and on the other, their stock-out.

For this purpose, first of all it is necessary to estimate the production lead time of a container (L), taking into account the working times of the single pieces and the time necessary for the transfer to the semi-finished products warehouse of the downstream department. Subsequently, it is necessary to determine the demand for components that will be expressed in this interval of time to establish the number of containers necessary to satisfy it. The formula to be used for this purpose is the following (Chase et al., 2008):

$$k = \frac{D \times L}{C}$$

where:
k = number of Kanbans;
D = demand of components in a unit of time;
L = production lead time measured in the same unit of time as the demand;
C = quantity of components in a standard container.

Should the demand for components be subject to a type of variability against which the company wants to protect itself with the establishment of safety stock (S, expressed in percentage terms), the formula changes as follows:

$$k = \frac{D \times L \times (1 + S)}{C}$$

14.4 Push/pull scheduling systems

14.4.1 Synchro MRP

Synchro MRP, developed in the 1970s by Yamaha Motor Corp., was created due to the recognition of the limits inherent in the push and pull scheduling systems (De Toni and Panizzolo, 2018). The former are based on the use of an MRP system that, in the presence of perfectly stable and known operating conditions, could guarantee the same advantages as a pull/just-in-time system. If, in fact, the production lead time of the components were perfectly known and reliable, the rate of compliance and the availability of the machines constant, MRP would allow for a scheduling able to minimize the stock of semi-finished products and guarantee dependable due dates. However, this does not take place, since the conditions mentioned above are rarely met. Moreover, the different departments of the factory seldom reach a good degree of coordination. This causes the accumulation of significant levels of **work-in-process**, at times further burdened by the use of safety stock established to deal with variability in the production processes.

Pull systems, on the other hand, propose an approach to scheduling that promises numerous advantages, including a significant reduction of stock, visual control of the load conditions of the departments, and a production control system that is simpler because it is decentralized. However, the use of pull systems based on Kanban requires the presence of numerous conditions that in reality are not always present. Operating with this method requires, among other things, a good leveling of production volumes, short unit setup times, a high level of versatility of machines and labor, and high levels of product conformance and availability of machines. For many companies, this has led to significant difficulties in the adoption of the Kanban method and at times inadequate results.

Over the years, hybrid approaches to scheduling have been developed, that are able to combine the characteristics of the push and pull systems. Synchro MRP is based on a combination of MRP and the Kanban system, and has been considered effective in contexts characterized by **low volumes and high variability** of products and processes. In this system, the completion of the product is planned through the Final Assembly Schedule (FAS), that in turn is exploded into various production orders through the MRP procedure. Each work center must therefore receive a dispatching list generated by the MRP. This constitutes the push component of Synchro MRP. The pull component regards the presence in the department of **two types of cards, Synchro 1** and **Synchro 2**, that resemble the withdrawal and production Kanbans. The former authorizes the movement of a container of pieces from the outgoing warehouse of the upstream department to the incoming warehouse of the downstream work center. The latter authorizes the production of a new container of pieces in a specific work center. The main difference between the classic Kanban system and Synchro MRP consists of the fact that in the latter, in order to launch a production activity two conditions must be met, which are:

- the presence of a Synchro 2 card in the department's board;
- the presence in the dispatching list of a production order generated by the MRP.

The mechanism of Synchro MRP is presented in Figure 14.2.

Consider a production system made up of two departments, A and B. The steps to be carried out are the following:

1. a container with 20 pieces of product Y, placed in the outgoing warehouse of department B, is withdrawn in order to fulfill a client order. The Synchro 2 card (S2-B) placed in this container is therefore detached and placed on the board of department B;
2. the operator of department B detects the presence of a Synchro 2 card (S2-B) in the department's board, relating to product Y. The operator checks that this card corresponds to a production order in the dispatching list of the MRP, and if it does, launches the production of 20 pieces of Y;
3. to start production, it is necessary for the operator of department B to withdraw the necessary material from the incoming warehouse, that is also placed in a 20-piece container. The Synchro 1 (S1-AB) card is removed from this container, and sent to the outgoing warehouse of department A;
4. in this warehouse, the Synchro 2 card (S2-A) is removed from one of the containers and the Synchro 1 card (S1-AB) is placed, that authorizes the movement of the container from the outgoing storage area of A to the incoming warehouse of B. The Synchro 2 card (S2-A) is then placed on the board of department A;
5. the department A operator detects the presence of a Synchro 2 card (S2-A) on the department's board. He/she checks that this card corresponds to a production order in the dispatching list of the MRP, and if it does, launches the production of 20 pieces.

An advantage of Synchro MRP is the ability to limit the accumulation of work-in-progress, because production is not launched if an order present in the dispatching list does not correspond to a Synchro 2 card on the department's board. Furthermore,

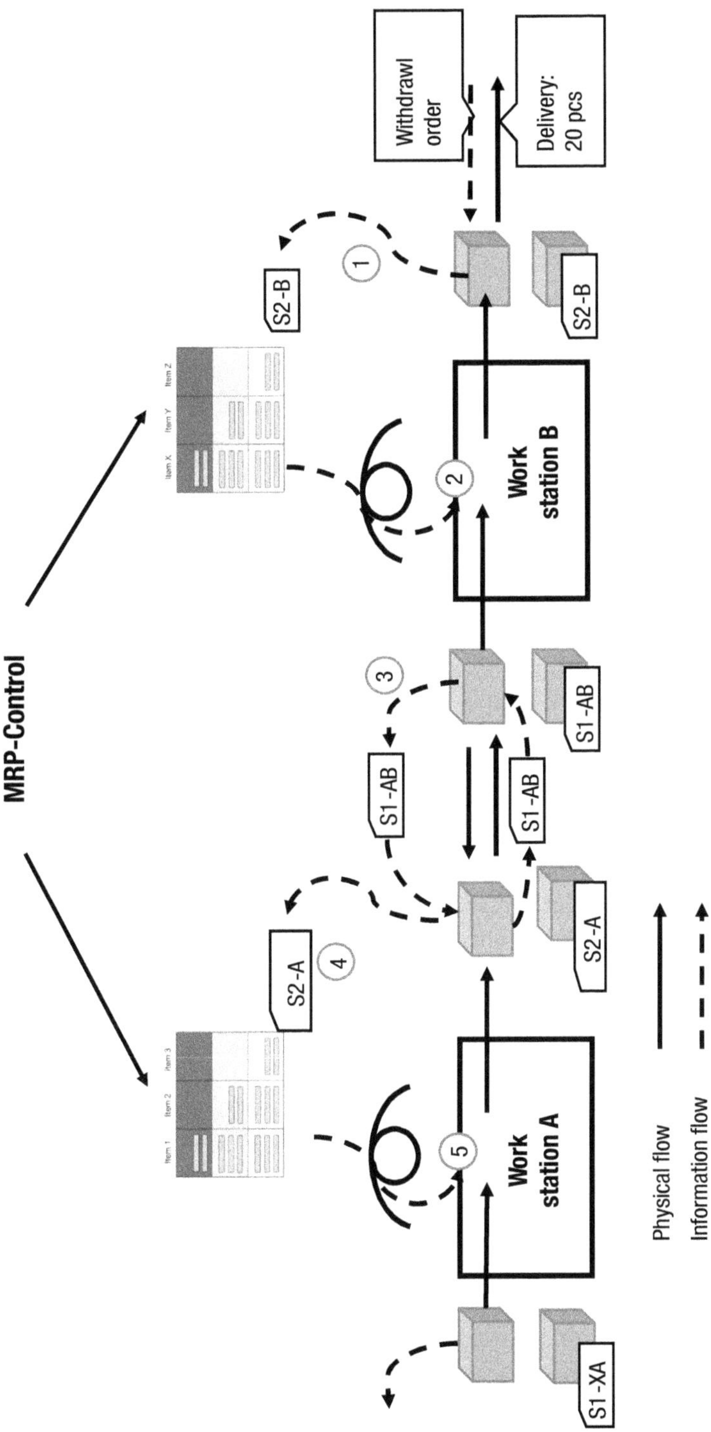

Figure 14.2 Synchro MRP: an example

to simplify the functioning of this system, the decision can be made to limit the MRP calculations to the most critical departments, i.e. those downstream, in which all of the components are generally processed. Since the production is "pulled" from these work centers, blocking the respective Synchro cards allows for aligning the production levels of the upstream departments with market demand.

14.4.2 Optimized Production Technology

Optimized Production Technology (OPT) is a software for scheduling production activities whose logic is based on the "**Theory of Constraints**" (TOC), developed in particular by Eliyahu Goldratt in the book "The Goal," published in 1984.

According to TOC, the achievement of adequate operational and economic performance in businesses can be compromised by the presence of what TOC calls "**constraints**" on the capacity to generate economic value. Although they can be found in all business activities, TOC concentrates in particular on the analysis of the constraints detectable within production processes that are the source of numerous pathologies, including low productivity of the system as a whole, accumulation of work-in-progress, long lead times and delivery delays. Ultimately, these inefficiencies make it impossible to fulfill orders, with evident consequences on the competitiveness of the company and its profitability.

According to TOC, the possibility to eliminate the causes of the inefficiency of production systems lies in the adoption of a production management system called **Drum-Buffer-Rope (DBR)**, implementable through the use of OPT.

The DBR system is the tool that allows for implementing TOC principles in the management of production processes, intervening on the inefficiencies due to bottlenecks. To better understand the DBR logic, consider the following example.

Figure 14.3 shows a production process that uses three different raw materials to make a product. RM1 and RM2 undergo work processes A, B, and C; RM3, on the other hand, is processed in D, E, and F. The two semi-finished products coming from C and F are subsequently assembled into a single finished product. In each phase of the process different production resources are used.

In analyzing the process, we can see that C, having the lowest throughput rate (100 pieces a day), is the bottleneck, or to use TOC terminology, the "constraint." As a consequence, C determines the rate at which the entire system generates finished products. If A and B continue to process semi-finished products by fully exploiting their production capacity, an excessive amount of work-in-progress will accumulate downstream of B. The same will occur upstream of the assembly department, where the parts coming from F accumulate at a much faster speed than the parts processed in C. If the production capacity of C is not sufficient to guarantee a production plan that satisfies demand in terms of volumes and delivery times, the system as a whole will underperform.

The implementation of TOC principles allows for, on the one hand, avoiding excessive accumulation of work-in-progress, and on the other, achieving better exploi-

Figure 14.3 The Drum-Buffer-Rope system

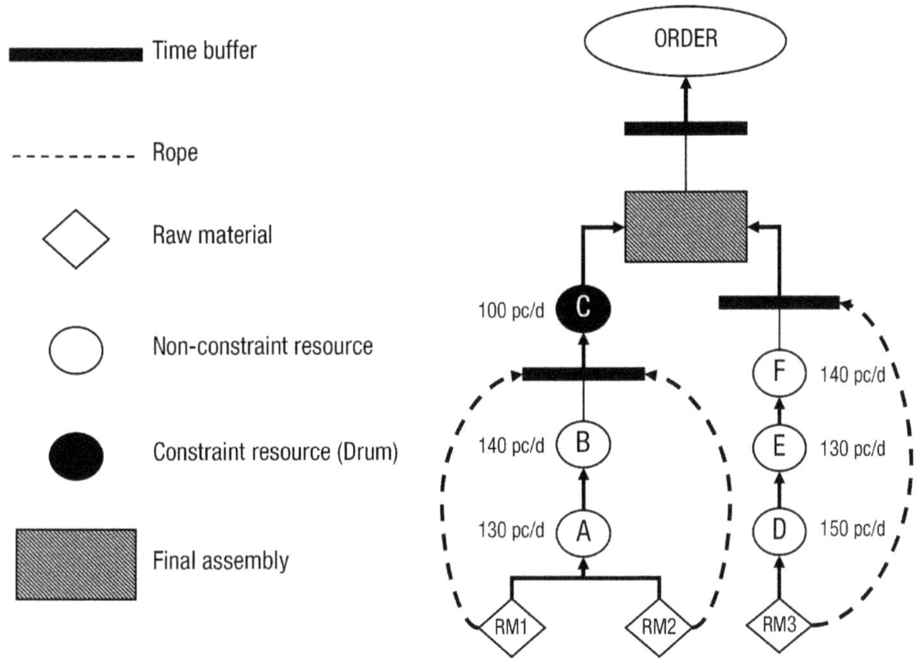

tation of the constraint. The solution proposed by DBR is based on three essential concepts. The first is the recognition of the existence of the constraint, that by determining the rhythm of production of the entire system, constitutes its "**drum**." In the example, the drum is process C.

The second element of the DBR system is the time **buffer**. Given the critical nature of the constraint with respect to performance of the entire process, it is necessary to create the conditions that guarantee its full exploitation. In the case illustrated in **Figure 14.3**, if demand is such as to absorb the production flow guaranteed by the constraint resource, a temporary stoppage of the same will cause a lack of sales, and ultimately, lost profits for the company. This situation can arise both in the presence of malfunctions or the absence of materials to process. Considering this, it is necessary to take action in order to maximize the availability of the bottleneck resource; but it is also necessary for it to be "protected" from the fluctuations of the throughput rate of the other elements in the system, that can entail the absence of material to be transformed. This implies the use of the time buffers, thus called because, in the DBR logic, it is more advisable to think of them as "stock" of time rather than stock of semi-finished products.

Various categories of time buffers exist. For example, the application of DBR to the production process shown in **Figure 14.3** entails, first of all, the creation of a buffer placed upstream of C (constraint buffer), to prevent the constraint from lacking parts to process following problems arising in the previous processing phases. The sizing of the buffer constraint is a problem for which a solution is provided by experience. In cases in which DBR has been implemented, it has been found that an appropriate constraint buffer must be able to cover a time horizon equal to three times the lead time of the semi-finished product processed by the constraint.

A second buffer is placed between F and the assembly department (assembly buffer). The reason for this choice is the need to prevent the parts produced by the constraint from waiting to be assembled due to the absence of a semi-finished product processed only by non-constraint. This would imply a suboptimal use of the bottleneck resource.

Lastly, a shipping buffer is created downstream of assembly, aimed at guaranteeing the fulfillment of the order within the planned due times, even in the presence of delays with respect to the production plan.

Once the constraint has been identified and the number and size of the time buffers have been defined, DBR develops the production plan. This implies only the planning of the constraint activities and the definition of the plan for entering the raw materials into the production system. The reason for this choice is to be sought in the considerations already illustrated concerning the centrality of the constraint with respect to the overall performance of the system. In fact, to guarantee adherence with the delivery deadline of the final products, it is sufficient to plan the activities of the bottleneck resource. Should the production capacity of the same not be sufficient to cover market demand, it will be necessary to prefer orders with the highest margin per hour of processing.

The plan for entering raw materials into the production system constitutes the third key element of DBR, which is generally referred to with the term **rope**. It represents the mechanism that "ties" the entire production process to the constraint, binding it to the overall production rhythms and volumes.

DBR is also implemented through the use of Optimized Production Technology (OPT), a software developed by the company Creative Output Ltd., that allows for drawing up the production plan based on the principles presented above, according to a finite capacity approach. The information available on the functioning of OPT is still very limited. Yet it is known that it has a modular architecture, whose essential elements are the following (De Toni and Panizzolo, 2018):

- **Buildnet**: creates an overall model of the production system ("engineering network") in terms of resources to be used and processes to be carried out, based on data such as sales forecasts, available stock, product bills of materials, work cycles, and available production resources;
- **Serve**: executes planning of the production activities in relation to the engineering network obtained with the previous step, in order to identify the bottlenecks in the system;
- **Split**: divides the production process into two sub-sets. One includes all of the constraint resources, the other the non-critical resources;
- **Brain**: schedules the production activities of the critical resources with a finite capacity approach, through algorithms that define the load for each resource, the

dimensions of the batches, and the start and end dates of the processing. This plan is an input necessary to schedule the activities of the non-critical resources.

OTP is based on **10 underlying principles**:

1. Avoid pursuing the complete saturation of non-critical resources. This can in fact generate a useless accumulation of work-in-progress, especially if a significant balancing is detected between the throughput rate of the non-critical resource and that of the constraint resource.
2. The optimal level of saturation of a non-critical resource depends on the throughput rate of the constraint resource.
3. It is necessary for the management to focus its improvement efforts on increasing the actual working time of the constraint resource, also through the reduction of the setup time.
4. The recovery of additional work hours for a non-critical resource does not constitute an advantage for the company; in this case, the use of frequent setups is not to be considered a source of inefficiency.
5. The constraint resource determines the throughput rate of the entire system and its operation is to be guaranteed also through the creation of adequate buffers.
6. The transfer batch must not necessarily be equal to the production batch size. If a non-critical resource upstream of the constraint is processing a batch of ample dimensions, it is possible to start transferring the already-completed pieces (transfer batch), to avoid interruptions in the activities of the critical resource.
7. The size of the production batch needs not necessarily be the same in the different phases of the process. It is advisable for it to be higher in the constraint, in order to minimize the time used for setups.
8. The production lead time of each batch depends on the availability of capacity and the prioritization logics adopted. Therefore, the two aspects should be assessed jointly.
9. It is not necessary to balance the production capacity of the different phases of a process, but rather its flow, that in turn must be aligned with demand.
10. The sum of the local optimums is not equal to the optimum for the system as a whole. To the contrary, it is necessary to pursue the optimization of only the constraint (i.e. maximize its throughput), because the increase of saturation of non-critical resources results in an excess of work-in-progress, i.e. an inefficiency.

15 Procurement Management

by *Giuseppe Stabilini*

15.1 The role of purchasing in business success

Business success cannot be separated from the ability to leverage a series of activities, processes, and skills entrusted to third parties, typically included in the concept of the "supply network." Regardless of the nature of their business, companies entrust phases of the value chain to suppliers that are able to manage processes more effectively or more efficiently than the companies could directly.

The correct management of suppliers and partners is thus critical in a logic of full integration between business choices and the multiple assets used. In this context, "**purchasing department**" is thus of fundamental importance, a center for monitoring skills and decisions relating to supply policies.

Suppliers contribute to business success in various different manners.

The primary aspect is certainly the impact that the goods or services purchased have on the **company's financial results**. Supplies directly affect the Cost of Goods Sold (COGS) and overheads, eroding profit margins. In addition, the payment management policies for supplies, together with inventory management, determine the company's Net Working Capital (NWC) requirement (Chapter 1). Even in less advanced companies, these two aspects guide the choice of suppliers.

However, supply choices cannot be separated from the **company's internal consumption processes**. In fact, the critical issues in terms of quality, times, and volumes determine specific needs in terms of level of compliance, speed and dependability, or flexibility of the supply (Chapters 4 and 5). An inefficiency in these aspects would not only impact operations, but would undoubtedly have negative financial effects, such as the need to rework products, 100% quality controls, excess inventory, or downtime caused by the lack of goods.

Two other aspects characterize the need for managerial supervision of supply choices in general. In the processes of **innovation or development of new products or services**, the technical departments need the support of the supplier to develop new solutions. The relationship must be carefully guided in order to avoid constraints in the management of the know-how developed (think of the intellectual property of designs or of particular technological solutions) or in the determination of the supply prices (think of the possible impact of incorrect volumes on purchase price calculation in the case of investments in specific assets by the supplier).

In the **marketing and sales processes**, many companies rely on the concept of "ingredient branding," that is, the positive influence on their customers of having a supplier with a particular brand or market position. The choice of the supplier must therefore consider commercial variables that affect price or sales volumes, rather than mere cost elements.

The role of procurement, as a function, and of purchasing decisions, is thus central to supporting the improvement of company performance. The ability to understand the **corporate strategy** and the context of both the market and the supply network must be the fundamental skill of each **procurement manager**.

15.2 Management processes and logics

The process of purchasing goods and services can be described as a series of phases and activities, which in a sequential logic guide decisions in order to satisfy companies' needs.

The process is typically described with the following phases (**Figure 15.1**):

1. **Origin of the requirement and definition of the Purchase Request (PR).** The peculiar characteristics of the need attributable to a business process (production, development, administrative management) are described and collected in a structured manner within the PR. The peculiar characteristics concern the specific requirements of the good or service, but also the logistical constraints, the terms of delivery or delivery, and any after-sales services.
2. **Planning the purchase.** The PR is assessed in light of the general purchasing policies of the company, for example with reference to combinations of expenditures, standardization of the article, and budget constraints.
3. **Search for potential suppliers.** The purchasing company selects a list of suitable suppliers to satisfy the needs. This phase includes the scouting activities of new suppliers or the selection of a group of companies within the company supplier register.
4. **Sending the Request For Quotation (RFQ) to suppliers.** The RFQ is sent to the identified suppliers, meaning the formal request to present a complete proposal of the technical, economic and service terms relating to the requirements expressed in the PR. Suppliers respond by sending their offers.
5. **Tabulation and analysis of the offers.** The offers of the suppliers who responded to the RFQ are tabulated within evaluation grids and analyzed in order to understand which offer and which supplier are deemed suitable to meet the company's needs.
6. **Negotiation.** The company starts the negotiation process with the shortlist of suppliers selected based on the RFQ. The negotiation concerns the various variables connected to the transaction, i.e. economic or technical aspects, those linked to the delivery or delivery times, the level of service requested, etc.
7. **Choice of supplier and definition of the agreement.** Based on the conditions proposed during the negotiation, the purchasing company identifies the best supplier (or suppliers, if it decides on a multiple allocation in order to satisfy) and defines

Figure 15.1 The purchasing process

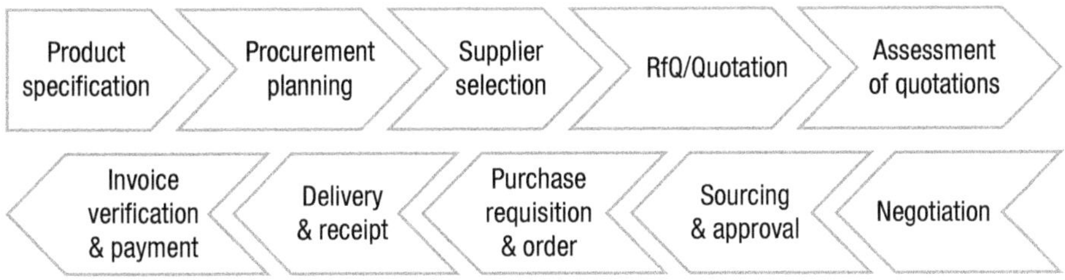

Source: adapted from Puschmann and Rainer, 2005.

the specific terms that regulate the relationship in agreement with the counterparty. The definition of the clauses of the relationship includes a declaration of intent of the substantial or formal buyer, with the stipulation of a purchase contract.

8. **Processing and issuing of the Purchase Order (PO).** The company processes the PO based on the conditions stipulated in the agreement and sends it to the supplier, for the satisfaction of the entire requirement or for a first part of it.
9. **Order fulfillment or service delivery.** The supplier will deliver the goods or provide the requested service. During this phase, control by the client company on the progress of the supplier activities and expediting actions may be envisaged in order to satisfy new needs that have emerged later (such as postponing delivery). In addition, a check can be made on the compliance of the services offered by the supplier with the agreements made.
10. **Reconciliation, payment, data storage.** In this last phase of the process, there is the closure of the administrative passive cycle, with the receipt of the invoice, the reconciliation with the PO and the transport document, if any, and the payment of the amount due. All data relating to the transaction is recorded in the company information system.

Several aspects make managing this process complex.

First, the needs of **different actors and business functions** converge in the process. Those who create the need often do not directly govern the negotiation phase. Contracts and payments are made by other functions. The receipt of the good or service involves different people and functions. This aspect can bring together different objectives, generating conflicts and inefficiencies.

Secondly, the process must govern both an objective of **effectiveness**, i.e. to monitor the performance of costs, services, and quality of the good or service purchased, and an objective of **efficiency and speed**, or develop the decisions and activities necessary to complete the process in a relatively short time. These two objectives are in clear conflict.

The solution to this **trade-off** must be sought in the ability of the purchasing department not only to carry out its activities with a greater degree of speed and/or automation, including in technological terms, but also trying to anticipate some phases of the process itself upstream of the birth of the requirement. Scouting and continuous auditing of new suppliers, preparation of standard contracts, framework agreements or conventions that decouple the moment of construction of the contract from the operational phase of order and delivery, and validation of free-pass supplies allow for obtaining both the monitoring of effectiveness (costs, quality) and process efficiency (short times between the emergence of the requirement and satisfaction of the same).

15.3 The organization and the purchasing process

From an organizational point of view, the decision on location of the purchasing activities and the relative responsibilities requires evaluating two particular choices:

1. the degree of **centralization or decentralization** of the procurement function;
2. the degree of **integration** in the decision-making process between the **procurement function and other stakeholders**, that are both internal (business processes) and external (upstream suppliers and downstream customers).

The placement of purchasing activities derives directly from the competitive and supply chain strategy that the company adopts. Businesses that require high flexibility, quick adaptability and the need for strong integration at the level of the **individual division or production plant**, push towards a strong decentralization of purchases, with advantages attributable to high integration and rapid decision-making.

Vice versa, in contexts where there is the need to search for purchasing economies, synergies in the management of relationships and where the negotiating power and volume scale allow for better control of the supply base, the purchasing department is placed in **headquarters structures**, with visibility and complete management of the expenditure of the various businesses/plants.

There are also **hybrid models**, that are highly appreciated in multinational corporations with different businesses and locations, which separate the various activities related to the purchasing processes, for example placing at the corporate level the definition of purchasing strategies for each spending category, supplier financial support and development policies, and the supplier vendor rating. Operational planning and order release/call off activities remain at the local level. In these cases, the Category Buyer represents a key role in the matrix with responsibility for defining the sourcing strategy at the corporate level, direct management of synergies that can be exploited at the central level, and coordination/control of purchasing activities also carried out at the local level, in an "end-to-end" logic.

The second choice concerns the level of integration between functions and actors in the purchasing process (**Figure 15.2**).

The **traditional and functional logic** provides for a low level of integration, with some phases of the process carried out by the internal customer and others, typically from "quotation" to "negotiation," carried out by the purchasing function. This mod-

Figure 15.2 **Purchasing organization and focus**

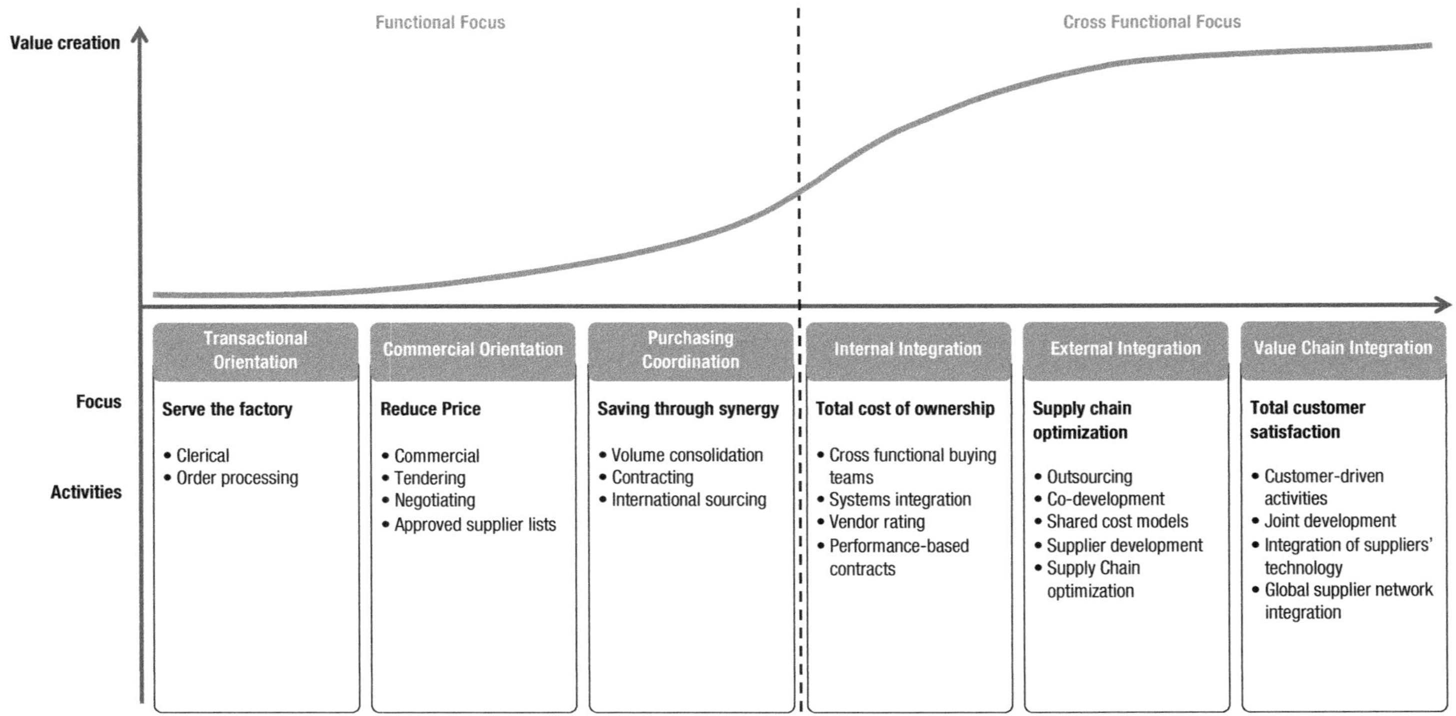

Source: adapted from van Weele, 2018.

el, widespread in companies where functional monitoring is a source of efficiency and simplicity, has several limits, especially with regard to the ability to monitor all the target and risk variables for which information asymmetries along the process do not allow for obtaining the best supply conditions.

For these reasons, many companies have adopted a **cross-functional logic**, with integration of all the functions involved throughout the process, both internally (internal customers) and externally (suppliers and customers).

Thus, Total Cost of Ownership models are introduced (Chapter 16) and Supply Chain Integration models are involved, which are aimed at understanding all aspects of the supply process and all the resources and assets, whether they are owned by the acquiring company or by the suppliers. This complete view allows for optimizing not only the company's economics, but also for recovering efficiency in processes typically not considered (processes belonging to the suppliers themselves). The integration of customers, such as distribution companies in retail chains or system integrators of equipment in the industrial sector, allows you to have a complete view of the network, also bringing the typical levers of demand management to the table.

15.4 Measurement of purchase performance

Measuring performance represents one of the key moments not only of reporting on the activities of the purchasing department, but also of guiding the decisions themselves.

The key measure is represented by "**savings**," i.e. the ability to influence company costs by reducing spending. Depending on the relevant product category, savings can be measured in terms of COGS or sales and general expenses. The presence of standard or budgeted costs allows to accurately measure the variance.

However, a financial vision must also accompany this economic vision, attributable to the effects of spending on corporate **net working capital**. In particular, the impacts on account payables deriving from payment terms and on inventories, determined by order policies and related volumes.

The complexity of supply relationships also requires introducing several **other measures** in order to complete the picture, such as:

- quality and claims;
- sustainability, people and environment;
- service, on time, in full, reliability;
- innovation and collaboration;
- risk and flexibility.

The set of measures is completed by other "**administrative**" ones, focused on internal processes, such as process lead times between PR and PO release or the expense value governed by the individual buyer.

From a financial point of view, the concept of "savings" has been widely criticized if the calculation considers only the purchase prices. In fact, there are several costs, associated with both the transaction (logistics or holding inventory costs) and the consumption of internal resources (time dedicated to control activities or material handling services) which should be considered in the general search for reduction of company costs.

The **Total Cost of Ownership** (TCO), fully described in Chapter 16, embraces this logic and summarizes all of the costs associated with the supply in a single economic value, both prior to the transaction (supplier audit and training), during the transaction (logistics, receipt and control, purchasing management) and subsequent to the same (defects, maintenance, miscellaneous consumption of materials or energy). The TCO allows for broadening the view to the "life cycle of the relationship," optimizing not only the expenditure, but also all the financial components involved in the supply, regardless of the moment or the function/budget associated with them.

The most advanced companies have started to develop advanced reporting models. An example is the "**Operative Total Shareholder Return**" (O-TSR) model, which introduces shareholder objectives (capital gains and dividends) in the evaluation of the purchase conditions, directly connected with the possibility for profit growth and cash flow growth for the company.

The model introduces three drivers of value creation (**Figure 15.3**):

1. sales growth;
2. margin improvement;
3. asset efficiency.

Each of these can be broken down into different metrics, which thus become a tool for tracing the value of purchase choices, in addition to simple structural sav-

Figure 15.3 Operative-Total Shareholder Return (O-TSR)

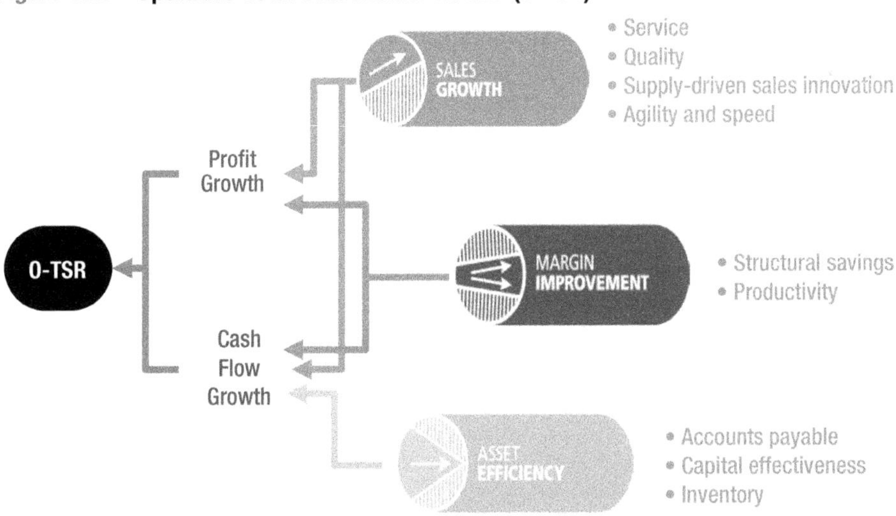

ings. Two aspects should be stressed. The model includes the factors of "agility and speed" that make it possible to work on revenues, an area typically not considered by those who usually only manage costs. Secondly, the efficiency axis incorporates the ability to reduce corporate assets, both NWC and fixed assets, through "buy" choices that reduce investments in the balance sheet. The concept of **Return on Net Asset (RONA)** is very close to CFO metrics.

15.5 Strategic Sourcing and Procurement Mix

Strategic sourcing represents the set of tools that enable the purchasing department to align procurement strategies and supply relationships with the objectives set by company processes. First, the correct understanding of the needs of the "internal customer" is sought, in terms of cost, quality, service and flexibility. Based on this understanding, Strategic sourcing promotes the systematic study of the environment, markets, products, suppliers, with the aim of continuously searching for the best supply alternatives.

As marketing studies and analyzes customers to identify their needs (whether expressed or latent) in order to guide the company's offer (for example, new sales methods, new products, etc.), **Strategic sourcing** studies and analyzes the internal customer to identify needs (in terms of quality, service and price) in order to search the supply market to find the product or service, or more generally the company, which is closest to these requirements. These are relationships between different subjects but characterized by the same information exchange. The complexity and breadth of a company's demand and supply market are therefore comparable, and the managerial tools used in their management must be equally complex and sophisticated.

Strategic sourcing, again in analogy with marketing in general, uses a set of levers, called the **procurement mix**, which are the variables that condition the objectives and procurement strategies and as a consequence, guide their decisions. The fundamental elements of the **procurement mix** can be summarized as:

- **product/service**;
- **price**;
- **communication**;
- **supply channels**.

Each "lever" allows you to analyze the supply relationship from a particular point of view. Furthermore, each lever includes a series of choices and tools for analysis that define the detailed strategies and guide the overall procurement objectives. The development of the strategies also requires the collaboration of the various corporate functions, in order to define common objectives to be achieved through specific choices.

15.6 The product/service lever

This section contains all the tools necessary to set up procurement policies consistent with the critical aspects of the goods or services purchased. The objective is to classify expenditures in terms of contribution to the competitive success of the company, and thus to define guidelines for the selection of suppliers and the type of relationships to be established.

An initial analysis useful for this purpose is the **ABC classification** of expenditures, also called a Pareto analysis. Uniform expense classes are identified, attributable to the various product categories. The classes are sorted in descending order and classified as "A," "B," or "C" according to their impact on the total. Class "A" is made up of all the categories, which together make up 80% of the total expenditures. Class "B" includes those that make up the next 15%. The remaining ones, forming the last 5%, are in class "C" (**Figure 15.4**). This simple analysis drives buyers to devote time and resources to class A codes with the aims to reduce the expense, focusing management attention marginally on low total impact codes. The three classes can also be usefully analyzed with respect to the number of PR, RFQ or PO issued for each product category, taking the number of these documents as a parameter to describe the absorption of the resources of the purchasing department.

A second useful tool is represented by the portfolio matrix. The matrix classifies the categories of purchases and outlines the key points for building the relationships with the sources of supply. The most used and appreciated by man-

Figure 15.4 The Kraljic matrix (portfolio matrix)

Profit Impact	Low	High
High	Leverage items **Competition** **(Price, Quality, Service)** Concentrate volumes Standardize needs Involve a large number of qualified suppliers	Strategic items **Partnership** Innovation and quality Collaboration and trust Co-development and investment sharing Business integration
Low	Non-critical items **Keep it simple** Process outsourcing to internal customer and suppliers Minimize TCO	Bottleneck items **Reduce risk** **Search for alternatives** **(Supplier, technical specification)** Other options: • Stock • Partnership • Make
	Low — *Supply Risk*	High

Source: adapted from Kraljic, 1983.

agement is certainly the matrix developed by **Peter Kraljic** (Kralijc, 1985). The matrix classifies the categories of expenditure into four groups, according to the following variables:

- importance of the good or service purchased in terms of added value by product line, economic impact compared to the cost of sales, impact on profitability, technical quality and quality perceived by the customer ("strategic" impact), on corporate competitiveness in general. For example, a raw material that alone represents an important share of the total cost of a finished product is certainly of high importance. This variable can be briefly described as "**profit impact**";
- complexity of the supply market (understood as adverse conditions for the purchase activity, risk implicit in the supply) in terms of number of alternative suppliers, characteristics of the offer (lack or abundance of availability of the good or service), pace of development of the technologies associated with the supply, presence of barriers to entry in the sector, cost and complexity of the supply logistics, patents or brand power of the supplier. This variable can be briefly described as "**supply risk**."

The placement of the expense category is based on qualitative analyses of the characteristics of profit impact and supply risk, while discussing the placement and subsequent approaches with all the company departments involved in the purchasing processes.

Thus, for each quadrant it is possible to identify the **categories** present and the most correct **supply policies** (**Figure 15.4**).

Non-critical materials. This quadrant includes all the materials or services that are not critical, i.e. characterized by both a low economic impact (low cost) and a low strategic impact (low impact both on the quality of the product/service provided, and on the customer's quality perception). These are typically commodities and standard goods or services; for this reason, they have low complexity from the supply market point of view (presence of numerous local supply alternatives). They are typically the **class C codes** of the ABC analysis. They generally represent from 5% to 20% of the total expenditure, despite being a very high number of items, around 80% of the total codes. Within this quadrant, for example, we find all stationery, indirect production materials (hand tools, lubricants), small parts, some direct materials, and cleaning and surveillance services.

Management should seek maximum **efficiency in the purchasing process**, precisely because the most significant cost item is the resources dedicated to the process, certainly not the price paid for the goods or services. The perspective to be considered is that of minimizing TCO. Efficiency must be sought in the study and construction of policies and purchasing processes with high management simplicity, decentralizing decisions directly to the user level. In recent years, the use of electronic catalogs or purchasing cards has made it possible to free the purchasing department of a series of administrative activities with low value added. In many cases, it is possible to delegate the management of material reorders autonomously to the supplier (practices such as Vendor Managed Inventory – VMI and Consignment Stock).

The possibility of standardizing the purchased materials, centralizing the purchasing process, allows for increasing the company's bargaining power. Consequently, the company selects a supplier with which to implement the purchasing process, with the aim of keeping to a minimum both the indirect purchasing costs (process) through management simplicity, and the direct costs (agreed price) by exploiting the leverage effect obtained by centralizing all the materials subject to exchange in a single negotiation.

Leverage materials. This quadrant includes all materials or services that enjoy a "leverage effect," that is, that are characterized by a high economic impact (their cost significantly affects the total cost of the product/service created) or strategic impact (direct impact on quality or on customer perception), but with a low complexity of the supply market (abundance of supply alternatives). They are typically class A codes, generally representing 45-50% of the total expenditure, although in terms of codes, they are not numerous (15% purchased items). In this category, for example, we find electric motors, electronic hardware, and some types of raw materials supplied on local markets.

Management must seek **competition** from alternative sources for the most effective means of obtaining the best possible supply. The high impact of the material or service on profits and the simultaneous presence of numerous suppliers allows for exploiting the dominant position of the company, making potential suppliers compete on the price, quality or service offered by the relationship (hence the leverage effect). The relationship can be short or medium-term, using spot agreements or open contracts for a not long period. The suppliers in this quadrant are "convenience partners," that is, subjects with which to maintain exchange relationships as long as they offer the highest performance compared to competitors.

Competition should be sought through **three drivers**:

- **increase in negotiated volumes**, bringing the contract to an attractive value for the supply market by both centralizing different business divisions and extending the terms of coverage of the contract (for example, annual agreements);
- **increase in the number of qualified suppliers**, that is, continuously searching for suppliers in the sourcing category and suitable for the specific supply needs, fueling direct competition;
- **standardization of the technical specifications**, intended as the ability to request a product or services from the market consistent with needs but as standard as possible, so as to be potentially offered by a larger pool of suppliers and eliminate elements of technological differentiation or exclusivity that reduce competition.

In some cases, a more stable relationship characterized by common objectives could make it possible to obtain greater advantages than the "competitive" approach. For example, the supplier could acquire skills during the relationship and suggest alternative solutions (material or technology changes, with a specific design) that are advantageous for the client company. In these situations, the strategy aims to move the material towards the "strategic" quadrant, establishing a very close relationship with the supplier, consisting common investments and technical and logistic integration. This strategy is certainly an exception, which can be found in rare cases, precisely because of the sacrifice by the client company of the initial situation, where there was high contractual power.

Strategic materials. This quadrant includes all strategic materials or services, characterized by a high economic or strategic impact, and with a high complexity of the supply market (few alternative sources). These are specialty materials (designed and manufactured to customer specifications), or which require high investments and technical skills from the supplier, or are purchased on markets where there are factors that limit the presence of alternatives. These goods are a very important part of the company's products and ensure the competitiveness of the company's products or service. They are class A codes, and they represent 45-50% of the total expenditure, although unlike the previous ones, they are not very numerous in terms of items. These include all direct materials built to specifications, such as proprietary chemical products, entire product systems, packaging, and special logistics services.

The high impact on profits and the scarcity of alternative sources forces the company to seek in the relationship the differential skills that can improve corporate competitiveness. The (formal or factual) **partnership** is characterized by joint investments, co-planning, and strategic and operational integration. The two companies benefit from the relationship, each according to their own role. In particular, in the supplier the client company finds a partner providing know-how and solutions. The supplier must be selected precisely based on its propensity to collaborate and follow the client company's industrial plan. The close link with the supplier can be the result either of a free choice by the company (the relationship requires high investments, impossible to replicate over many suppliers), or of a monopolistic situation (supplier patent or differential skills). In this second case, the "lock-in" effect is configured, where the partnership strategy clashes with the commercial strategies of the supplier itself.

The pre-established partnership may not offer the expected performance. To escape the lock-in effect represented by the relationship, the client company may have an interest in simplifying and standardizing the object of the exchange, in order to be able to deal with a less complex market, where to take advantage of the leverage effect.

Bottleneck materials. This quadrant includes all the so-called "bottleneck" materials or services, typically specialties with a low economic and strategic impact, but with high complexity of the supply market (few alternatives, due to technical skills and specificity or high personalization of the product or service). These materials are often found in this quadrant due to the presence of a supplier company with an established brand, which sets constraints on supply choices. While representing a marginal share of items and expenses, these materials are characterized by the need for effective management in order to minimize the risk of stock-out which would have a negative impact on production performance (hence the name of "bottleneck").

The **search for alternatives** is certainly the most suitable strategy to free the company from the risk of supplying these materials. The alternative must be sought both at the product level, eliminating, simplifying or standardizing the specifications of the requirement, and at the market level, continuing to scout new sources or trying to develop new suppliers. Both strategies would bring the material into the non-critical materials quadrant. If this solution is not practicable, the allocation of these supplies to companies with which partnership relationships have been undertaken could reduce the risk. If it is not possible to replace these materials or services, characterized by high risk regarding supply, the best strategy is aimed at reducing the risk of stock-out, through the formulation of long-term contracts with an emphasis on quality and reliability, asking the supplier to ensure the continuous availability of the material, and if necessary, to keeping safety stock very high.

The formulation of the management strategy for each product category purchased represents the path that guides the intervention plan and the actions that allow for continuously aligning the purchasing relationship at the optimal point. In fact, as stressed in the specification of the strategies, the client company opens up a series of alternatives, each with its own opportunities and risks, which must be managed thanks to different information and decisions.

The Strategic sourcing activity will therefore be configured differently depending on the type of material and strategy that the company intends to pursue.

Finally, the Kraljic matrix also guides the **performance measurement system**. Indeed, it is clear that each sourcing category needs an appropriate system of measures for the strategy implemented and the benefits sought. Leverage materials can be measured, for the most part, with savings metrics, while non-critical materials will need to calculate a TCO. Strategic materials must be observed in terms of revenues and value/gross margin created, while bottleneck materials need measures of "damage" (related to risk) and not of the purchase price.

15.7 The price lever

This section is aimed at considering the economic variables of the relationship as a tool to guide the behavior of the supplier and of the relationship more in general. The price, and more generally the TCO, can guide the supply relationship in two distinct ways:

- the **determination of the purchase price**, or fixed price or bonus / malus ratios;
- the **cost structure of the supplier or the cost breakdown analysis structure**.

Some supply situations allow you to predetermine the price of the product or service purchased. Specifications are available, clear and complete. The supplier has a clear view of operating costs and generated margin. In these cases, a **fixed price** is the best choice, to protect the interests of both parties.

Vice versa, if the technical specification cannot be clarified a priori, or some aspects of the service, such as particular elements of performance, may vary over time and affect the downstream processes, the purchasing departments prefer to enter a "variable" price, or a bonus-malus system. The price paid will be directly linked to an underlying metric, linked to performance, which may increase or decrease the consideration paid. These contracts, called "incentives," make the orientation of the relationship clear, with a directly link to the agreed-upon price.

In some cases, where uncertainty is greatest, the contract is even constructed at a "**cost plus**" price, that is, by acknowledging that the supplier will cover all the resources used, subject to authorization by the client company.

The analysis of the supplier's cost structure allows for going beyond the price and reaching a level of knowledge of the formation valuable for both the buyer and the relationship itself.

First, the analysis allows us to understand the **nature of the cost components**:

- fixed or variable costs;
- direct or indirect costs.

These distinctions and the importance of the various cost items highlight how particular conditions, such as the volumes traded or the planning period (short to medium term), can become effective negotiating levers or not. For example, a cost structure characterized by significant direct fixed costs will be very sensitive to volumes and asset saturation. Conversely, a preponderance of variables will allow the supplier to be able to exit the agreement without suffering particular economic damage.

Furthermore, the characteristics of the resources employed in the supplier's production structure can affect the characteristics of the relationship itself. Strong exposure to a particular raw material listed on international markets, or a supply chain operating in areas with different currencies, can be elements of risk if the contract sets a fixed price. This knowledge guides the possible construction of prices indexed to raw materials and exchange rates or hedging policies in order to isolate and eliminate risk.

Finally, knowledge of the cost structure also makes it possible to recognize the value underlying the supply relationship, and as a consequence, to identify a correct price for the same. As indicated in **Figure 15.5**, the price could be linked to a margin per unit of use of critical assets (contract A) or a mark-up on costs (contract B). In the first case, the changes in the costs of raw materials are reflected only in the price, while in the second case they also increase the margin.

The considerations on the price paid show that it is necessary for a buyer to be able to govern both the economic/financial aspects of the relationship, but also to technically and operationally understand the production processes used by the supplier. This level of sharing often arises through **"open book" relationships**; in other cases, it comes from a buyer with significant experience in the supply industries.

Figure 15.5 **Purchase agreement for cocoa cream**

Contract A

Ingredient	Quantity/Ton	Price (€/Ton)	Standard Cost (QxP)
Cocoa	0.50	2,275.50	1,137.75
Sugar	0.20	550.00	110.00
Vegetable fats	0.30	1,150.00	345.00
Processing	0.50	200.00	100.00
Margin	1.00	300.00	300.00
Total			**1,992.75**

Contract B

Ingredient	Quantity/Ton	Price (€/Ton)	Standard Cost (QxP)
Cocoa	0.50	2,275.50	1,137.75
Sugar	0.20	550.00	110.00
Vegetable fats	0.30	1,150.00	345.00
Processing	0.50	200.00	100.00
		Sub-Total	**1,692.75**
Margin	17.72%	300.00	300.00
Total			**1,992.75**

Source: adapted from Morelli et al., 2018.

15.8 The communication lever

This section collects the set of policies aimed both at promoting the corporate image among potential and active suppliers and at improving the exchange of information between the various parties. The choices regarding which information to exchange with the supplier and which means to use for the information transfer fall within this "lever" of strategic sourcing.

The "**supplier's day**" events organized by the company for its entire fleet of suppliers fall within the communication lever. These events are precious opportunities both to present business plans to partners and to reward the best suppliers. An award has a value that goes far beyond recognition. It allows the client company to retain suppliers and reward the ability to offer excellent services, stimulating competition. Furthermore, it represents an element to support the growth of the supplier itself, which thanks to the use of the "qualified reference" connected to the award can activate new customers or have access to new industries, improving both his own economic/financial situation and facilitating technological exchange between different sectors and companies, all elements of value for the same client company.

The communication lever also includes the **planning methods** shared with the supplier and the underlying economic commitments. One of these is represented by collaborative planning programs (**Figure 15.6**). In particular, operational collaboration and research through integrated processes of greater flexibility and speed leads

Figure 15.6 **The rolling plan in procurement**

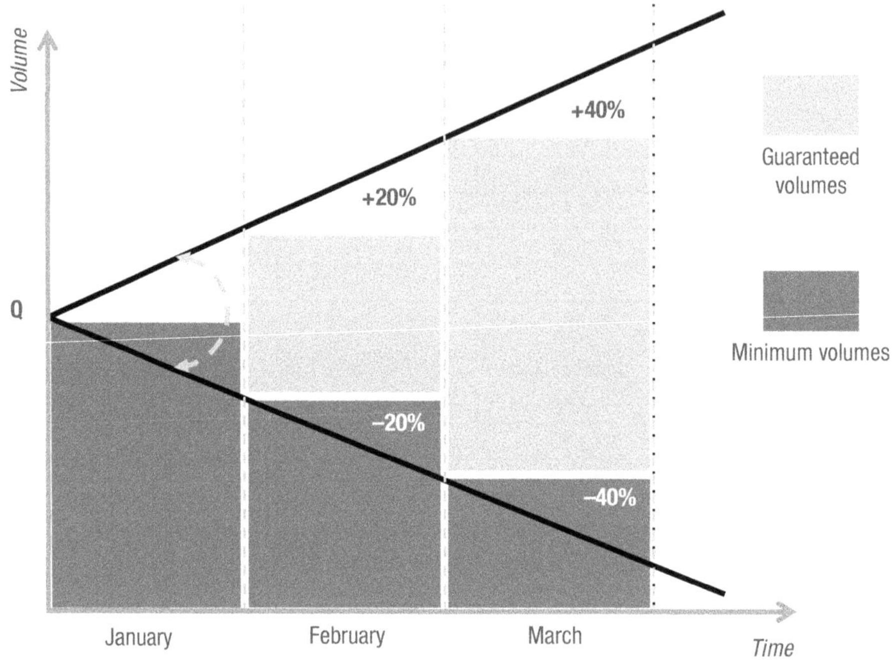

to sharing with the supplier not only the short-term order plan, but also predictions about the future trend of the demand for materials or services. The implementation of "**rolling plans**" allows for giving greater visibility to the supplier on the trend of production consumption. With reference to the example in Figure 15.4, in an order the purchasing department states in an order a requirement for month 1 of a quantity Q of material, the subject of the transaction. At the same time, on the basis of the expected production requirements, the supplier is asked to guarantee the production of certain quantities (Q + 20% and Q + 40%) for the following months 2 and 3. In this way, the purchasing office "books" production capacity of the supplier and ensures continuity in the supply flows. Against this guarantee, the supplier is assured of the commitment by the client company to purchase an amount equal to Q-20% and Q-40%, again for months 2 and 3. This share represents a demand certain for the supplier, allowing business continuity and assessments of any development plans for its production capacity. The width of the angle α is a function of the variability of the needs of the client company and the bargaining power of the counterparties.

Order communication processes and systems also belong to the communication lever. Regarding the processes, as already stressed, the company should try to optimize the information transfers, implementing lean procedures able to trace the relevant information but also to absorb the shortest possible time. The IT systems for order management and transmission also allow small businesses to make operations efficient, reducing manual interventions to a minimum and transferring data to the supplier at limited marginal costs.

In recent years, communication processes have merged into platforms called "**supplier portals**." These systems bring together all the supplier relationship processes in a single environment. This provides the scouting and supplier qualification processes, the offer request and contract construction platforms, and the planning and release of orders, up to the payment and control of the services provided.

The evolution of information systems to support the purchases has allowed for the development of so-called "**digital procurement**," that is, an environment where the buyer's activities (expenditure analysis, performance control) are integrated with the budget processes, to detect needs and audit suppliers, up to the release of orders and the management of the passive cycle. These systems are capable of both making activities more efficient and improving the effectiveness of decisions and the integration capacity of partner suppliers.

15.9 The supply channels lever

The last "lever" of strategic sourcing concerns the strategic choices for setting up the supply channels. In particular, the tools aim to structure the supplier base in order to obtain the maximum possible performance in economic, quality, service and ease of management terms. Proper management of supply channels is one of the most strategic components of strategic sourcing. The setting of these choices directly influences the performance of suppliers, and more generally, of supply relationships. In addition, these strategies embrace a series of problems that compete both at the top of the company and at the internal functions level, forcing every decision to be constructed and shared with a multifunctional team of people.

The interventions analyzed below concern:

- the choices for **allocating supply quotas**, which are part of the broader assessments regarding strategic sourcing;
- the **supplier selection and evaluation process** (Chapter 16).

The traditional approach to the supply market suggests the simultaneous use of multiple sources of supply, in order to create a competitive arena where to exploit the bargaining power of the purchasing company. This structure allows for avoiding dependence on a single source, to ensure continuity of supply and to maintain constant competitive pressure. However, any form of collaboration or integration is precluded, prerequisites for common investments aimed at improving the quality of the product or service and of the relationship in general. Consequently, the allocation options allow for obtaining evident benefits, but at the same time, they expose the company to risks that cannot be underestimated. In order to govern these choices, it is necessary to specifically define the objectives that the client company wants to maximize in the relationship and choose the most consistent alternative allocation of supply quotas.

With reference to the objectives, the considerations on the optimal number of suppliers to which to allocate the needs cannot therefore ignore an in-depth analysis of:

1. **The needs of the client company**, with reference to the critical performance expected from the supply relationship. This includes the specifications of both the product and the logistical process associated with the supply;
2. **The characteristics of the supply market**, in terms of concentration, technologies, critical issues, and volatility.

The use of **centralization and concentration on the single source** to seek purchasing economies, greater cooperation by the supplier, the simplification of quality management, logistics and deliveries, and the minimization of costs related to tooling, molds, or constraints on specific investments, suggest the choice of a single source of supply. Furthermore, even the presence of specific know-how or patent barriers, or a very small order, require the use only one source.

Maintaining a **high level of competition between sources** makes it possible to stimulate the continuous search for the best performance, to which to link the most substantial quantities of purchases, while using comparison methods (benchmark control of the performance) to maintain the possibility of continually innovating the group of suppliers, which evolves over time by including new companies and eliminating the less performing ones, aims to reduce the risk that extraordinary events (strikes, accidents or natural disasters) could constitute a cause of production stoppage. This entails ensuring independence from single suppliers, with the search for a high level of elasticity, understood as flexibility in volumes, guaranteed by the allocation (or reduction) of production quantities into small portions over multiple sources. All of these factors suggest a multiple supply model, with multiple suppliers active at the same time.

However, both solutions conceal numerous **risks**. On the one hand, the relationship with a single supplier can degenerate towards dependence, due to the impossibility of changing the source without losing the unrecoverable costs of the investments made. On the other hand, the relationship with multiple suppliers accentuates the supplier's propensity for opportunistic behavior in the short term.

The definition of the strategy for allocating supply quotas thus moves within a spectrum of choices ranging from single supply, where only one supplier manages 100% of the company's needs, to multiple supply, or where there is the simultaneous presence of multiple suppliers to whom portions of the company's total needs are allocated. Among these are a series of **hybrid solutions** that try to mediate the advantages, but also the risks, which characterize the two extremes (**Figure 15.7**).

Sole sourcing. The entire business requirement relating to a product family is purchased from a single supplier. The choice in this case is obligatory as this is the only one present on the market, due to patents, regulations or know-how that no other company is able to reproduce.

This is the most unfavorable strategy for allocating supply quotas for the client com-

Figure 15.7 The alternatives for allocating supply quotas

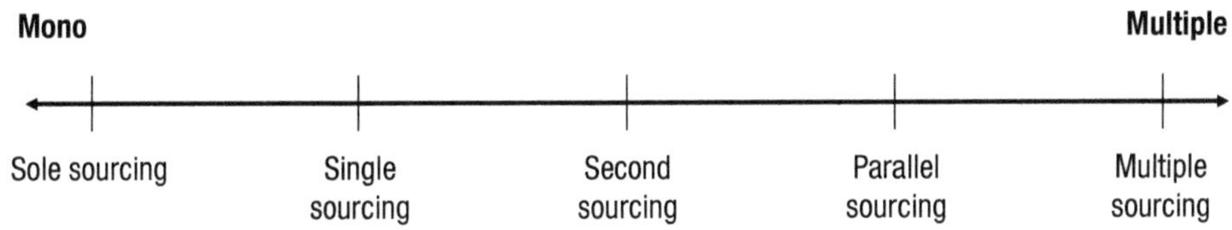

pany, since there is essentially no choice, with high contractual power on the part of the supplier. The relationship is completely guided by the supplier, which imposes its own strategic and commercial choices. It follows that, in order to align the needs of the client company and the services offered, it is necessary to work to involve the supplier in the client company's business plans, if necessary to the point of sharing the value generated. Conversely, if the supplier imposes its own choices, the viable alternatives remain a general adjustment of the purchase policies (for example, with orders of volumes and times dictated by the commercial terms of sale) or the search for and possible development of an alternative source (with a different technology or a new subject).

Single sourcing. The entire business requirement relating to a product family is purchased from a single supplier. On the market, however, there are alternative sources, comparable to the one chosen in terms of the level of performance offered.

The choice to build the relationship with a single source derives from an autonomous decision of the purchasing company, aimed at reaping advantages linked to economies of purchase for larger volumes or collaboration on the part of the supplier. The agreement often has formal foundations (a framework contract or partnership), the relationship is medium to long-term, resulting in high stability, and the exchange goes beyond the goods or services concerned, involving the production systems or the innovation process. The integration of processes is aimed at reducing uncertainty and increasing the degree of coordination. The parties are willing to make and share investments in the relationship, seeing lasting benefits. For example, there is the delegation of quality control to the supplier, according to shared sampling specifications, with consequent free pass acceptance of the materials.

The greatest risks of single sourcing are attributable, first of all, to a high level of dependence and the impossibility of both controlling the performance of the supplier (there is no benchmark), and of changing the source in a short time (high switch costs). These situations actually protect the supplier, who could implement opportunistic behavior to the detriment of the relationship's performance. Second, a single source limits flexibility. In fact, an excessive increase in the volumes required would only affect the production capacity of the supplier, that could potentially be unable to increase its production capacity in a short time. Furthermore, in situations of high dependence of the supplier on the client company (percentage of supplier turnover generated by the relationship), the reduction in volumes could also lead to economic/financial tensions, with negative effects on the supplier's stability in the medium-term.

In conclusion, this choice of allocation of supply quotas is consistent with high investments in know-how and specific technologies, confidentiality, the search for involvement of the counterparty, and integration of logistics/production flows.

Second sourcing. The business requirement is divided between two suppliers, that are comparable in terms of skills, processes, and performance. A majority share of the requirement is allocated to the first supplier, while a marginal share to the second. This is a solution derived from single sourcing, which, while maintaining the collaborative approach given to the relationship, tries to reduce the risks associated with the single supply. In fact, although the "second" supplier receives very small quotas, it performs three very important functions:

- it allows you to have a second source of supply to which to allocate peaks of needs, which cannot be absorbed by the main supplier, allowing for greater flexibility in supply volumes (elasticity);
- it maintains the possibility for the client company to lower the supply quota (or even close the relationship) in the event that the performance of the "first" supplier is not consistent with expectations;
- it provides a continuous benchmark to evaluate the performance of the "first" supplier.

This structure is adequate in situations where opportunistic behavior by the supplier occurs or as a correction of a previous single sourcing strategy that presented critical issues.

Parallel sourcing. The business requirement is divided between several suppliers, who work in parallel on similar codes within families of products or services. In this case, each supplier exclusively manages the single code (or groups of codes) of the family, the merchandise class, or the type of service.

This is a hybrid system, halfway between single supply and the multiple supply. This model is characterized by the presence of two or more companies, similar to each other in terms of capacity, technologies and know-how, which supply the same company as a "single" supplier of components or services comparable to each other. This structure maintains competition between the sources, in form rather than in substance, maintaining a high performance of the relationship, allowing for both the control and the replacement of the supplier in a short time. The supplier is offered a medium-term, collaborative, and stable relationship.

This choice is adequate when there are limited setup costs for the single source, there is the need to make specific investments in the relationship, and there are purchasing economies and there is a desire to create a medium to long-term relationship.

Multiple sourcing. The business requirement is divided between several suppliers, who work simultaneously on the same product or service. The strategy is to constantly maintain multiple alternative suppliers with whom to have non-preferential relationships and to allocate supply quotas in relation to the performance provided by them.

The large number of suppliers allows for continuous monitoring of quality, cost, and service performance, and constantly stimulates competition between companies. In addition, it limits dependence on sources and reduces the risk of interruptions to supply flows. The disadvantages lie in the division of volumes, which entails greater inefficiencies for the supplier due to smaller batches and more frequent setups, which in turn affect the client company in terms of loss of purchase economies and any price increases. In addition, there is a need for continuous monitoring of performance, in order to stimulate competition and allocate volumes based on the performance provided.

The use of multiple suppliers also limits the possibility of activities carried out in a joint and coordinated form. The expected duration of the relationship is not always sufficient to ensure the recovery of the efforts made in the activities carried out together with the supplier, and lastly, to evaluate the costs related to the management and control of numerous sources. This solution is appropriate for the purchase of standard materials and commodities, where the costs of starting the relationship (technical alignment of the supplier and related investments) are limited.

From the analysis of the pros and cons of the different alternatives, it can be seen that there is no optimal solution that allows for high performance and limited risks on all the variables considered. The choice of the allocation strategy must be set for each product category purchased, considering the impact it generates on the company's operations from a production and logistics point of view, and the value generated within the company's product or service. **The trade-off** between the search for integration and collaboration, on the one hand, and cost and competition on the other, forces us to identify an often intermediate position between the extreme alternatives presented.

Risk assessments relating to the supply structure also refer to the "supply channels" lever.

The analysis maps suppliers on a Cartesian level by comparing the level of dependence between the parties (**Figure 15.8**).

- **Customer dependency ratio** expresses the level of dependence of the customer on the supplier, directly attributable to the strategic sourcing policy adopted (number of suppliers). It is calculated as the number of suppliers present for each category/family of products and services purchased. The threshold could be indicated at 2/3 suppliers.
- **Supplier dependency ratio** expresses the level of dependence of the supplier on the client company. It is calculated as the value ordered by the customer company out of the supplier's total turnover. The threshold could be set at 30%/40%.

The low/low quadrant includes suppliers who do not create or have any kind of dependency. This quadrant is called opportunistic and flexible supply, as neither party suffers from excessive damage created by the counterparty's decision to close the relationship. It is the quadrant that best describes the supply relationship, or the outsourcing of activities, products, and services to a market with which you have a short-term relationship.

The high/high quadrant gathers the suppliers who create and have a high degree of dependence. It is a stable and strong relationship, where in essence there is a de facto partnership. The criticality of the relationship is limited precisely by the mutual interest of the parties in collaborating in order to support both businesses. The strong and exclusive relationship of dependence will likely lead to a merger between the two companies.

The most critical quadrants are the others.

The high/low quadrant includes suppliers in high-risk sourcing, that is, that indicate a high dependence of the client company, but a lack of dependence of the supplier. In many cases, they can be small and medium-sized enterprises and companies that buy products or services from multinationals. The size of the value in order and the interests of the multinational can lead to the decision not to continue supplying the items, causing considerable difficulties both in the short and medium-term.

The last quadrant includes the suppliers in a condition of lock-in, that is, that generate low dependence but have a significant dependence on the client company. These are typical situations in certain sectors (fashion or furniture), where the client company has control over the brand and product design, while the supplier works 100% for this client only. The low dependence of the customer, which still has different supply alternatives, clashes with the lock-in and responsible sourcing effect, i.e. the need to continue the relationship independently of other opportunities due to the risk of being called on to respond for abuse of dominant position, with the consequent responsibility for the economic and financial conditions of the supplier.

Figure 15.8 **Risk management in purchasing**

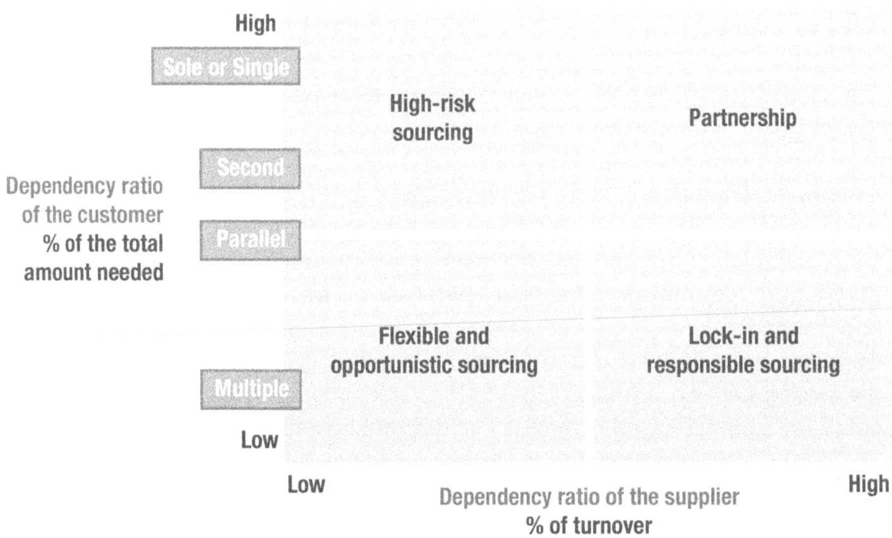

16 Process of Selecting and Evaluating Suppliers

by *Giuseppe Stabilini*

16.1 Introduction

The growing importance assumed by suppliers within various business activities and their role in pursuing the strategic objectives of the client company require the development and structuring of **effective systems to support decision-making** in purchasing activities.

The correct assessment and management of supply channels are among the strategic components within the levers of the procurement mix typical of strategic sourcing activity.

The setting of the logics and the selection criteria for suppliers represents the moment of **synthesis between different visions**. The requirements of the internal process, used for the products or services provided, the monitoring of economic and financial objectives, and the need for flexibility, such as innovation, converge at the moment of partner selection.

The management of the **selection and evaluation of suppliers** therefore assumes importance in both strategic and operational terms. From a strategic point of view, evaluating a supplier means first understanding what the company's needs are in relation to the supply market, what capabilities you want to identify in the supplier, what type of relationship you want to establish, and what time horizon will link the two companies.

From an operational point of view, the selection and evaluation process allows for the choice of suppliers who will have exchange relationships with the company, or directly influence the performance of Quality, Cost, Time, and Flexibility that the suppliers themselves will be able to offer, with a consequent impact on the effectiveness and efficiency of the various business processes. This impact will thus have to be managed both in the short term, determining any actions to improve the performance of the relationship, and in the long term, with the choices of reallocating supply quotas and possible selection of new sources.

The selection process must also ensure that every choice is made with **full transparency** and not only protecting the interests of the company and the shareholder, but also allowing the market to express its full potential through competition. Transparency also serves to protect the purchasing department, eliminating corruption and collusion between the buyer and the seller.

The use of **techniques and methods** for making purchase decisions therefore constitutes a strong need on the part of the purchasing department, both to govern the complexity of the system and to avoid incorrect decisions.

A methodological and rational approach to the selection and evaluation choices of suppliers makes it possible to make the selection process more effective, through a correct focus on the real problems, or indeed on the objectives pursued, a greater ability to evaluate all the relevant criteria, and a more precise systematization of the model that regulates the decision.

Conversely, the same approach can improve the efficiency of the decision process, by facilitating automated comparisons and analyses of the information collected, the possibility of using decision models and processes already used in the past, the elimination of factors not relevant to the purposes of the decision process, and the simplification of the communication activities related to the decision, indicating the justifications associated with it, both within the company and with suppliers.

Finally, in recent years, the visibility of customers and end consumers on the supply chains and the sharing of sustainability objectives has led to greater attention in the assessment and design choices of a company's supply base, to protect not only operations, but also the brand and the strategic positioning of the company itself.

16.2 The supplier selection and evaluation process

The process of selecting sources of supply can be schematically represented through four distinct phases:

- **definition of the problem**, or rather specification of the objectives to be achieved through the identification of the supplier. This phase includes a clear definition of the specifications relating to the material or service needed and a definition of the characteristics that the supplier must have;
- **definition of the selection criteria**, i.e. identification of the critical variables which, defined on the basis of the problem in question, will guide the choice of the supplier;
- **qualification of the most suitable suppliers** in relation to the particular needs of the supply. This phase, also called *vendor selection*, has the aim of identifying a restricted circle of suppliers (*vendor lists*) from among which to select the best source. The output can consist both of a reorganization of the suppliers into homogeneous groups (*sorting*), and of an actual ordering of the same (*ranking*);
- **final choice**, or precise identification of the supply source or sources that will begin the transaction relations with the company, the number suggested by the defined sourcing strategy.

These four phases can refer, in general terms, to three **purchase situations** relating to the degree of novelty of the material or service to be purchased:

- purchase of a new material or service (**new task**). The good to be purchased is completely new and thus there is no past experience for the company. There are no suppliers available or they are not known. High level of uncertainty about the specifications of the asset and a decision that requires different skills;

- modified repurchase of a material or service (**modified rebuy**). The good is new, but it is purchased from known suppliers, or conversely, the good has already been purchased in the past but new suppliers are sought. Medium level of uncertainty about the specifications of the asset and decision that requires a limited spectrum of skills;
- unchanged repurchase of a material or service (**straight rebuy**). The good does not present uncertainty from either the point of view of the specifications or that of the companies able to supply it. In general, this is a simple reallocation of supply quotas to active suppliers.

For the first two situations, i.e. those characterized by a medium/high degree of purchase novelty, a complete selection and evaluation process of the suppliers is set up, which involves the production of relevant information (market analysis, definition of the characteristics of the object to exchange, specific objectives of the relationship), the subsequent selection of potential suppliers, and finally, on the basis of the criteria identified, the final choice. The process collects information in order to judge the potential of the new supplier with whom to establish the relationship, thus producing a selection and evaluation "ex ante" of the potential supplier.

The third situation, that of unchanged repurchase on an already-known market, allows for bypassing the initial stages of the process, already having available an appropriate amount of information to be able to proceed directly to the final choice. The evaluation process is based on the information produced during the repeated transaction relationships with suppliers. This assessment is defined as "ex post" and the result is a repetitive process put in place both to monitor the actual consistency of the service offered by suppliers with respect to contractual agreements, and to provide information on the potential decision to reallocate supply quotas.

16.3 The depth and width of the selection and evaluation

The problem of defining "what" we intend to evaluate assumes first of all that the nature and objectives of the supply relationships are specified, definable through the traditional tools of strategic sourcing with the collaboration of the internal customer. In fact, the degree of complexity of the supply, the logistical integration pursued, the complexity of the material or service, the impact on the internal process and the continuity of the relationship, direct the assessment towards a different observation amplitude.

Thus, two variables can be identified that describe the **type of assessment**:

- **the type of relationship (depth)**;
- **the object of the evaluation (width)**.

In general terms, three different depths of relationships can be identified between the client company and the supplier:

- **no integration**, when the relationship is based on the mere short-term exchange of goods or services, with the client company that controls the quality of the different supply lots or individual service deliveries. In the case of exchange of goods, the client company protects itself from the variability of the service level of the supplier through the creation of security stocks;
- **integration of processes**, when the relationship envisages an alignment of the processes of the two companies, with common efforts in order to reduce inventories, avoid duplication of quality controls and free pass supplies, and an agreement that lasts for a medium to long-term time horizon;
- **partnership**, when the relationship goes beyond simple integration and involves joint investments in research and development, logistics, information systems, etc. in order to optimize the business relationships between the parties, both in the short and long term.

At the same time, the object of the evaluation, which expresses the extent of the evaluation, can consist of different levels of investigation:

- **at the product or service level**, namely the supplier output level. The observation focuses on the traditional performance variables of the supplier, such as the price, quality, and conformity of the products, delivery times, the deliveries dependability, and the level of logistical service and flexibility (see Chapters 4 and 5);
- **at the level of operating processes**. At this level, the assessments refer to the organization and management of business processes that offer the product or service being exchanged as an output. The focus is on the ability of the supplier to keep all the critical variables under control, and the stability and repeatability of the process, in order to ensure a quality-compliant output (good or service) over time;
- **at the company system level**. At this level, the assessments that, in addition to the previous ones, consider the strategic potential of the supplier, consider the different ways in which the supplier can contribute to the business of the client company, such as with technological development capacity and affinity of strategic orientation. The goal is to understand how much the supplier will be able to maintain or increase its level of competitiveness in the future.

By crossing depth and breadth, three **classes of suppliers** can be identified, with three different evaluation models (**Figure 16.1**).

The activity of the **normal supplier** (vendor) is characterized by an simple and regular exchange of typically standard goods or services. The company controls and evaluates the supplier only on the basis of the output produced and the performance offered, evaluating the conformity of the same with the specifications defined in the agreements. The client company protects itself from any variability in the supplier's performance by creating safety stock. The levels of performance recorded thus concern the simple transaction price and quality and service compliance.

The **integrated supplier** is characterized by a high level of coordination of processes with the client company. The evaluation conducted on the process and the

Figure 16.1 **Depth and width of the assessment: the types of suppliers**

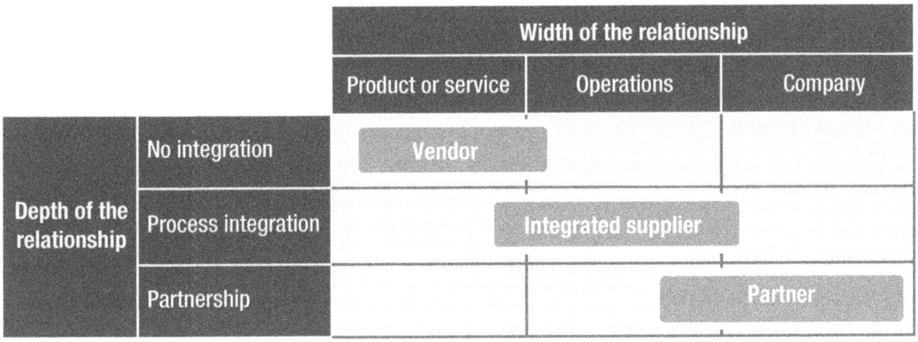

quality control system makes it possible to avoid checks upon acceptance of the goods, limit the stock of materials, and reduce the accessory costs related to the transaction, in a Total Cost of Ownership (TCO) logic.

The **partner** is characterized by a very high level of operational and business integration. Compared to the previous case, the object of the assessment also extends to a directional audit, in order to understand the medium to long-term strategic choices and the willingness of the supplier to support investments in the relationship, for example with common investments in assets or technological development projects.

16.4 The integrated vision of the process

The effectiveness of the supplier selection and evaluation process must be guaranteed by an **integrated approach**. The supplier that comes into contact with the company must be guided and managed in three distinct phases:

1. **supplier scouting and qualification**;
2. **source to contract**;
3. **supplier performance management**.

The process can be represented as a **funnel containing filters**. The first stages are aimed at very large populations of companies, where the mesh of the filters is very large in order to allow for collection and contact with numerous business opportunities.

Throughout the process, the number of suppliers decreases, just as the filter closes its mesh in order to observe characteristics and measure more detailed and specific elements of performance.

Each phase therefore assumes its own management objectives and logics, which differ significantly from the previous or subsequent phases. In the past, companies focused mainly on the last phase, sometimes not even with absolute control, in order to oversee the more strictly operational aspects of the relationship. The upstream phases often regarded direct control of the buyers and their own experience. The increasingly critical nature of the upstream phases, due to both the search for the best sources, and to traceability and transparency, has required a complete revision of the processes, thanks also to the use of new IT solutions and new information providers.

Finally, the ensuing discussion focuses on first-tier suppliers. The need to monitor supply chains requires that these assessments also applied to entire supply chains, acting directly or by requesting the collection of the same information throughout the supply chain from suppliers or third-party partners. In this case, transparency and traceability, regardless of the degree of distance from the main company, assume fundamental importance in order to safeguard global characteristics and performance. The entire process is described in **Table 16.1**.

Table 16.1 The phases and characteristics of the selection and evaluation process

	Supplier scouting and qualification	Source to contract	Supplier performance management
	Expand the database of potential suppliers to offer numerous alternatives	Select the pool of suitable suppliers and, from this, identify the best source (s)	Monitor the performance delivered and the supply relationship
Database	Registered suppliers database	Qualified supplier database	Active supplier database
Information sources	Public and private databases	Audits and visits to the supplier Request for Information Request for Quotation	Private databases and management information systems modules (ERP or vertical procurement)
Information provider	Company (Procurement and other functions) Supplier Third parts	Company (Procurement e other functions) Supplier Third parts	Company
Object of evaluation	The company as a company name or business division	The company as a company name or business division The single plant or organizational unit of the project	The service provided in comparison with the order The supply relationship
Degree of automation	Low if integrated collection processes are not structured	Very low, both for the specificity of the information and for the methods of collecting it	Very high Some manual additions
Phase output	List of suppliers with the main features	The best supplier (s) compared to business needs	"Supplier Scorecard" report with all the services at a glance
Critical issues	Populating database for new suppliers Constant updating of information	Storage of relevant information Time and resources to manage the activities	Completeness of the database Systems integration

Supplier scouting and qualification

The first phase of the process feeds the firm's ability to operate in the broadest possible competitive context. The goal is to have a **database with as many suppliers as possible**, organized by macro-family of products or services offered.

The information collected in the "registered supplier database" is aimed at understanding if the company has the characteristics to be able to participate in a possible negotiation. For this reason, company information is stored: company name, type of company, balance sheet data, and main ratios. In addition, business information is added, such as the types of products or services, the locations and activities carried out (development, production, after-sales services, for example) and the main references.

The process of collecting and uploading information to the database can be carried out in different ways.

- The purchasing structure itself collects the information when it comes into contact with the supplier (websites, company documentation, direct meetings) and supervises the loading of the same according to the preset, proprietary scheme. Despite having the significant advantage of a full system governance, this method has significant limits, as all the cost and resources are dedicated exclusively to the purchasing organization.
- The supplier can independently upload information on an "open" system. This mode allows for delegating the costs of the process (costs that become commercial investments for the potential partner) and requesting the population of a database that can also be deep and detailed. Obviously, the verification of the correctness of the information released remains the responsibility of the purchasing company, which will put in place phase control systems, from the start or subsequently.
- Some service providers, rating agencies, have developed their own business model based on the ability to collect and manage information at the individual company level. These firms integrate company and balance sheet data with summary indicators such as firm size and reliability, probability of bankruptcy, and non-fulfillment of payments; indicators that often integrate judicial information and news about the supplier.

Firms that are more structured and more attractive than potential suppliers have developed proprietary systems integrated with their ERP that allow data to be collected from different sources, thus using all the methods illustrated, obtaining the benefits of each.

The output of the phase is an **organized list of suppliers**, which can be quickly and efficiently consulted by different entry keys (supplier, product category, geographic area, main characteristics), in order to rapidly identify the pool of suppliers to be involved in a purchase project according to the particular characteristics required. The list may also present "rough" judgments, giving particular suppliers priority in the selection. Furthermore, if suppliers have had relationships with the company in the past, the database also presents information on past performance.

Source to contract

The second phase of the process is activated by a specific purchase request. The phase involves two elements. The first relates to the selection of the most suitable supplier pool with which to start the dialogue. The pool can be larger or smaller depending on the purchase strategy. The second is the actual selection of the supplier or suppliers with whom to enter into the purchase contract.

The dialogue with suppliers aims to collect all the information in a "**qualified supplier database**" to allow for a better understanding of the potential of the individual company, and based on the choice model, to support the final choice. This phase is the real funnel of the process, where we move from a situation of full opening to the market to one of closing in on a few relationships, with the drafting of the agreement.

The information covers all aspects of the supply: operations, such as production capacity and constraints, quality, service, such as delivery times and flexibility, as well as economic aspects. The breadth of the variables often requires that different corporate functions be involved, such as the applicant body itself.

The construction of the information framework can be done in different ways.

- **Audit and site visits to the supplier's facilities**. These visits are very useful both to collect and verify the critical characteristics sought, and to understand the supplier's general management and management framework. Although effective, the visits require considerable investments both from an economic point of view and in terms of dedicated resources. For this reason, the suppliers to be involved must be carefully selected, both for the critical nature of the relationship and for the competitive impact of having a new possible source.
- **Request for Information or Quotation**. The company formally asks suppliers to indicate their supply conditions in view of a simple market analysis and dialogue (RFI – Request for Information) or an actual negotiation aimed at a contract (FRQ – Request for Quotation). While efficient, this method exposes the company to a risk of information manipulation, which requires careful verification and control.
- **Third party certification**. Many companies entrust this phase, like the previous one, to third parties specialized in the auditing and evaluation of suppliers. In this phase, however, the staff in charge of the activity benefits from considerable experience in the specific product category or production technology, allowing not only for outsourcing the phase, but also for governing the process with know-how sometimes superior to that of the buyer company, and with faster execution. The focus of these activities makes it possible to operate on certified standards and models, also responding to requests from company and supply chain accreditation bodies (for example, ISO or Sustainability standards). In many situations, the au-

dit is also accompanied by operational support in the construction of the specific purchase technique or, ultimately, of the actual commercial negotiation.

These methods are combined with purchasing practices, consistent with the product category of the expenditure, and the criticality or more or less competitive negotiation strategy adopted.

The output of this phase is the **choice of the supplier or suppliers** with whom to enter into the agreement, or the contract or the individual purchase order. The information and reports produced must be filed and made available to give structured and precise feedback to the sources that have not been chosen, both to allow for alternative solutions if there are problems with the chosen partner, and to support similar decisions in the future. Many companies allow the information and judgments collected to remain under the control of only the buyer, without institutional sharing. This practice entails the risk not only of inefficiency in the activities, but also of loss of the know-how generated if the person changes role or company.

Supplier performance management

The third and final phase of the process oversees performance control and provides a structured summary of the entire cycle of relations between the purchasing company and the supplier. The objective is to have full control of the performance achieved in the supply of products and services, fully supporting both replacement projects in the supplier base and support for the improvement and development of the supplier company.

The "**Supplier Performance Scorecard**" report summarizes the entire cycle of relationships between the supplier and the company, providing in a single document a holistic vision of the quality of the supply relationship, and comparing it with the expectations or performance of other competing suppliers, to quickly allow for making strategic decisions to reallocate supply quotas or evolve the supplier base (**Figure 16.2**).

The information for the report is automatically provided by the company information systems, constructing the "**active suppliers database**." The order and performance data provided, from different points of view, is processed in standard mode in order to propose summary indicators. To these are added to the qualitative feedback collected with surveys or the direct judgments of the various internal customers. The Supplier Performance Scorecard allows for rapidly and completely interpreting the quality of the supply source, supporting every consequent decision.

The abundance of data and the need to have a real-time vision makes it necessary to build data management and collection processes that are as comprehensive as possible, open to external and interoperable databases. This means sacrificing the level of detail, that can still be recovered with ad hoc analysis projects.

Figure 16.2a Example of "Supplier Performance Scorecard" – SAP Ariba software solution

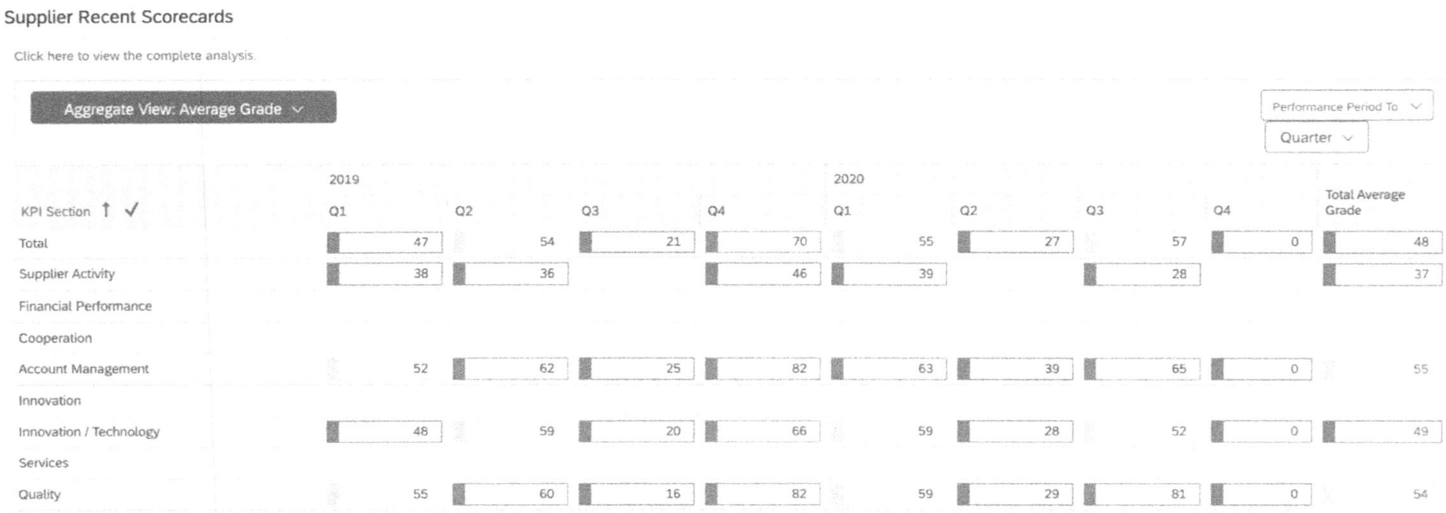

KPI Section	2019 Q1	Q2	Q3	Q4	2020 Q1	Q2	Q3	Q4	Total Average Grade
Total	47	54	21	70	55	27	57	0	48
Supplier Activity	38	36		46	39		28		37
Financial Performance									
Cooperation									
Account Management	52	62	25	82	63	39	65	0	55
Innovation									
Innovation / Technology	48	59	20	66	59	28	52	0	49
Services									
Quality	55	60	16	82	59	29	81	0	54

Source: SAP Ariba software solution.

Figure 16.2b Supplier Performance Trending

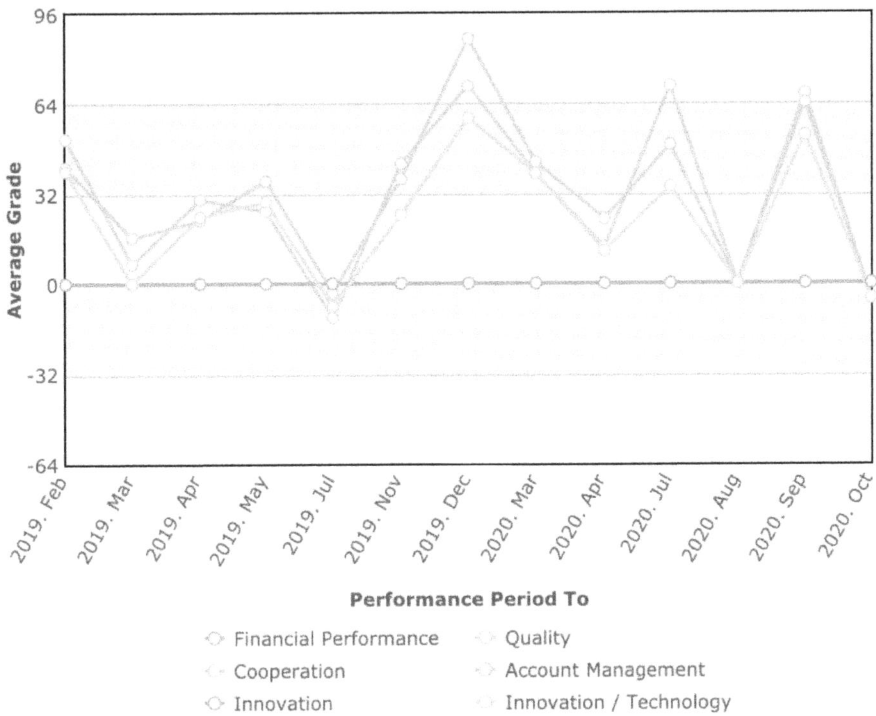

Source: SAP Ariba software solution.

16.5 The approach to evaluation

An important aspect in managing the whole process is building a **correct approach in supplier evaluation**.

The first choice concerns "who" should have the burden of evaluating the supplier. In the simplest hypothesis, it is the purchasing department itself with the reference buyer, that expresses opinions on the supplier. This choice is efficient and appropriate in simple or low-critical contexts. It is more appropriate to also involve the internal customer, or the person who receives the product or service. This actor has a full view of the problems related to the goods or services offered, during both the selection and check phase, and therefore integrates the assessment with aspects more related to the business or the technical and operational requirements.

It would be more correct to involve the **whole organization**. In fact, many aspects of the relationship touch on different functions and processes, multiplying the points of contact and therefore the observable performance between supplier and customer. Think of technical aspects evaluated by development functions, such as after-sales maintenance services, evaluated by maintenance or after-sales. The involvement of several people highlights the need to organize a multi-function selection and evaluation process, integrating many subjects. The complexity of collecting judgments will increase, but the result is a clearer and more focused partner analysis. If the supplier service is provided to external parties, such as a delivery logistics service to end customers, they are involved in the evaluation process.

The second choice concerns the **nature of the judgment** assigned to the characteristics or performance of the supplier. System design offers three options.

The simplest is represented by the expression of a **score**, for each criterion involved (a numerical rating) capable of summarizing the judgment. Although it is easy to implement, independent of the characteristic to be explored, and flexible, this option offers no "objective" basis of measurement and requires manual data entry. Furthermore, it leaves the score to be assigned to the discretion of the subject involved.

An evolution is represented by a **transparent survey-based model**, where specific characteristics with sequential levels are associated with the numerical score, characteristics that guide the compiler in assigning a score corresponding to the observed situation. Although it is an improvement over the option previously illustrated, it cannot be automated.

The best method is certainly represented by a **quantitative metric**, capable of collecting information and reworking it through a mathematical algorithm to indicate performance clearly and punctually. The quantitative system offers an objective view of performance, with great capacity for synthesis, but can only be implemented if structured data is available.

In **Figure 16.3** an example of the application of the three options in judgement method.

The intersection of the subjects involved with the nature of the evaluation criteria makes it possible to classify the supplier evaluation model (**Figure 16.4**).

"**Short-sighted**" models only involve the personnel of the purchasing function with subjective criteria. The myopia is given by the inability to fully observe the performance and to do so with subjective judgments.

The "**revenge**" models involve the whole organization, but they do so on a basis of subjective judgments that often do not allow for reading the real performance. Even a slight disservice can lead to different negative judgments according to the sensitivity of the individual.

The "**self-reference**" models use objective criteria but only on variables of direct interest to the purchasing function. Thus evaluations emerge on costs, payment terms, and commercial conditions in general. In this case the vision is also partial and connected only to classic expenditure variables.

The reference model is the "**360° panorama**," that combines both the involvement of the entire organization in contact with the supplier, and a complete set of objective criteria on different characteristics of the service provided. Although this last approach is better from various points of view, the need to combine efficiency and criticality of the product/service purchased suggests applying more simplistic models where the relationship with the supplier does not have a significant impact on company performance.

Figure 16.3 **Application of the three options in judgement method**

KPI	Judgment method	Data collection	Processing	Ending result
On Time Delivery 1-10 scale	Numerical rating	Employee A. Rating = 8 Employee B. Rating = 6 Employee C. Rating = 10	Average rating = (8+6+10)/3 = 8	8
	Survey methods	Employee A. Employee B. Employee C. **Rating System** 10 = Always on time 9 = In the majority of cases on time 8 = On time with some delays 7 = Some delays 6 = …	Average rating = (7+6+8)/3 = 7	7
	Quantitative metrics	(see PO table below)	Average delay = (4+10+7)/3 = 7 days	7 days

Survey methods – data collection grid:

	1	2	3	4	5	6	7	8	9	10
Employee A.							X			
Employee B.						X				
Employee C.								X		

Quantitative metrics – data collection:

PO ID	Actual delivery date	Agreed delivery date	Actual – agreed (days)
PO_1	21/03/202x	17/03/202x	+4
PO_2	30/03/202x	20/03/202x	+10
PO_3	12/03/202x	5/03/202x	+7

Figure 16.4 **Evaluation methods**

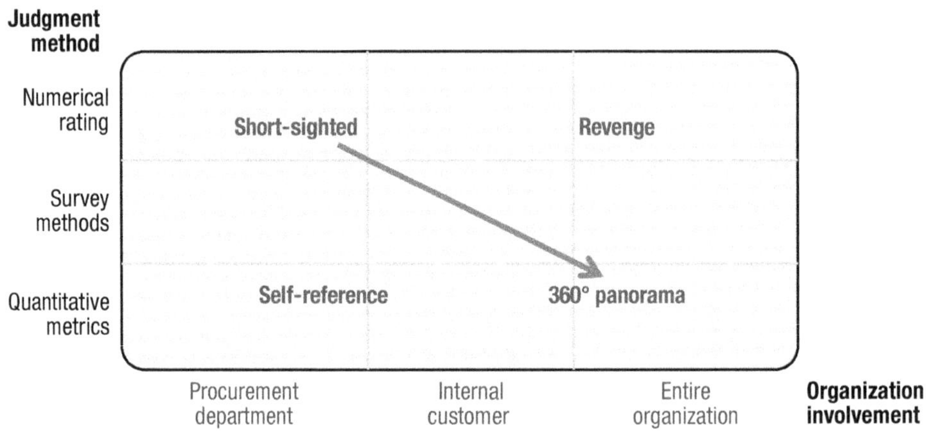

16.6 The ex-ante evaluation, potential

The ex-ante evaluation aims to analyze a new source of supply in order to start a transaction relationship with it with certain characteristics. These assessments are fundamental to fully conduct the "Supplier scouting and qualification" and "Source to contract" phases described in paragraph 16.4.

The greatest criticality of the source selection and evaluation process lies precisely in the fact that the "**potential**" of the supplier must be assessed, that is, the performance that can be provided by the supplier is derived by observing a series of indicators that should describe the performance offered in future transactions.

The analysis thus focuses not on the object of the future transaction but on the organization as a "factory" of the product or service itself. By detecting characteristics, skills, and process quality, a reasonable description of the possible performances that the supplier will be able to provide is obtained. In some contexts, such as professional services or activities that make extensive use of supply markets or of outsourcing, the evaluation should include numerous input resources, such as the characteristics and skills of management or second-tier suppliers.

The **information** therefore regards the company and the competitive context, such as:

- economic and financial situation;
- design capacity and technological know-how;
- characteristics of the production system;
- QA system and recognized certifications;
- previous experience in the sector;
- references from other companies;
- composition of turnover (markets, customers).

Depending on the type of product/service offered, details on production systems and critical resources for the business can be added to this information, identifying inputs and other relevant elements. The nature of the observed elements does not allow for the construction of quantitative indicators, but the analysis is limited to defining a level of qualitative performance (high, medium, low; positive, neutral, negative; poor, sufficient, good, excellent) or a type present/not present (on/off, used for ISO certifications) that is however sufficient to allow comparisons to be made between the different alternatives.

The final choice can be made through the use of one of the methods illustrated below.

16.7 The ex-post evaluation, control tools

The ex-post evaluation for the unchanged repurchase (**straight buy**) has the objective of identifying, form among the suppliers currently used by the company, the restricted group or the individual company to which to allocate portions of needs or the totality of the purchased items. This information is analyzed in the "Supplier performance management" phase of the process described in paragraph 16.4.

This assessment for the final selection phase uses all the information obtainable from the **control system** of the active suppliers. In fact, the structuring of information collection is practically the same whether a final choice has to be made or whether it is collected with the aim of monitoring the performance provided.

The most commonly used metrics concern:

- **financial conditions**, such as the price charged and terms of payment extensions;
- **quality level** expressed as conformity of the product or service with respect to expectations. In many cases these indices also evaluate the various severity levels of defects differently;
- **reliability of the volumes and completeness of the order**, i.e. compliance of the delivery with the volumes and materials or services requested;
- **timeliness, or speed**, quantified by the extension of the time interval between the date of issue of the order to the supplier and that on which the goods are available;
- **dependability, or punctuality**, measured in terms of temporal deviation from an expected or agreed-on delivery date;
- **flexibility**, evaluated in terms of the ability of the supplier to adapt its service (like a change in volumes or delivery) to changing conditions and unexpected requests from the client company.

The levels of performance are compared with both the levels defined by the contract and those of similar suppliers with two objectives: to evaluate the conformity of the supply and to identify the best suppliers from different points of view. However, these will be accompanied by evaluations on the company and on processes (typical of the ex-ante evaluation) which will allow for the overall monitoring of the relationship, in both current and potential terms.

The first step in the construction of the system of choice is represented by the **production of the data**, that is to say, by the detection of the service provided in the single supply.

For each criterion, one or more **indicators** must be constructed which, using the information from the single supply/order, calculate the level of performance achieved. There is no correct total number of indicators, but it is useful to reiterate that, especially in control contexts, a large number of indicators will require numerous surveys, a more complex database, and a less intuitive synthesis and final choice process. For this reason, many companies have chosen to limit the number of indicators to a small group (for example, 5 metrics), focusing them on truly important aspects of performance and making the calculation process still effective but relatively simple.

In the construction of the indicator it is necessary to identify the following fields of the hypothetical record layout:

- **index name**: assign to each indicator a "name" that expresses what the indicator intends to measure. The name must express the object of analysis clearly and intuitively;
- **the calculation formula**, indicating the necessary data and the type of calculation;
- **the codomain**, i.e. the maximum and minimum values released by the formula (for example, from 0% to 100%). Where possible, the indicators should have a similar codomain, so as to facilitate the reading and the subsequent application phase of the final choice methods;
- **excellence**, i.e. the maximum value obtainable by the "perfect" supplier (for example 100%). Here too, there is a tendency to standardize the values of excellence of the various indicator towards a single limit (for example 100 or 0).

These indicator will then be applied to the single transaction, that is, to the single material or service code purchased for each supplier. As anticipated, the existence of all the information necessary for the calculation of the indicators within the business information system makes it possible to automate this activity, which otherwise would be too expensive in terms of resources absorbed.

The final choice can be made through the use of one of the methods illustrated below.

16.8 The methods of selection and evaluation of suppliers

The choice of the suitable supplier or suppliers to support the company business occurs at various points in the evaluation process.

In the "Supplier scouting and qualification" phase, management must choose the group of suppliers with whom to activate the dialogue for the presentation of offers. In the "Source to contract" phase it is necessary to identify the most suitable source with which to sign the agreement. In the "Supplier performance management" phase, the services of the active suppliers are processed in order to verify compliance with the contractual agreements or renewal of the supplier fleet.

This moment of "choice" of the best source, regardless of the phase, aims to provide the client company with an analytical but also a synthetic evaluation of the supplier with respect to a series of criteria identified.

Take for example the data shown in Figure 16.5.

The three suppliers are assessed on the basis of three criteria: economics, quality and lead time. In order to identify the best supplier it is necessary to consider both the level of performance achieved and the importance of each criterion. By way of example, if criterion 1 "economics" were the most important, the choice would probably be provider C. Conversely, a high importance of the criterion 3 "lead time" would lead to choosing provider A.

Referring to this case, if within the variability of the performance we also want to accept suppliers with performance on one or more criteria that is low, but compensable through the excellent performance of other criteria, we will use **fully compensatory logic**. Using a "simple average" or "weighted average" as a synthesis formula reproduces this logic exactly. Conversely, we assume that all the criteria have a minimum acceptability threshold, identified for reasons of simplicity and shown in the graph with a dotted line. In this case, it is noted that supplier B has a performance below the minimum threshold for criterion 1 "economics." In this case you should apply a logic of non-compensation (**non-compensatory logic**), or poor performance on one of the detected and observed selection criteria, that however entails the rejection of that supplier. The synthesis must therefore be made considering Boolean logic (values 1 or 0) in order to exclude unsuitable suppliers in the synthesis process.

Figure 16.5 **Supplier evaluation example**

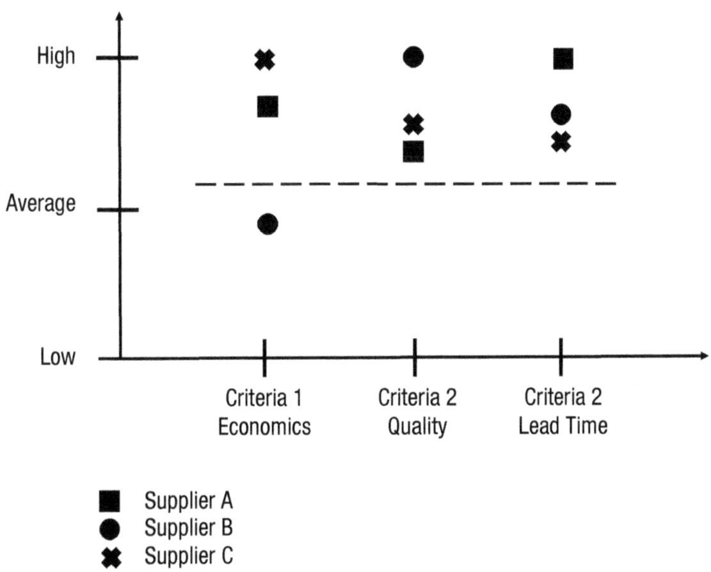

Knowledge not only of the logic to be applied (compensatory or non-compensatory) but also of the mathematical characteristics of the synthesis models applied are fundamental for the correct evaluation of suppliers. Cases are not rare in which the search for transparency and traceability of the process obliges the buyer to make the evaluation model explicit using more or less complex mathematical models. Such models could incorporate a valuation logic different from that intended by the buyer, leading to a transparent process, but one which indicates that they are better than the other supply options when they are not.

A second important point in the management of the process is represented by the determination of the **importance of the various evaluation criteria**. At the management level, the attribution of importance is associated with a "weight" that will directly influence the final judgment on the supplier. This decision must comply with the following principles:

- **all functions must be aligned on the assigned weights**. In fact, since the overall performance of the supplier (and its supply lots) are directly influenced by the assigned weights, these will direct the supplier towards the satisfaction of certain relevant indicator, even at the expense of other less important ones. For example, a supplier assessed mainly on the basis of the number of rejects in the batch supplied could tend to be systematically late, but with a complete and qualitatively perfect batch, thus satisfying the Quality function, but with major problems of stock management or production;
- each criterion assumes a **different importance** depending on the **type of material or service** involved in the supply. Among the goods purchased there will be products, components, or services where the delivery dependability is primary, and others where the price is the relevant variable, even at the expense of the level of service;
- each criterion assumes **different importance** depending on the **type of organization of production of the purchasing company**. Taking only the Service indicators, for example, in a production system that works in JIT mode with suppliers, the timeliness and speed of deliveries are very important, unlike management based on MRP where, vice versa, dependability (punctuality of delivery) is the preponderant service.

In order to analyze the **evaluation and choice models** of the suppliers most used in the company, the following are illustrated:

- **Categorical method (beauty contest)**;
- **Analytic Hierarchy Process (AHP)**;
- **Total Cost of Ownership (TCO)**;
- **Linear weighted average model (Vendor Rating)**.

16.8.1 Categorical method ("beauty contest")

The **categorical method** is certainly the **simplest and most efficient model**, but also the least transparent. The selection of the supplier is made on the basis of a list of criteria, the so-called "categories," on which the evaluator is called to express a qualitative opinion.

The criteria are listed in such a way as to describe all the relevant aspects in the supply relationship. This method is also called J.M. Keynes's "beauty contest," in reference to the concept of allocating resources to the subject that best responds to the characteristics desired by the promoter of the operation.

For each criterion, a **"positive," "neutral," "negative" judgment** is expressed (**Figure 16.6**).

The evaluation is made, therefore, by comparing the supplier performance both on the basis of historical data or previous experience, and by evaluating the characteristics of the supplier and the offer presented by it.

The same list of criteria is used to evaluate all supply alternatives, thus building a complete picture. By observing the performance summarized in the tables for each supplier, the evaluator proceeds to choose the best alternative.

There are many **positive factors**. This method is very simple in its application, allows for a wide range of flexibility on the variables to be observed, and allows for the expression of judgments by the figures belonging to the different corporate functions. Furthermore, the case-by-case application of both compensatory and non-compensatory logics is very simple.

In contrast, the method does not provide adequate tools for the final choice, which is based on a purely subjective decision. All the assessments relating to the trade-offs between positive and negative judgments and the summary rules for the final choice remain implicit and difficult to interpret and communicate.

The application of the categorical model is suited to two particular contexts:

Figure 16.6 Example of evaluation with categorical method: quality section

Judgment Macro Criteria: Quality	Negative	Neutral	Positive
• Certifications			√
• SPC		√	
• "5S"	√		
• Six Sigma	√		
• ...			

- for the selection of **professional service providers**, where the choice of the supplier is based not only on technical aspects, but on intangible elements such as trust, collaboration skills, people skills, and quality guarantees of the services offered. In these cases the "subjective" aspects are so broad that it is unnecessary to overcomplicate the model;
- for the **selection of a group of suppliers** with which to proceed in subsequent selection phases. The efficiency of the model allows for a screening of the supplier base in a short time and with limited resources, selecting a pool of companies with which to continue the dialogue in order to make the final choice.

16.8.2 Analytic Hierarchy Process (AHP)

The Analytic Hierarchy Process (AHP) (Saaty, 1980) allows for simplifying all the numerical calculations typical of the various other methods. The AHP in fact makes it possible to obtain the quantitative synthesis data through the expression of qualitative judgments in pairs of criteria which, through simple analyses, are transformed into mathematical information.

From a logical point of view, the method allows for identifying the "**weight**" of the various criteria and the "**performance**" of the supplier on each of them.

The first step in applying the method is the **hierarchical deconstruction** of the problem. In the case of supplier evaluation, this means exploding the final judgment of the source in the series of criteria and indicators that determine it (**Figure 16.7**). For simplicity, in the following example we only consider aggregate level 1 indicators. The steps described can be applied in the same way to the different levels below.

Each criterion is compared with the rest in pairs, in order to determine which one takes on a higher importance, or a greater weight. The scale for expressing such judgments is represented in **Figure 16.8**.

Let's assume we compare criterion A (first of the pair) with criterion B (second of the pair). If criterion A is more important, the scale ranges from a maximum of 9 to

Figure 16.7 Supplier evaluation criteria

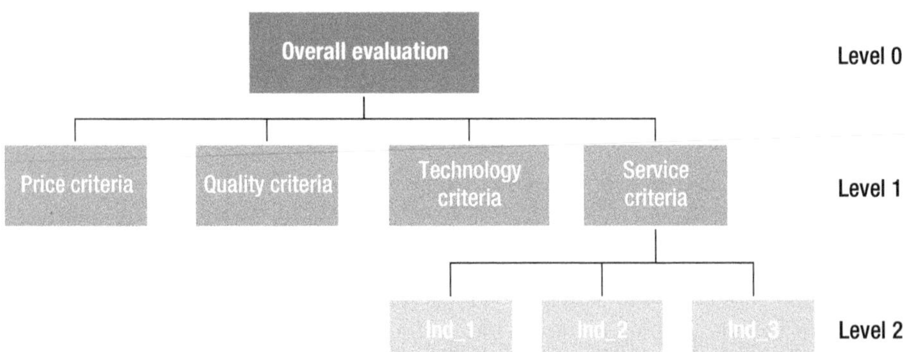

Figure 16.8 Evaluation scale

Criteria A — Criteria B

9 ◄ - - - - - ► 1 ◄ - - - - - ► 1/9

Judgment if Criteria A > B	Score
Absolute importance of A	9
intermediate values	8
Demonstrated importance of A	7
intermediate values	6
Essential or strong importance of A	5
intermediate values	4
Weak importance of A	3
intermediate values	2
Equal importance (A = B)	1

Judgment if Criteria A < B	Score
Equal importance (A = B)	1
intermediate values	1/2
Weak importance of B	1/3
intermediate values	1/4
Essential or strong importance of B	1/5
intermediate values	1/6
Demonstrated importance of B	1/7
intermediate values	1/8
Absolute importance of B	1/9

a minimum of 1, or equivalence between criteria. Conversely, if criterion B is more important, the scores range from 1 (equivalence) to 1/9 (maximum).

In order to facilitate the comparison of all possible pairs it is necessary to build a **matrix** where the criteria are shown in rows and columns. The crossing cells express all possible combinations (**Figure 16.9**). The cells on the diagonal will neces-

Figure 16.9 Original and adjusted matrix

Original Matrix	Criteria			
	Price	Quality	Technology	Service
Price	1	3	4	1/5
Quality	1/3	1	2	1/7
Technology	1/4	1/2	1	1/8
Service	5	7	8	1
Sum by Column	**6.58**	**11.50**	**15.00**	**1.47**

Adjusted Matrix	Criteria				Weight AVG by raw
	Price	Quality	Technology	Service	
Price	0.15	0.26	0.27	0.14	**0.20**
Quality	0.05	0.09	0.13	0.10	**0.09**
Technology	0.04	0.04	0.07	0.09	**0.06**
Service	0.76	0.61	0.53	0.68	**0.65**

sarily have a score equal to 1 (equal importance), being the pair formed by the same criterion.

The evaluator must **fill in the original matrix**, being careful to express judgments that are consistent as a whole, or that correctly reflect the importance to be attributed to each element of the selection. Once the original matrix has been completed, the system calculates in the **adjusted matrix** the relative importance of each cell with respect to the sum of the column. The last step is determined by the calculation in the adjusted matrix of the **average per row**. This value expresses the weight of each criterion. Note that the sum is normalized and equal to 1.

The **second part** of the AHP method allows you to attribute **a score to the services offered by suppliers**. The procedure is the same.

An original matrix is built for each criterion, placing the names of the suppliers in a row and in a column. In the example, three alternatives are hypothesized, identified with S1, S2, and S3.

Each supplier is compared with the others, using the same scoring scheme illustrated in Figure 16.8, building four original and adjusted matrices. The judgments and scores will make it possible to obtain the performance of each supplier relating to each criterion (last column on the right) (Figure 16.10).

The **last step** of the AHP is to **combine the weights with the performance**.

Figure 16.11 shows the suppliers in the rows and the criteria in the columns. Each crossover cell multiplies the supplier performance by the individual criterion with the weight of the criterion itself.

The multiplication "weight and performance," which is the weighted average of the performance, allows for obtaining the final score for each provider. The sum of the scores in the line allows for obtaining the **total score** achieved by each supply alternative. In the example, supplier 2 is the best of the group, thanks above all to its excellent performance on the "service" criterion, the most important criterion where S2 is clearly the best.

The **advantages** of using the AHP can be traced back to the following points:

- **simplicity in assigning the score**, given that each assignment is made by comparing pairs of objects (indicators, suppliers), reducing complexity compared to other approaches that envisage a systemic and simultaneous method;
- **ability to synthesize complex conditions into scores**, such as the comparison of different technologies proposed by suppliers. In these cases, it is not necessary to build metrics or normalize performance, preference is given to one or the other element directly;
- **ability to rationalize the importance of indicators** through the construction of a hierarchical vision of the criteria.

There are, however, **several unfavorable aspects** to the use of AHP. The major criticism of the AHP concerns the subjective process (judgments) of assigning weights to the indicators. Any evaluation requires subjective steps and often the construction of complex metrics not only lacks transparency in regard to subjective steps, but could even lead to mathematically incorrect solutions.

Figure 16.10 The evaluation of suppliers' attitudes

Price	Original matrix S1	S2	S3	Adjusted matrix S1	S2	S3	Price Performance AVG by raw
S1	1	1/4	1/9	0.07	0.06	0.08	**0.07**
S2	4	1	1/3	0.29	0.24	0.23	**0.25**
S3	9	3	1	0.64	0.71	0.69	**0.68**
Sum by Column	**14.00**	**4.25**	**1.44**				

Quality	S1	S2	S3	S1	S2	S3	Quality Performance AVG by raw
S1	1	2	8	0.62	0.63	0.53	**0.59**
S2	1/2	1	6	0.31	0.32	0.40	**0.34**
S3	1/8	1/6	1	0.08	0.05	0.07	**0.07**
Sum by Column	**1.63**	**3.17**	**15.00**				

Tecnology	S1	S2	S3	S1	S2	S3	Tecnology Performance AVG by raw
S1	1	2	1/8	0.11	0.17	0.10	**0.12**
S2	1/2	1	1/9	0.05	0.08	0.09	**0.08**
S3	8	9	1	0.84	0.75	0.81	**0.80**
Sum by Column	**9.50**	**12.00**	**1.24**				

Service	S1	S2	S3	S1	S2	S3	Service Performance AVG by raw
S1	1	1/3	5	0.24	0.23	0.33	**0.27**
S2	3	1	9	0.71	0.69	0.60	**0.67**
S3	1/5	1/9	1	0.05	0.08	0.07	**0.06**
Sum by Column	**4.20**	**1.44**	**15.00**				

Figure 16.11 AHP summary table

		Criteria						
		Price	Quality	Technology	Service	Total score	Supplier	Rank
Suppliers	S1	0.014	0.055	0.007	0.173	**0.249**	S1	2
	S2	0.051	0.031	0.004	0.432	**0.519**	S2	1
	S3	0.139	0.006	0.047	0.041	**0.233**	S3	3

Furthermore, AHP is a **fully compensated method**, so it is necessary to include only suppliers who have all the minimum characteristics required in the selection, perhaps by applying the categorical method for screening to the large pool of suppliers.

Finally, the method requires **repeating most of the assessments** if a new element is added to the criteria or supplier matrices, lengthening the process.

In conclusion, the AHP method is appropriate in two particular situations:

- in a context of **conflict in the company** where the method assumes the role of supporting the reduction of contrasts between functions by organizing the analysis and decision process in a linear manner;
- in **complex supply lines**, with different subjective variables, such as the ability for collaboration or trust, which can be efficiently processed through the comparison in pairs, providing structurality and transparency that would otherwise remain in the background.

16.8.3 Total Cost of Ownership (TCO)

This model is based on an all-inclusive view of the economic stakes at hand in the supply relationship. The **total cost of ownership** consists of determining the economic value of the supply as the sum of all the economic or cost variables in existence during the relationship.

Each supplier is evaluated not on the price, but on the basis of the **total costs generated on the client company system**: all costs which, although not directly caused by the supplier, are connected to the construction of the relationship, to the transaction between the companies and, to the management of the activities once the relationship with it has been started, up to the termination of the relationship itself.

This model also incorporates the typical **accounting principles** of Activity-Based Costing and the financial principles of Net Present Value, having to both allocate to the ratio the indirect costs of various cost centers, and to discount future flows of expenditures in the entire life cycle of the supply relationship.

For this reason, this approach can be useful in specific situations:

- **when the purchase project is complex** and embraces not only a multiplicity of products or services to be purchased during the relationship, but anticipates the same having effects on average over a long period of time, forcing the company to consider not only the existing expenses in the decision period, but all costs that will emerge in future periods (life cycle concept);
- when the elements of evaluation of the supplier and of the supply are legible due to **certain economic effects or estimated on the basis of a plausible scenario**. Since the TCO is a sum of costs, the model is not able to manage variables that are difficult to evaluate in economic value, such as the level of trust or the ability to collaborate.

The TCO model is consistent with **two particular categories of expenditure**. In both cases, the purchase price is an important component but it does not fully describe the TCO. In detail:

- the purchase of **non-critical materials and services** (non-critical items), where the economic variable to be monitored should not stop at the purchase price of the good or service, but must include all costs that the company sustains for asset management or use of the service, including any allocation of human resources (and related costs) in the management of administrative and control processes;
- the purchase of **investment assets**, such as plant and machinery. The length of their life-cycle makes it necessary to evaluate spare parts and repair costs, maintenance, as well as energy absorption, the need for support from human resources, and productivity. The enhancement of all these costs, as well as the analysis of possible alternative scenarios (full or partial saturation of the asset), allow for having a single reference with respect to the total costs.

The **complexity** of calculating the TCO requires a structured approach capable of bringing out all the possible costs to be included in the calculation model. A useful reference is represented by the following scheme (**Figure 16.12**):

- **pre-transaction components**;
- **transaction components**;
- **post-transaction components**.

The **pre-transaction costs** are the costs prior to the transaction and concern all the activities carried out to define in detail the characteristics of the requirement and to research and subsequently qualify the source of supply. In many cases these activities, in particular the qualification of the supplier and the management and updating of information within the supplier register, are not recorded as costs but are considered extraneous to the costs associated with the transaction, almost as if they were a fixed cost for the company.

Figure 16.12 Total Cost of Ownership: a conceptual framework

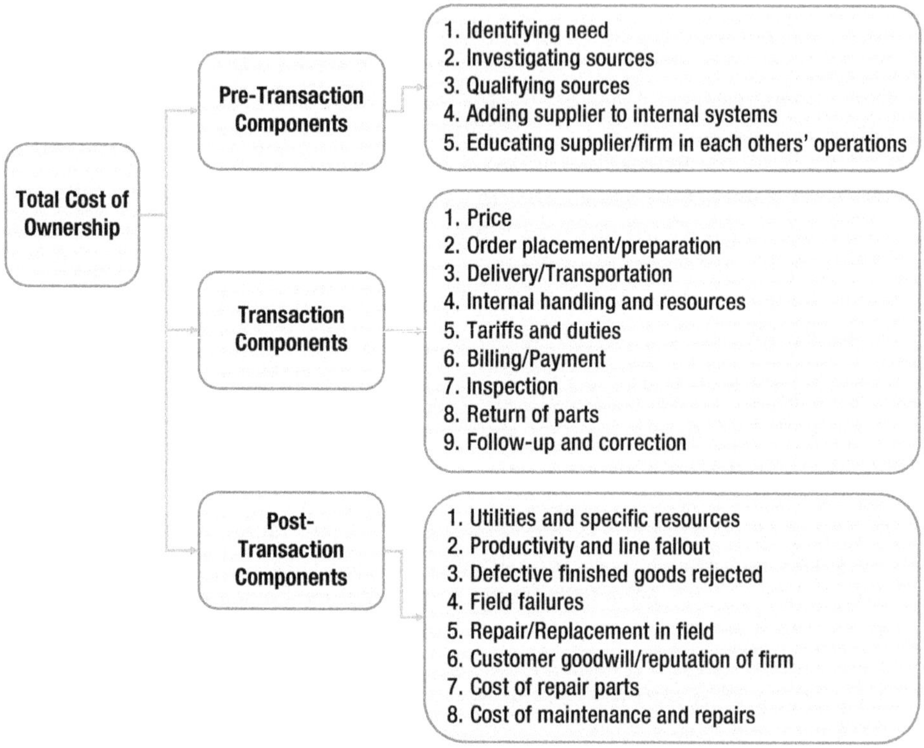

Transaction costs are the costs associated with the core activities of the purchasing process. In particular, the purchase price is the main cost item, even if the other activities carried out in this phase may, in some cases, take on significant percentages of the total cost. In many cases some of these costs, such as those related to transportation logistics, can already be incorporated in the supply price. In the case of low-value services or materials, the costs associated with personnel dedicated to the administration, management, and control of supplies must emerge.

Post-transaction costs are all costs associated with post-transaction activities, when the purchased good or service is used in internal company processes. The more the cost occurs in a period following the transaction, the more there is a tendency on the part of many companies not to refer it directly to the purchase choice. In many cases, these costs are considered as irrelevant in the purchase decision, or to be charged to other company functions, which are not however attributed to the purchasing department. Consider, for example, the consumption of utilities by an industrial plant or the cost of spare parts.

Many of the items listed are certain or identifiable a priori. Others, such as maintenance interventions, require the construction of a scenario to which all future items should refer. The robustness of the assessment requires the comparison of alternative scenarios, showing each time the best solution and the deviation from the basic scenario, an indicator of sensitivity to the uncertain variable.

As an example, **Figure 16.13** presents a comparison between two alternatives of supplying an item of direct material. The example refers to the "Item X" and all the variables in the scenario are certain. It can be seen that despite the more competitive price proposed by supplier B and supplier C, the TCO of the three options available is almost similar due to some cost components, such as the holding cost and positive financial income generated by longer payment conditions, which balance the higher unit price proposed by supplier A.

Continuing to refer to this example, however, we stress that the economic evaluation of the TCO does not make it possible to consider the possible impact of the

Figure 16.13 **Examples of TCO application**

Item	ITEM X
Suppliers (name)	A; B; C
Stock holding cost (%/year)	12%
WACC Interest rate (%/year)	5%
Annual Demand (units/Y)	5,000
Demand Type	Constant

	Supplier A	**Supplier B**	**Supplier C**
Unit Price (€/unit)	20.00	19.45	19.80
Order Quantity (unit)	1,000.00	3,000.00	3,000.00
Payment terms (days)	60.00	–	120.00
Ordering Cost (€)	100.00	100.00	50.00
Purchase cost (€)	20,000.00	58,350.00	59,400.00
Holding cost (per time-order)	240.00	2,100.60	2,138.40
Financial Expenditure (+) or Income (-) (€ per order)	–166.67	–	–990.00
Ordering cost (€)	100.00	100.00	40.00
TCO (per order)	20,173.33	60,550.60	60,588.40
Unit cost (TCO / Order Q)	20.17	20.18	20.20

Purchase cost (€)	= Unit_price × OrderQ
Holding cost (per time-order)	= Unit_Price × (OrderQ/2) × Stock Holding Cost % × (OrderQ/Demand)
Financial Expenditure (+) or Income (-) (€ per order)	= - Unit_Price × OrderQ × (Interest_Rate × Paym_Terms/360)
Ordering cost (€)	= Ordering Cost

supply choice on company assets, in particular on Net Working Capital. In fact, the purchase lot of 3,000 units generates a liquidity requirement which is approximately 3 times higher than for supplier A, further aggravated by payment conditions which reduce the payables accounts. Therefore, while the TCO method has the advantage of constructing a complete picture of the costs, on the other hand, **it does not consider any other objectives of the related supply**, such as the impact of finance on which the method is not able to offer any support.

In conclusion, TCO is an excellent tool for gaining **greater awareness of the impact of procurement choices**. It can be used both when selecting a new supplier and for monitoring the performance of active suppliers. In addition, through its complete vision of the company processes, TCO offers an operational comparison table between the different company functions, increasing the involvement of the different areas of responsibility and making it possible to search for economies and efficiencies beyond the traditional purchase price.

16.8.4 Linear weighted average model (Vendor Rating)

The linear weighted average model is certainly the **most popular approach** to evaluation. It is also called the **Vendor Rating** for its ability to score the various sources of supply. The calculation process summarizes **two key variables**.

On the one hand, there are the **weights** given to the various selection and evaluation criteria. Weights represent the importance of the criteria themselves. Typically, they are expressed as a percentage (for example, 10% or 30%) whose sum is equal to 100%, allowing you to view their relative importance. They are determined both by direct attribution and through the use of the AHP method (the part relating to the weights of the indicators).

On the other hand there is **performance**, the levels of performance of the single suppliers with respect to each criterion. This can be performance describing both a characteristic (as in the case of an ex-ante evaluation) and performance actually achieved. In the second case, the performance derives directly from an evaluation metric built on real data. The metric typically monitors performance in terms of economics/price, logistics and service, quality and flexibility. Many of these variables are in fact governed quantitatively by contract or by purchase order and are compared with the service actually provided.

Since the metric is expressed in a specific unit of measurement, the model foresees a point of **normalization of performance**.

For example, the supply lead time could be expressed in days or weeks, the quality in parts per million (PPM – Part Per Million), and the price based on a percentage deviation from a historical reference price or market price.

Please refer to Chapters 4 and 5 for more information on metrics.

Some operational aspects of measuring performance.

- Performance detection is often carried out on an **individual order basis**, comparing actual performance with expected performance. In the case of repeated orders from the same supplier, when calculating the performance in the period (a quarter, for example), the dimensions of the purchase orders must be considered (such as the volumes ordered in the case of purchase of goods), thus calculating a weighted average.
- The services must be recorded both in terms of "**average**," descriptive value of the service rendered in the various orders, and in terms of variability, or of "**sigma**" of the performance, an indicator of the degree of dispersion of the service rendered compared to the average. This second point is important and often underestimated. Many managerial and logistical decisions encounter difficulties due to a lack of constancy in the supplier's performance (think of lead times or punctuality) which leads to the need to consider this aspect as well in the evaluation metrics.

Once the performance has been calculated, it must be reported on a unique scale, for example from 0 to 10 points or from 0 to 100 points. This operation is called "**standardization**." There are different mathematical ways of normalizing the values, which however do not consider the managerial value of the judgment to be given at the different performance levels. For this reason the suggestion is to adopt a non-linear normalization, constructing performance intervals that make it possible to transform any data into a reference on a single scale. Figure 16.14 provides an example relating to the process of normalization of the Quality performance metric.

Both the individual metrics and the final rating will be expressed with a score based on the same measurement scale.

The final judgment for each supplier is called the **Global Vendor Rate** (GVR) and is calculated as:

$$GVR_{Supplier\ X} = \sum_{i=1}^{n} Performance\ M_i \times Weight\ M_i$$

Figure 16.14 Example of normalization

Supplier Alfa
Quality

Performance measured = **120 PPM**

Normalized performance =

$(((\mathbf{120} - 100) / (200 - 100)) \times (85 - 95)) + 95 =$
$((20 / 100) \times -10) + 95 =$
$(0.2 \times -10) + 95 =$
$-2 + 95 = 93$

PPM	Normalized score
7	100
100	95
200	85
500	80
1000	50
> 1000	0

Or:
(((data – Higher_Perf.) / (Lower_Perf. – Higher_Perf.)) × (MIN_Score – Max_Score)) + Max_Score

Where:
Supplier X = name of the individual supplier
M = evaluation metric
n = number of metrics provided by the evaluation system
Performance M = measured value on the single metric
Weight M = weight relative to the single metric

The synthesis can also be done in two steps. We can first calculate the macro performance GVR (price, quality, service, for example), so as to have a summary framework by type of service rendered.

Subsequently, by attributing a weight to the individual macro performance GVR, the final performance of the supplier is obtained.

$$GVR_{Supplier\ X} = GVR_{Price} \times Weight_{Price} + GVR_{Quality} \times Weight_{Quality} + \\ + GVR_{Service} \times Weight_{Service}$$

Based on the **comparison of the various GVRs** for the suppliers involved, management will make the final choice, whether to award a supply contract (Source to contract), or to manage the active supplier base, with the best winners and the worst excluded from new relationships (Supplier performance management).

16.9 Use of information generated in the supplier selection and evaluation process

The information generated during the selection and evaluation process can be used not only in reference to the **choice of the partner**, but also to start **improvement projects** relating to the supplier base.

In the phases of Supplier scouting and qualification and Source to contract, the characteristics and competitiveness of the companies involved allows for defining a management strategy for the product category.

In the event of a substantial **alignment of the companies**, the management can seek and encourage competition through the continuous search for new suppliers. Furthermore, by carefully evaluating the type of relationship to be established through the TCO, it can be understood whether the best strategy is competition (seeking better conditions in the market) or collaboration (identifying the source that is most reliable and ready, to reduce not the purchase conditions but the total management costs of the category, the TCO).

In the event of **polarization of the pool of suppliers**, with a few excellent suppliers compared to other mediocre ones, the strategy should be both to search for new alternatives, in order to identify companies useful for fueling a more competitive context, and to define investment plans with particular suppliers in order to eliminate the causes of lower performance or define growth paths with them that can lead to a general improvement of their offer.

In both cases, the information must not remain tied to the contingent choice alone, but as indicated, should feed a **medium-term strategy** conducted by the purchasing department.

During Supplier performance management, the **performance of suppliers** (hereinafter referred to with GVR) is compared to the **purchase volumes assigned** to the different companies in the same product group. By crossing the GVR values obtained by each supplier with the quantities assigned to them and defining the intervals ordered by decreasing values (for example, through the ABC curves), it is possible to evaluate the **consistency** between the assignment policies followed by the procurement function and the services provided by the supplier companies (**Figure 16.15**).

It can be seen that there is an area, arranged diagonally, along which all the suppliers should be located. If we adopt a policy of bundling purchases on one or a limited number of suppliers, then they should all be placed within the upper left quadrant. If there is a policy of assigning increasing volumes to different suppliers, based on the services provided, we should refer to the exception of the situation of suppliers who are outside the gray area, who need an appropriate reorientation. Only in the case of a supply market characterized by a monopoly could high volumes assigned to a supplier with poor performance be accepted.

In the example in **Figure 16.5**, suppliers 4, 2 and 5 are positioned correctly: supplier 4 is assigned significant volumes thanks to its high overall performance, supplier 5 is entrusted with residual quantities due to its modest performance, while supplier 2 is characterized by modest performance and is entrusted with modest, or residual, volumes.

The positions of suppliers 1 and 3 must definitely be reviewed. Supplier 3 attracts low volumes of activity even though the services offered are high. At this point, the actions can be directed towards increasing the supply quotas, that is, a shift to the quadrant occupied by supplier 4 (evaluating the possibility with respect to its production capacity), or its marginalization, against other suppliers with the same performance.

Figure 16.15 Global Vendor Rating and procurement policy

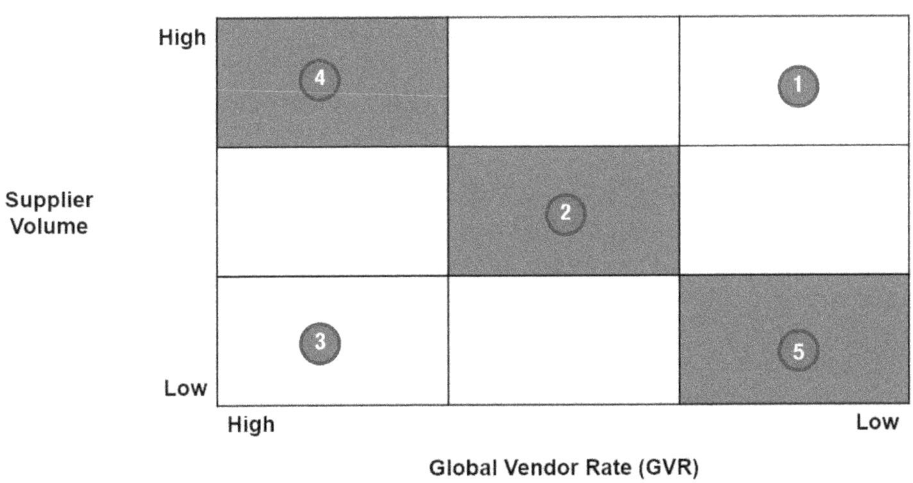

Conversely, supplier 1 is characterized by modest performance, but by high allocations. Here too, the solutions are oriented either towards its exclusion or its recovery. If you want to recover it, it will be necessary to renegotiate the contractual terms in order to align them with the performance provided. Alternatively, in order to marginalize it, there will be a tendency to gradually reduce assignments in order to marginalize the relationship.

Finally, the **GVR represents the supplier performance evaluation system**, and precisely for this reason, **it must be shared** with it, both as regards the results and for the calculation schemes used. In order to guide the behavior of suppliers, the company must periodically send a report of the services provided by the supplier (every three or six months), possibly accompanied by trends and blind comparisons with the main suppliers of its own class. This communication, in addition to creating greater commitment in the supplier, also has beneficial effects on the performance itself. In addition, this information, completed by an internal audit of the supplier, allows for correctly orienting the improvement and development projects of the supplier, highlighting the performance gaps and the areas that require investments.

16.10 Conclusions

In conclusion, it is thus evident that the success, or failure, of the supplier selection and evaluation process is fundamentally due to the **correctness of the information** summarized in the indicators and the criteria used for the synthesis and subsequent final choice. If, on the one hand, the **construction of the metrics** is not difficult to implement, the presence of a single database, or interfaced databases, is however critical, in order to automate the collection and initial analysis of the information. This phase would in fact be too time-consuming and resource-consuming to be carried out manually for all supplies, such that the result of decision-making based on the systemic analysis skills of people would be obtained.

Furthermore, the **reliability of the data** and the **tracking of any changes** decided on in the relationship (such as the reprogramming of a delivery date) must be ensured in order to produce information that describes the potential performance in the most realistic way possible, or that provided by the supplier. Failure to comply with this condition leads to the production of reports that cannot be used for making decisions.

The **involvement of the "internal customer"** is also fundamental. Only those who have expressed the need are able to define the correct level of performance expected by the supplier. The attribution of the "weights" to be assigned to the metrics cannot be an activity carried out independently within the purchasing office, but must result from the work of a cross-functional group comprising all the parts/company functions involved in the supply, which therefore have clear what they expect from it. The definition of weights that are not consistent with the business goal often leads to frictions between functions that certainly do not facilitate the provision of the service by the supplier (who does not know which part to satisfy).

The **involvement of suppliers** must be total at all stages of the process. The supplier must know exactly what are used to evaluate its work, thanks to which it can direct its behavior and improve the service offered. In addition, periodically, the supplier must be able to access summary reports regarding its results and those of its competitors, in order to understand the logic of assignment of the volumes purchased by the company. In fact, while on the one hand the GVR allows for evaluating the supplier, on the other it should link high performance to high supply quotas. Without this direct link between "result and reward," the GVR allows for control of the supplier, but removes the incentive to improve its performance.

Finally, the structuring of processes and methods for the selection and evaluation of suppliers guides the construction of the procurement department performance. Indeed, the system for evaluating the activity of the buyers cannot ignore the quality of the sources of supply offered to internal customers.

From an operational point of view, the supplier selection and evaluation process generates information for the **supplier scorecard**. This document fully photographs all relevant information relating to each supplier. It summarizes both the company characteristics and the performance deriving from active relationships, comparing the data both in an historical logic and positioning it with respect to the other suppliers by product category. The document, produced and updated automatically, allows the buyer to make the data collection and processing phases more efficient; phases that are essentially carried out by the information systems, fully supporting the decision-making processes typical of the management of supply relationships.

Figure 16.16 summarizes the design principles of the evaluation system for suppliers and the supplier scorecard.

Figure 16.16 **Building a supplier scorecard: principles**

1. **The measurement system allows for scoring flexibility.** Introduction of new criteria. Change in the weights.
2. **Internal customers evaluate supplier performance.** Ending user involvement. Evaluation in real-time with the use of the product/service supplied.
3. **Scorecards are reviewed and acknowledged by suppliers' top managers.** Formal meeting on a predefined timeline.
4. **Suppliers with more than one location receive multiple scorecards.** Team or human resource-driven services.
5. **Scorecards include cost-based measures whenever possible,** applying a TCO approach.
6. **Scorecards are updated in real-time.** Automatic updating, calculation, and report production
7. **The measurement system separates the critical few from the marginal many.** Ability to focus on the most relevant criteria and events.
8. **The metrics database allows for user flexibility in retrieving and displaying data.** Data analytics and data browsing features embedded in the system.
9. **The measurement system provides early-warning performance alerts.** Ability to give a signal to management in case of performance warning.
10. **Suppliers can view and compare their performance online.**
11. The measurement system is benchmarked against best-practice companies.

Source: adapted from Trent, 2010.

17 Lean Management, Total Quality Management, Six Sigma

by *Valeria Belvedere*

17.1 Lean Management

Lean Management is the set of logics, practices and tools for improving processes for the production of goods and provision of services, developed thanks to the experience gained above all in Japanese companies, and at Toyota Motors in particular, which since the 1950s has developed what is now known as the "**Toyota Production System**" (Womack et al., 1990). From the moment of its conceptualization to date, Lean Management has found wide implementation in both service and manufacturing companies due to its ability to guarantee the simultaneous improvement of all production and logistics performance measures, and thus to allow company operations to be ready to deal with new competitive challenges (Akmal et al., 2020; Bortolotti and Romano, 2012; Holweg, 2007).

This approach is based on the constant search for "**waste**" (also called *muda* in Japanese), that is, all the improper uses of company resources that do not generate "value" from the customer's perspective (Rother and Shook, 2003). The waste that can arise in production processes can be traced to the following categories:

- **Overproduction**: this occurs in particular in the presence of technical constraints that do not allow for production volumes to be aligned with the actual needs of the market. The excess quantities are sometimes not reusable because they are subject to phenomena of physical and/or commercial obsolescence, which cause a rapid reduction in the value of the product or even its physical deterioration. On the other hand, in cases where it is possible to allocate these quantities to satisfy a future market demand, they must be stored in the warehouse in the form of stocks of products which, in turn, are the cause of further waste, including the space to be used for storage, the electricity required for heating/cooling the premises, etc.;
- **Inventory**: especially in situations where the production planning process is managed on a forecast basis, it is a structural characteristic of the company to hold stocks of finished products, suitably sized to meet demand. However, especially in sectors characterized by short product life cycles and a wide product range, the demand forecasting process can be characterized by low accuracy, which does not allow for easy alignment of the stock to market demand. Therefore, in the event that the sales forecasts are higher than actual demand, the stock of products is oversized, generating an unnecessary use of space for storage and resources used for producing the good (especially materials, electricity, and manpower);

- **Overprocessing**: this phenomenon consists of carrying out processes aimed at giving the product some function or performance that the customer does not need (and for which he/she would not be willing to pay). This generally results in too tight tolerances – for example related to the variability margins of some dimensional characteristics of the product, or in the presence of unnecessary attributes, as in the case of certain finishing treatments on components of the product that the customer cannot see;
- **Defects**: the production of parts not meeting design specifications constitutes waste, by definition. In fact, these parts may need to be reworked, aiming to correct the defect, while in other cases they must even be discarded. In both cases the company incurs costs which are the direct effect of the waste associated with the defects;
- **Waiting times**: industrial plants are often subject to downtimes, due to multiple causes. This determines the non-use of production resources which are theoretically available, but not in a position to operate, for example due to lack of orders from customers or downtime due to breakdowns and maintenance;
- **Transport**: the transport of materials (within the production site or between different companies) represents waste because during transport the product does not undergo useful transformations from the customer's perspective; consequently, it should be considered a *muda*;
- **Motion**: as in the case of goods, unnecessary movements of production workers are also wasteful, resulting from inappropriate process design or inefficiencies that force staff to frequently travel along the production lines.

In the lean philosophy, the elimination of these types of waste makes it possible to simultaneously achieve the improvement of various elements of production performance, which can be linked to efficiency, product quality, and the logistic service delivered to the customer. In addition, since Lean Management assigns a prominent role to the workforce as a driver of process improvement, the use of this management philosophy is also accompanied by an improvement in working conditions of factory personnel, mainly thanks to job enrichment and the improvement of safety standards in the factory.

The attention placed on the elimination of waste, on the one hand, and on the enhancement of human resources, on the other, has led to the belief that the lean philosophy is also useful for improving sustainability performance, thanks to the beneficial effect that its techniques can produce on the environmental and social "waste" that is associated with industrial production processes (Genc and De Giovanni, 2020).

17.1.1 Lean principles

In order to achieve the elimination of the types of waste peculiar to production processes managed according to traditional approaches, Lean Management indicates the need to reorganize company's operations according to the following principles (Womack et al., 1990; Rother and Shook, 2003):

- **value**: first of all, it is necessary to distinguish between activities that create value from the customer's perspective and those that do not add value. The latter must be removed, as they are the origin of numerous types of waste, such as the use of resources for unproductive purposes and the lengthening of production and delivery times to the customer;
- **value stream mapping**: this is a process of mapping production and logistics activities, aimed at providing a representation of the current state ("as is") that helps to identify the main inefficiencies, to be eliminated through the adoption of tools and approaches typical of Lean Management;
- **flow**: once the not-value-added activities have been eliminated, it is necessary to ensure that the production system operates without interruptions, which can be induced by various circumstances. For example, unexpected downtime due to breakdowns, sudden unavailability of materials due to supplier delays, and inaccuracies in materials handling processes lead to inefficiencies and waste, in particular waiting and unnecessary accumulations of work-in-progress inventories. From the perspective of Lean Management, it is necessary to identify and implement practices and solutions that allow processes to proceed smoothly and without interruptions;
- **pull**: once the processes have been redesigned according to the principles stated above, it is necessary to make sure that each phase of the process takes place according to the pull logic, that is, only if an order from the customer has been received. In the case of the final step of the production process, the customer is by definition external, while for the others it is represented by the production department immediately downstream. This principle implies that nothing can be manufactured (finished or semi-finished products) in the absence of demand;
- **perfection**: Lean Management postulates the need to seek continuous improvement, also called **kaizen** in Japanese, through a systematic process of analysis of value flows and research and elimination of inefficiencies that may arise in the performance of production and logistics activities.

Over time, numerous tools, organizational solutions, and methodologies have been developed with which to implement the principles of Lean Management (Bicheno and Holweg, 2016). The main ones are illustrated in the following sections. The *kanban* system, which represents one of the best-known tools of Lean Management, is described in Chapter 14 as a shop-floor scheduling system.

17.1.2 Value stream mapping

"Value Stream Mapping" is the mapping method used to describe the "**current state**" ("as is") and the "**future state**" ("to be") of the process under observation. More specifically, the former represents the way in which the process is currently managed, accompanied by information useful for identifying the main inefficiencies and therefore directing improvement actions. The "future state," on the oth-

Figure 17.1 Value stream map standard symbols

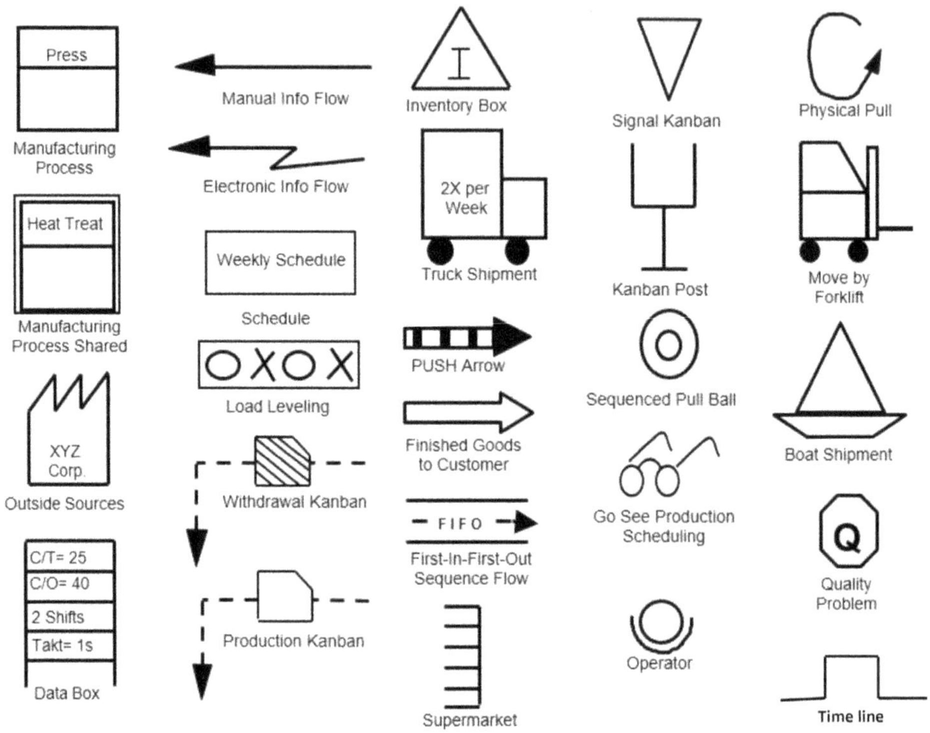

Source: adapted from Rother and Shook, 2003.

er hand, is the description of the process as it would appear after the adoption of improvement actions aimed at eliminating (or reducing) current waste. The "value stream map" (current and future) is normally represented using standard symbols, shown in **Figure 17.1**.

In the context of a manufacturing company, the value stream map is the graphic representation of the activities into which the production and delivery process is divided. These activities are usually divided into two categories:

1. **Value-added activities**, which the customer is willing to pay for, because from his/her perspective, they create "value." These include, by way of example, the transformation of the product, such as the cutting of a fabric to obtain a garment;
2. **Not-value-added activities**, which the customer considers as such because they do not contribute to modifying the functional characteristics of the product. This is the case of shipping and handling, as well as quality control, at the end of which the good remains unchanged, from the customer's perspective.

Box 17.1 shows an example of a "current state" value stream.

Box 17.1 "Current state" value stream: an example*

Jay is a little bakery that produces two types of cookies, one with chocolate and the other with nuts. The production process consists of four stages, endowed with one operator each:

- **Mixing ingredients**: the amount of time necessary for the operator to process a batch of 900 cookies is equal to 20 minutes. When shifting from one type of product to the other, a cleaning activity lasting 20 minutes is carried out to prevent allergies;
- **Rolling and cutting**: once the dough has been prepared, it is rolled out on a table. This activity lasts 5 minutes per batch (900 cookies). Then, the operator gets the cookies using a cutter. This cutting operation requires 1 second per cookie. To clean the cutter 10 minutes are necessary. The worker places the products on a tray where 90 cookies can be put (1 minute/tray) and moves the tray to the cooking department (30 seconds/tray);
- **Cooking**: the oven can process up to 10 trays in parallel. One minute is necessary to place a tray into the oven. The cooking process lasts 15 minutes. Once the cookies have cooled down (5 minutes), they are removed from the oven and brought to the packaging department (1 minute/tray). Chocolate and nuts cookies cannot be cooked together. Once cooking has been completed, a cleaning activity is carried out that lasts 20 minutes;
- **Packaging**: the operator places cookies into the jar (12 cookies per jar). To complete one jar 20 seconds are needed.

Along the process, the following stock of work in progress and finished products has been reported upstream each stage:

Stage	Chocolate	Nuts	Total in batches
Rolling and cutting	4,800 cookies	2,400 cookies	8
Cooking	3,600 cookies	1,800 cookies	6
Packaging	3,000 cookies	2,400 cookies	6
Finished products	360 jars	240 jars	8

To conduct the analysis of the current state ("as is") of this company and draw its value stream map, the single batch (900 cookies each) is adopted as unit of measurement in all stages of the process, including the packaging department.
First of all, the cycle time of each operation must be computed as follows (m/b stands for "minutes per batch"):

$$Cycle\ Time_{(mixing)} = 20\ m/b_{(processing)} + 20\ m/b_{(cleaning)} = 40\ m/b$$

$$Cycle\ Time_{(roll\&cut)} = 5\ m/b_{(roll)} + 15\ m/b_{(cut)} + 10\ m/b_{(clean)} + 10\ m/b_{place} + 5\ m/b_{move} = 45\ m/b$$

$$Cycle\ Time_{(cooking)} = 10\ m/b_{(place)} + 15\ m/b_{(cook)} + 5\ m/b_{(cool)} + 10\ m/b_{(remove)} + 20 m/b_{(clean)} = 60\ m/b$$

$$Cycle\ Time_{(packaging)} = 20\ seconds/jar = 25\ m/b$$

To compute the lead time necessary for the work in progress to be processed in the next stage, it must be multiplied by the cycle time of the downstream stage. Computations are as follows:

$$Lead\ Time_{(between\ mixing\ \&\ rolling)} = 8\ batches \times 45\ m/b = 360\ m$$

$$\textit{Lead Time}_{(between\ rolling\ \&\ cooking)} = 6\ batches \times 60\ m/b = 360\ m$$

$$\textit{Lead Time}_{(between\ cooking\ and\ packaging)} = 6\ batches \times 25\ m/b = 150\ m$$

In order to measure the amount of time finished products must wait into the warehouse before being delivered to the client (stock coverage), the available stock of finished products (600 jars, equal to 8 batches) must be divided by the customer demand. Assuming that the latter is equal to 1 jar per minute, the stock coverage of the finished products is equal to:

$$\textit{Stock Coverage}_{(finished\ products)} = \frac{600\ jars}{1\ jar/m} = 600\ m$$

On the basis of this analysis, the Value Added Ratio can be computed which is equal to the total processing time divided by the total lead time:

$$\textit{Value Added Ratio} = \frac{40\ m + 45\ m + 60\ m + 25\ m}{360\ m + 360\ m + 150\ m + 600\ m} = \frac{170\ m}{1470\ m} \cong 0{,}12$$

Assuming that Jay can fully exploit its production resources, with an uptime equal to 100%, its "as is" value stream can be represented as in Figure 17.2.

Figure 17.2 "As is" value stream

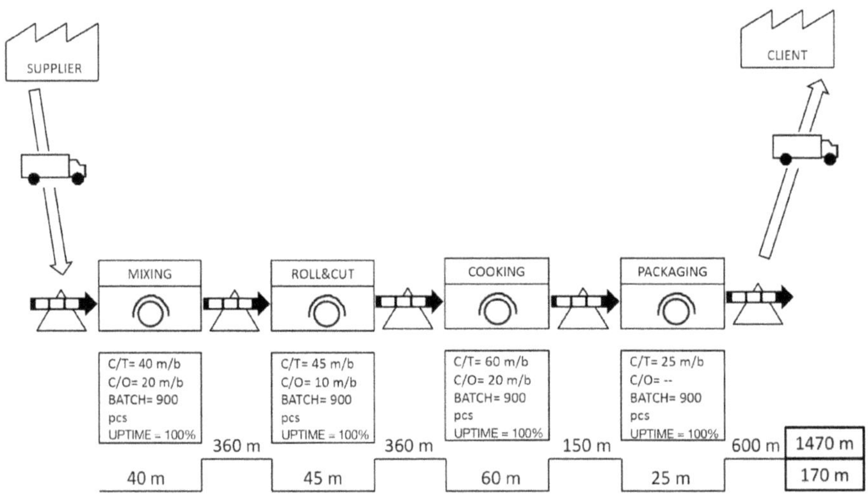

* This example is adapted from the exercise "Cookie Please" written by Andrea Antognazza, SDA Bocconi School of Management.

17.1.3 Standardization of work cycles and redesign of the layout

According to one of the key principles of Lean Management, production must be managed in a pull logic. In order to align the production output with customer orders, and operate in conditions of high productivity at the same time, it is necessary to adapt the resources involved in the production lines to actual demand. Therefore, should it be necessary, it must be possible to integrate the manpower employed on a line with additional employees who allow for increasing production capacity quickly, and vice versa. For this to happen, it is necessary to have multiskilled operators, that is, workers able to carry out a wide variety of tasks in order to allow for their easy reallocation within a production line.

To achieve a high degree of employee flexibility, companies must conduct training programs for production workers, in order to make them capable of carrying out a wide range of activities. This is accomplished through programs for the rotation of tasks and standardization of work cycles. The latter, in particular, implies the identification of efficient methods of carrying out operations and their transposition into standard procedures quickly accessible to all operators, in order to allow for the alternation of processes on different products.

To achieve greater workforce flexibility, it is also necessary to redesign the layout of the production line. The type of layout that best meets this need is the **"U-shaped" line**, shown in Figure 17.3.

The advantage of this type of layout (also known as "**cellular manufacturing**") consists of the fact that production employees work in close contact one with each other, which enables a mutual learning process and skills extension.

Therefore, if, for example, it is necessary to produce a smaller quantity of output in the presence of a decline in demand, it is possible to reduce the number of operators employed on the line, expanding the tasks of those who remain. The example in Figure 17.2 represents a situation in which 3 operators carry out the tasks of a line divided into 8 stations.

The U-shaped layout also implies the specialization of each line on a product family that requires the same type of processing, in order to ensure greater efficiency in the use of production resources.

Figure 17.3 **U-shaped layout**

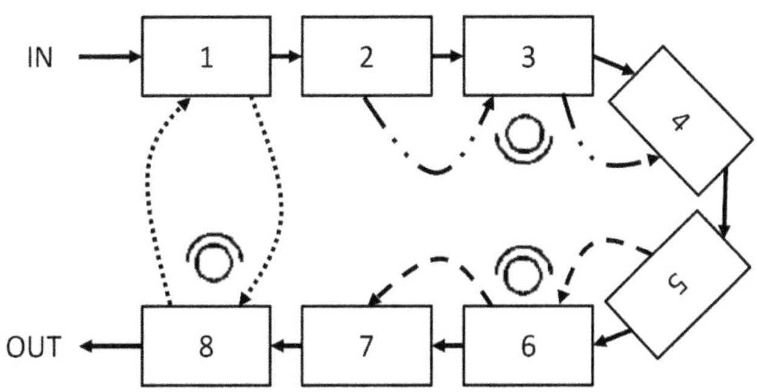

17.1.4 Workload balancing and "mixed model" production

The use of production lines specialized by product families can expose the company to the risk of excessive variability in demand and therefore to concrete difficulty in sizing the production capacity of the departments. In order to overcome this difficulty, it is necessary to achieve **heijunka**, a Japanese word that indicates the need for leveling the type and quantity of production over a fixed period of time in the last step of the production process. The focus on this phase is due to the fact that it has the role of "pulling" the production of the upstream departments, determining their workload.

This involves drawing up a monthly production plan that, based on the overall volumes to be manufactured, levels production on a daily basis. This way of operating allows for achieving a stable workload (and therefore, setting a stable production capacity) for each department. In this context, daily production is planned according to the so-called "**mixed model**" approach.

To understand the rationale behind the mixed model logic, suppose a company needs to produce a total volume of 100 units in the next month (Grando, 2018). It is also assumed that the company produces only three different finished products (A, B, C) and that the overall sales are divided as follows: A 50%, B 25%, and C 25%. In this case, producing according to the mixed model logic would mean planning the production of these three items on a daily basis as follows: A, B, A, C. In other words, the company chooses a daily production mix that follows market demand. Scheduling with the mixed model logic produces an important benefit for the customer: if the company produces all the items with high frequency, it is able to guarantee a better response time to the market.

The company also achieves a direct advantage from the use of the mixed model logic. In fact, if large production batches were produced, a considerable accumulation of stocks of finished products would be observed, which generates numerous forms of waste. On the contrary, by operating with a mix that follows the pattern of demand, inventories remain at much lower levels.

Figure 17.4 compares a daily production plan elaborated according to the mixed model logic to one which, all things being equal, is based on the traditional logic of large batches. Against a total monthly production of 2,000 pieces, the first plan entails daily production of a total volume of 100 pieces (assuming that the month is made up of 20 working days), divided into the three types of items in order to replicate the mix of demand (50% A; 25% B; 25% C) and guarantee the availability of all types of products on each single day. The second plan operates with large fixed batches of 100 pieces, whatever the item. Therefore, in this case it takes 3 days for the company to produce the entire product range.

A department capable of operating according to the mixed model approach must have a production pace perfectly aligned with that of sales. This pace represents what is called **takt time**, calculated as follows:

$$Takt\ Time = \frac{Time\ available\ for\ production\ in\ a\ day}{Customers'daily\ demand} \qquad [1]$$

Figure 17.4 Mixed model production (plan A) vs. Production plan with large batches (plan B)

Plan A				
Item	**Monthly production**	**Daily production$_{(day\ 1)}$**	**Daily production$_{(day\ 2)}$**	**Daily production$_{(day\ 3)}$**
A	1000 pc	50 pc	50 pc	50 pc
B	500 pc	25 pc	25 pc	25 pc
C	500 pc	25 pc	25 pc	25 pc
Total	**2000 pc**	**100 pc**	**100 pc**	**100 pc**

Plan B				
Item	**Monthly production**	**Daily production$_{(day\ 1)}$**	**Daily production$_{(day\ 2)}$**	**Daily production$_{(day\ 3)}$**
A	1000 pc	100 pc	0 pc	0 pc
B	500 pc	0 pc	100 pc	0 pc
C	500 pc	0 pc	0 pc	100 pc
Total	**2000 pc**	**100 pc**	**100 pc**	**100 pc**

The mixed model technique affects only the department where the last step of the production process is carried out, while the production plan of the upstream departments is determined by using the kanban system, described in Chapter 14.

Finally, it must be considered that in order to operate according to the mixed model approach, it is necessary to have the possibility to retool the systems very quickly, to ensure that the cost related to the numerous setups does not make the use of this logic inconvenient. We will discuss this topic in the following section.

17.1.5 Setup reduction

As already pointed out in the previous section, an essential condition for producing according to the pull logic is that setup times be short. This condition allows for achieving two important benefits, linked to the possibility of reducing the size of production batches and improving the level of saturation of production capacity.

Total setup time can be cut in two different ways. The first consists of reducing the total number of setups carried out over a certain period of time, by designing the products according to the principles of product modularity (Baldwin and Clark, 2000). In fact, if the degree of variety is reduced during the design phase, especially at the component level, the conditions are created to re-size the mix of items to be produced and, therefore, the number of setups.

The second solution consists of cutting the unit setup time, trying to reduce the interval during which the machine is stopped in order to start the production of an item different from the previous one. In this case, given a certain mix of products to be made, the objective pursued is to limit as much as possible the time during which the equipment is stopped for tooling operations. In this regard it should be remembered that the setup can be divided into two components:

- **internal setup**: the set of operations for preparing a new process, carried out with the machine stopped;
- **external setup**: the set of operations that can be carried out while the machine is running.

It is necessary to look for organizational and technical solutions that allow, first of all, to reduce the internal setup operations, making sure that they can be performed while the machine is working. Furthermore, by observing the way the setup is carried out, it is possible to identify unnecessary steps, which can be eliminated, and others that can be sped up or simplified, for example by ensuring the rapid availability of the tools used or by replacing some components of the machine with others that are easier to use. Lastly, once the new configuration of the setup has been defined, it must be translated into a written procedure so that it represents the new standard for carrying out the activity, which operators must follow.

The techniques used to reduce the length of setup operations are called **SMED – Single Minute Exchange of Die**. If these interventions make it possible to reduce the unit time setup in such a way as to make it substantially irrelevant, it is possible to produce according to the so-called **one-piece-flow** approach, i.e. generating a sequence of products different from one another, as may be required in a mixed model line.

17.1.6 Total Productive Maintenance

One of the conditions for producing according to the pull approach is high efficiency of machines, necessary to guarantee a reliable production flow that can satisfy demand in terms of timing and quantities. This condition can be compromised by the occurrence of breakdowns in the system and the consequent extraordinary maintenance activities.

Total Productive Maintenance (TPM) aims at maximizing plant performance through maintenance plans based on the principle of preventing breakdowns and delegating ordinary interventions to the machine operators (Nakajima, 1988). This approach, which has spread since the end of the 1980s, seemed particularly suitable for industrial contexts where production is managed according to pull logics, where the ability to respond promptly to market demand without drawing on the stock of products and semi-finished products requires high plant performance, to be sought in the elimination of downtimes related to failures and unscheduled maintenance, and in the improvement of machine efficiency and the reduction of defects.

To evaluate the effectiveness of the implementation of TPM, **Overall Equipment Effectiveness** (OEE) is calculated, which has already been illustrated in Chapter 8.

The experience gained in the companies that have adopted TPM proves that OEE increases thanks to improvement in all its determinants. More effective maintenance management allows for reducing downtime due to breakdowns and extraordinary maintenance, delaying the obsolescence of the machine by guaranteeing higher levels of efficiency, and finally, reducing the defects and yield drops characteristic of machines affected by frequent breakdowns.

All of this makes it possible to have effective production lines, able to respond promptly to customer requests (whether external or internal), in compliance with the agreed-on product quality standards.

17.1.7 5S

The expression "5S" refers to the system for eliminating waste and improving productivity based on the correct organization of workstations. This system arises from the observation that numerous errors, inefficiencies, and even accidents involving employees are generated by the presence of objects or materials that are not necessary for performing the activity, by the irrational and disordered arrangement of what actually is necessary for carrying out value-adding operations, and more generally, by the absence of standardized routines that allow operators to work quickly, safely, and efficiently.

To achieve this last result, the 5S method suggests the following improvement path:

- **seiri** (sort): this activity is aimed at eliminating from the workstation everything that is not necessary for carrying out production activities, through the correct description of the processes and the identification of the input materials and pieces of equipment to be used;
- **seiton** (set in order): selected materials and equipment should be arranged and organized in such a way as to make them clearly visible, in order to avoid waste related to unnecessary search efforts. This result is often achieved through the creation of shaped tables that accommodate the components to be assembled, making it possible, on the one hand, to check the completeness of the kit necessary for the completion of the processing, and on the other, to quickly identify the pieces to be assembled. Similar solutions, based on visual management, are used for the organization of tools and pieces of equipment;
- **seiso** (shine): the cleaning of the workstation is a necessary condition to prevent equipment breakdowns and malfunctions, accidents at work and product defects that may derive from contamination or contact with other substances and materials, such as oils, shavings, dust, etc.
- **seiketsu** (standardize): this principle concerns the need to identify standard work procedures, to be applied repeatedly to ensure high levels of efficiency;
- **shitsuke** (sustain): it is necessary to promote the pursuit of excellent practices in organizing workstations and carrying out production activities, in order to guarantee the consolidation of the new procedures and the benefits they bring about.

The 5S method is widely used in all those organizations that intend to adopt lean principles. Its implementation requires only a modest allocation of resources, but produces rapid and significant results, which concern the efficiency of the machinery, the scrap rate, downtimes due to breakdowns, accidents, and the amount of space occupied by the production departments.

The combination of these benefits means that the adoption of the 5S method is often a prior step necessary for the implementation of more sophisticated practices, in particular Total Productive Maintenance.

17.2 Total Quality Management

Total Quality Management (TQM) is a managerial approach that aims to achieve long-term business success and profitability, to be obtained through customer satisfaction (Feigenbaum, 1991). This implies the ability to manufacture products characterized by a high degree of quality, to be understood from a dual perspective. On the one hand, the customer who buys an item implicitly expresses the expectation not to find any defects or non-conformities with respect to the design specifications. On the other hand, the decision to purchase (and subsequently repurchase) a particular product also depends on its ability to "surprise" the customer by offering aesthetic and functional characteristics that are not taken for granted. Therefore, to ensure the development and production of a good quality product, it is necessary to involve several business functions. For example, the phenomenon of product defects affects the Operations Department, which is responsible for the production of items compliant with the specifications provided by the Design Department.

However, the ability to give the product distinctive characteristics that make it different from those offered by the competitors depends on the effectiveness of the new product development process, which in turn entails the involvement of numerous other functions, such as Marketing, Design, Process Engineering, Quality Assurance, and Logistics.

This highlights two characteristic aspects of TQM. The first concerns the need for the involvement and commitment of the whole organization in the pursuit of quality. The second concerns the inherent complexity of the concept of quality, which is well highlighted through the numerous definitions proposed over time by the leading scholars in the field of quality management:

- conformance to specifications (Crosby, 1984);
- predictable degree of uniformity and dependability at a low cost with quality suited for the market (Deming, 1989);
- fitness for use (Juran, 1988);
- bundle of commercial, design, production, and maintenance characteristics that allow a product to meet customer expectations (Feigenbaum, 1991);
- satisfaction of customer needs (Ishikawa, 1986).

Each of these definitions highlights different characteristics of quality. For example, Crosby's approach is functional to the assessment of defects, understood as failure to comply with the design specification. The other authors, however, despite the variety of definitions proposed, highlight the need to adopt the customer's perspective in order to judge the quality of the product, which may vary according to the needs and subjective motivations of the client.

The basic principles of TQM can be traced back to the following (Dale et al., 2013; Feigenbaum, 1991):

- **Customer Centricity**: as already pointed out, according to the TQM approach, the success and long-term profitability of the company depend on the ability to offer the customer products that can satisfy his/her needs, which can be explicit and implicit. Furthermore, it is necessary to apply the same principle to "internal customers," or to the various entities of the organization that interact with each other, ensuring the satisfaction of their requests. Over time, this approach can lead to the improvement of the overall performance of business processes, which determines customer satisfaction.
- **Focus on processes**: according to this Japanese approach, companies with an orientation toward long-term success are characterized by the ability to focus their improvement efforts on processes. This principle can lead to significant changes in quality management. For example, according to a short-term approach, to avoid the delivery of defective products to the customer, quality control can be carried out on all production volumes. While on the one hand this minimizes the risk of immediate customer dissatisfaction, on the other, it does not allow for the causes of defects to be identified and removed, and thus the company will continue to incur high quality costs. On the contrary, focusing on processes means that the company will adopt a prevention-based approach, aimed at identifying the main causes of defective products (for example, excessive complexity of the product resulting in the assembly errors, obsolescence of machinery, poor training of production workers, etc.) and at adopting appropriate corrective actions. In the medium to long term, this will make it possible not only to minimize the quantity of defective products delivered to the customer, but also to cut overall quality costs, thanks to the reduction of scraps and the downsizing of Quality Control activities.
- **Continuous Improvement**: in the TQM approach, process improvement should be pursued systematically, through the constant search for problems to be tackled and the planning and implementation of appropriate corrective actions. The method adopted to achieve continuous improvement is called Plan-Do-Check-Act (PDCA), developed by Deming. According to this approach, depicted in **Figure 17.5**, the improvement of processes (and the related results) is the outcome of a systematic planning activity (plan), implementation of corrective actions (do), control of results (check) and implementation of new interventions (act), defined on the basis of the feedback received in the previous phase. It follows that, especially in companies endowed with a Quality System certified according to ISO standards, whose rationale is based on the PDCA cycle, the performance measurement

Figure 17.5 **Plan-Do-Check-Act (PDCA) approach**

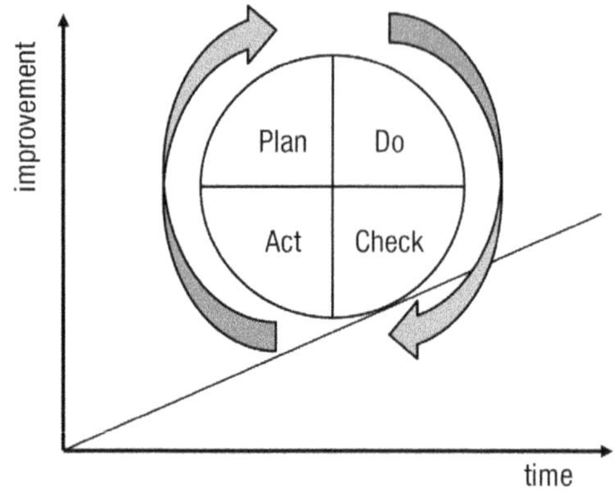

process is particularly important, since it constitutes the essential requirement for assessing processes and planning actions.

- **Staff involvement**: the focus on continuous improvement determines the need for a high degree of staff involvement. The experience of many companies demonstrates that only those who carry out specific tasks on a daily basis, implementing procedures and using the necessary inputs (for example, materials and machinery along the production lines), can identify specific opportunities for improvement and propose appropriate corrective actions. To facilitate this process, an "ideas box" is often set up in companies where the principles of TQM are adopted, where all employees can submit their proposals for improvement, that are subsequently evaluated in order to identify those worthy of being implemented based on the degree of relevance of the problem addressed and the probability of success of the intervention. Generally, in situations where this tool is adopted, a bonus is also awarded to the employee who has proposed the most deserving project. In order to make the involvement of personnel effective, however, not only is a general awareness of the issue of quality necessary, but training programs are also needed that allow workers to learn the principles of TQM and the operational tools with which to implement those principles.
- **Top Management Commitment**: while the involvement and training of personnel allow processes to be improved in a bottom-up manner, the most significant results of TQM are achieved when the organization is able to focus its efforts on real improvement priorities through a top-down process. This implies a high commitment by the management, which must include specific quality objectives in the strategic planning process, to be pursued over the long term.
- **Data Analysis:** the continuous improvement of processes must start from the analysis of objective data, which makes it possible to conduct a check-up on the current state and drive the identification of the main criticalities. Therefore, in companies that adopt the TQM approach, the tools of statistical process control, illustrated in section 17.3.1, are widely used.

The spread of TQM during the 1980s and 1990s led, among other things, to the institution of awards aimed at rewarding the organizations that most successfully implemented this methodology. The best-known are the Malcom Baldrige National Quality Award and the EFQM Excellence Award.

The **Malcolm Baldrige National Quality Award** (MBNQA) was established in 1987 in the United States by the National Institute of Standards and Technology in order to promote the spread of TQM and reward the most deserving American companies. The MBNQA is awarded annually and includes awards for six different categories:

- Manufacturing;
- Service;
- Small business;
- Education;
- Healthcare;
- Non-profit.

The criteria by which companies are evaluated fully reflect the key principles of TQM, namely:

- **Leadership**, understood as the ability of top management to guide the organization in the application of the principles of TQM;
- **Strategy**, which refers to the capacity of the organization to draw up and implement strategic plans;
- **Customers**, with a focus on the company's ability to create lasting relationships with its customers;
- **Measurement, analysis and knowledge management**; this dimension concerns the use of data in order to highlight areas of potential process improvement;
- **Workforce**, i.e. the extent of the effort made by the company in training and involving staff;
- **Operations**, to be understood as the ability to design, manage, and improve key processes;
- **Results**, understood as the performance achieved by the company (also in comparative terms with respect to competitors) in relation to: i) products and processes; ii) customer satisfaction; iii) employees; iv) leadership and governance; and v) financial, strategic, and market results.

The **EFQM Excellence Award** was established in 1991 by the European Foundation for Quality Management, with the same aims as the MBNQA, and provides annual awards for the following categories of companies:

- Private large companies (over 1,000 employees);
- Private small to medium-sized companies (less than 1,000 employees);
- Large public companies (over 1,000 employees);
- Small to medium-sized public companies (less than 1,000 employees).

The assessment leading to formal recognition is based on an audit conducted in accordance with the EFQM model, which over the years has undergone various updates. According to the latest version, released in 2020 (see **Figure 17.6** in this regard), the assessment of the degree of excellence achieved by the company must be based on three important questions:

1. What is the purpose of the company? Why does it intend to pursue a particular strategy? (**Direction**)
2. How does the company intend to pursue its aims and implement its strategy? (**Execution**)
3. What are the results achieved and what are the objectives for the future? (**Results**)

In this case as well, there is overall consistency between the logic of the model used and the principles of TQM, according to which the achievement of excellent results depends on the ability of the company to develop action plans focused on the most important objectives and to put them in place through appropriate initiatives.

Figure 17.6 **EFQM model**

Source: ©EFQM 2021. The EFQM Model is a registered trademark of EFQM.

17.3 Six Sigma

Six Sigma is a methodology developed in the 1980s, which pursues excellence in organizations through the systematic reduction of production waste (Pande et al., 2000). Its name derives from the Greek letter sigma (σ), which in statistics refers to the **standard deviation** of a variable. In **Figure 17.7**, one can observe that, if the probability density function of a variable resembles a Gaussian (or normal) distribution, 99.9997% of the observations fall in the range equal to 6σ. If this variable refers to a feature of the product, x_0 is the nominal value set by the Design Department, and is also the average value that is likely to be found in the sample of items that undergo quality control.

Production processes compliant with the requirements of Six Sigma are characterized by output with a range of variability around the nominal value that is very low. In particular, according to Six Sigma principles, perfection is reached when the range of tolerance chosen by the Design Department can contain six times the σ of a process. In these cases, the conformance rate is equal to 99.9997% and the scrap rate is so low as to make it no longer useful to measure it in percentage terms. Therefore such a unit of measurement is replaced with a measure in "**parts per million**" (ppm). **Table 17.1** shows the relationship of the different levels of σ (i.e. the amplitude of the range of tolerance expressed as multiples of the sigma in the process) with conformance rates and ppm.

Six Sigma is associated with several improvement methodologies. The best-known and distinctive is the DMAIC, which is the process improvement approach discussed in Chapter 3. Additional tools used in organizations adopting Six Sigma (among others) are the following:

- Statistical process control;
- Design of experiments;
- Failure mode and effect analysis;
- Quality function deployment.

Figure 17.7 Density function of the Normal (Gaussian) distribution

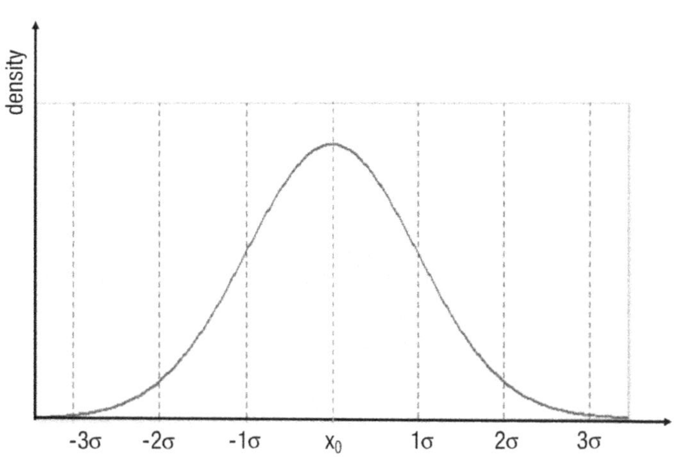

Table 17.1 **Relationship among sigma, conformance rate and ppm**

Sigma	Conformance rate	PPM
1	30.9%	690,000
2	69.2%	308,000
3	93.3%	66,800
4	99.4%	6,210
5	99.98%	320
6	99.9997%	3.4

Source: elaborated from Pande et al., 2000.

17.3.1 Statistical Process Control

Statistical Process Control (SPC) has its origins in the work of Shewhart, conducted at Bell Laboratories, and is aimed at the identification of **"out of control" production processes**, that are characterized by actual values beyond the range of tolerance (Shewhart, 1931). During the Second World War, these techniques were used by US industry to guarantee adequate quality standards for military equipment, and subsequently were adopted in the civilian field by numerous Japanese companies, which have based their success in part on an innovative approach to quality management.

Over the last few years, following the spread of Six Sigma, SPC has received renewed interest at an international level, due to the consistency of its objectives with the philosophy underlying Six Sigma, aimed at reducing process variability.

SPC, in particular, aims to ensure control of processes through the use of statistical tools, which provide useful information to plan suitable actions for reducing variability (Oakland, 2008). The deviation of actual values from nominal values, can in fact derive from several causes. In some cases, the inputs can determine different outcomes downstream of a transformation process. This is the case of some food products (such as wine or olive oil), in which the characteristics of the raw materials are influenced, among other things, by the environmental and climatic conditions of the place of collection, and as a consequence determine significantly different yields depending on the period, delivery batch, storage conditions, etc.

Further causes of variability are attributable to the equipment used in the production processes. Sometimes an inadequate maintenance program can produce different outcomes in the characteristics of the product. Think of the cylinders of rolling mills, which inevitably wear out over time and require interventions and adjustments, or even the dyeing processes of clothing, in which the lack of accuracy in cleaning the machine from one batch to the next can cause even small contaminations that compromise color stability. Obviously, machine malfunctions can also be a cause of high variability, especially if there is a drift in the values recorded in a certain time interval.

Sometimes, the variability of a process derives from the operator's skill and degree of experience, as well as from the conditions of the workplace. If the premises are poorly lit, not ergonomic, and not equipped with the materials and tools necessary for the operator to carry out the production activities, the degree of process stability will tend to decrease, as will the conformance rates.

High variability is a pathology that a company must try to eliminate, or at least reduce. For this purpose, SPC proposes a set of tools based on statistical methods, which allow for highlighting any critical issues and guiding the company in their removal.

Suppose that a sample of n pieces produced in a certain period of time have been subjected to quality control and that for each of them the measurement x of a certain characteristic has been detected (think, for example, of the weight of a jar of canned food, the thickness of a sheet, etc.). The values corresponding to this measure will be identified as follows:

$$x_1, x_2, x_3, \ldots, x_{n-1}, x_n$$

First, the **simple mean** of the detected values must be computed, which makes it possible to position the performance of the process with respect to the range of tolerance defined in the design phase. Secondly, to assess the variability of the process with respect to the mean, the **range** (i.e. the difference between the highest and lowest recorded values) and the **sigma** (or standard deviation) will be calculated. With respect to the single sample of n pieces tested, the three indicators are calculated as follows:

$$\bar{x} = \frac{\sum_{i=1}^{n} x_i}{n}$$

$$R = x_{max} - x_{min}$$

$$\sigma = \sqrt{\frac{\sum_{i=1}^{n} (x_i - \bar{x})^2}{n - 1}}$$

These elaborations are functional to the construction of the main instrument of SPC, namely the **control chart**. In this regard, see **Figure 17.8**.

The individual samples are shown on the horizontal axis according to the time sequence in which they are checked. The vertical axis, on the other hand, has the same unit of measurement as the observed feature (for example, millimeters if it is a thickness). On the vertical axis, the mean $\bar{\bar{X}}$ is reported (equal to 100 in the example in **Figure 17.7**), computed on the $\bar{x}$ values recorded in the k samples analyzed. It is obtained as follows:

$$\bar{\bar{X}} = \frac{\sum_{j=1}^{k} \bar{x}_j}{k}$$

A process is considered to be under control if all the $\bar{x}$ values fall within the interval equal to 6 times the sigma of the observed phenomenon, centered on the mean $\bar{\bar{X}}$.

Figure 17.8 Control chart: an example

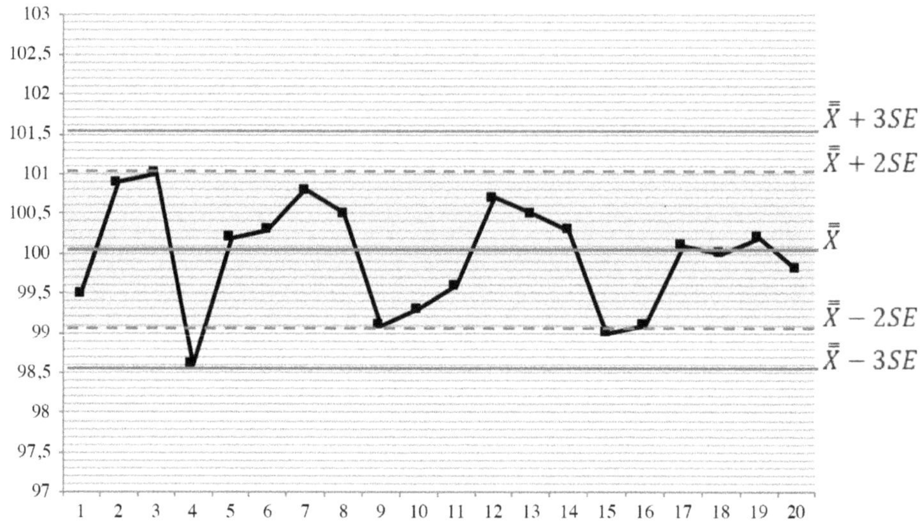

Since in this case the data under observation is not that relating to the individual pieces subjected to quality control, but to the averages detected in each sample, the sigma is replaced with the **standard error**, calculated as follows:

$$SE = \frac{\sigma}{\sqrt{n}}$$

If a process is stable, assuming that the population values are distributed according to a normal curve, more than 99% of the means $\bar{x}$ should fall within the range $\bar{\bar{X}} \pm 3SE$, with a strong concentration around $\bar{\bar{X}}$, and only one case out of forty would be placed in the area outside the range $\bar{\bar{X}} \pm 2SE$. The example shown in **Figure 17.7** assumes a mean equal to 100 and SE equal to 0.5. Therefore, the threshold values are 98.5 ($\bar{\bar{X}}$ – 3SE), 99 ($\bar{\bar{X}}$ – 2SE), 101 ($\bar{\bar{X}}$ + 2SE) and 101.5 ($\bar{\bar{X}}$ + 3SE), respectively. The data shown in **Figure 17.7** thus represents a situation of stability of the process. These considerations immediately highlight the information potential of a control chart. In fact, the probability that the means found on two consecutive samples will be positioned immediately beyond the interval is so low ($\frac{1}{40} \times \frac{1}{40} = \frac{1}{1600}$) as to make the event worthy of attention, because it is likely connected to the onset of a pathological situation.

More generally, the presence of trends (positive or negative) or values systematically higher or lower than the mean $\bar{\bar{X}}$, even if included in the interval $\bar{\bar{X}} \pm 3SE$, is clear evidence of situations that require an analysis aimed at ascertaining the possible presence of criticalities. The study of the level of stability of the process is also crucial for the calculation of the **process capability**, which expresses the ability of a process to produce objects that comply with the design specifications. To do this, suppose that the range of tolerance set for a given product feature is $x_0 \pm \varepsilon$, where x_0

is the nominal value. The width of this range is equal to 2ε. During the design phase, it is useful to first check whether the machinery used in production will be able to guarantee compliance with this tolerance, through a range of variability of the results that cannot be greater than 2ε. If the process is under control, with a standard deviation equal to σ, more than 99% of its products will fall within the range of 6σ. Therefore, it is possible to carry out the aforementioned check in the design stage by calculating the following process capability index (*Cp*):

$$Cp = \frac{2\varepsilon}{6\sigma}$$

In cases where this indicator is greater than 1, the range of tolerance is wider than the process variability range; therefore, at a first analysis, there should be no problems in the production phase. On the contrary, if the value obtained is less than 1, it is possible to state that too narrow margins are being defined in the design stage and that the process will inevitably be characterized by the systematic production of defective parts. In this regard, see **Figure 17.9**. This situation, also known as overspecification (Coman and Ronen, 2010), usually results in unsatisfactory economic results, due to the high cost of manufacturing products with excessively tight tolerances, for which the market does not recognize value.

In reality, even when Cp is greater than 1, some difficulties in meeting the specifications can occur as in **Figure 17.10**. The graph represents a situation where the process variability range (6σ) is narrower than the range of tolerance. However, the values reported on the process outputs are distributed according to a normal curve that has a mean equal to $\bar{x}$, which is greater than the nominal value x_0. Therefore, the process is more precise than required, but with values concentrated around a mean that is different from the target.

The use of the Cp indicator is therefore not sufficient to identify situations of this type. It is necessary to use a second indicator, **Cpk**, which considers not only the

Figure 17.9 Process capability

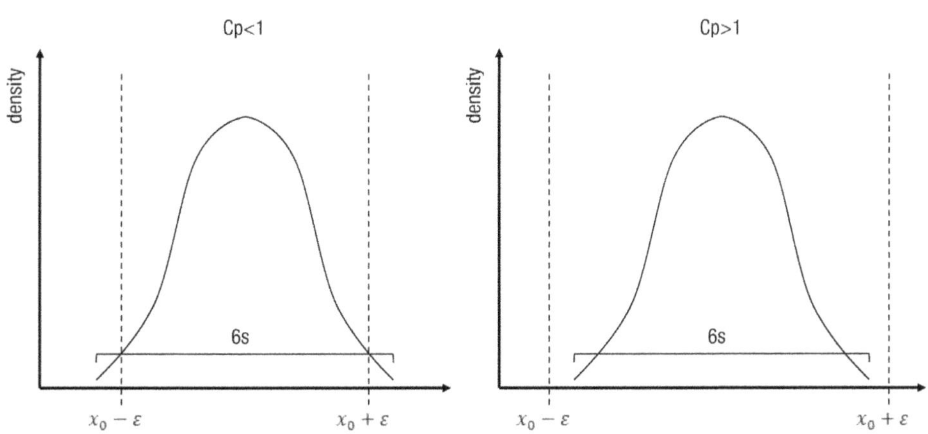

Figure 17.10 Process capability in non-centered processes

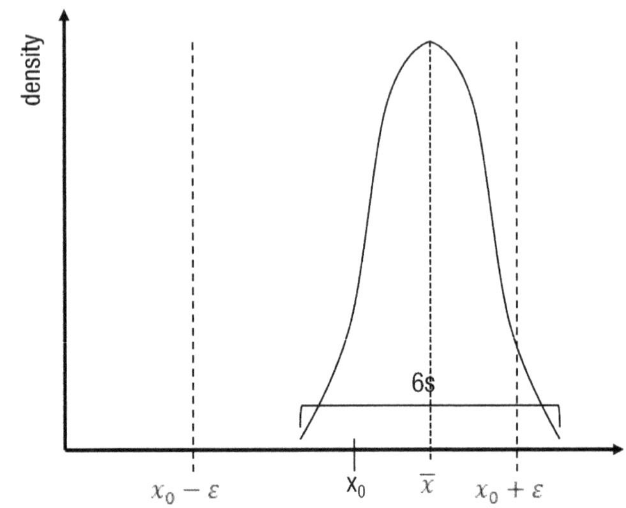

variability of the process, but also its centering. Cpk, in particular, is equal to the lowest of the two following indicators:

$$Cpk_u = \frac{(x_0 + \varepsilon) - \bar{x}}{3\sigma}$$

$$Cpk_1 = \frac{\bar{x} - (x_0 + \varepsilon)}{3\sigma}$$

A Cpk lower than 1 highlights the fact that the variability of the process and its centering are such as not to allow compliance with at least one of the two extremes of the range of tolerance. In this case, the solution to the problem may lie not only in the reduction of the variability of the process, but also in its different centering, which can sometimes be easily obtained only through the calibration of the system.

17.3.2 Design of Experiments

Design of Experiments (DOE) is a technique aimed at designing the experiments to be conducted on the product in order to identify the factors that affect variability more than others. DOE, as implemented today in business, can be traced back to the work of Genichi Taguchi, who during the 1990s developed an innovative approach to the concept of quality (Taguchi and Clausing, 1990). According to Taguchi, non-quality is the loss a product causes to society after being shipped to the client. Assume that for a given product feature, in the design stage, it has been decided that the nominal value should be x_0 and that the range of tolerance is $(x_0 - \varepsilon; x_0 + \varepsilon)$. In the event that a value outside this range is observed, the product must be discarded, reworked, or downgraded, incurring a cost equal to C. If, on the other hand, the observed value falls within the range, although not coinciding with the nominal value, according to the traditional logic called "**zero defects**," inspired by Crosby, the cost incurred by the company is zero. Consider in this regard **Figure 17.11**. Taguchi, on the other hand, claims that in reality a zero cost situation occurs only when the observed value is equal to x_0. Otherwise, the product will present an anomaly with respect to an ideal situation defined in the design phase, and therefore, will give rise to a cost, which will grow more than linearly as per the deviation between the observed value and x_0.

Taguchi suggests that the objective to be pursued is not the generic aim of conformance with specifications, but the centering of the mean value actually detected on the nominal value and the minimization of variability around it. For this purpose it is necessary to identify all of the factors that can affect variability, quantifying their impact.

DOE is a branch of applied statistics that deals with planning, conducting, analyzing, and interpreting controlled tests aimed at identifying:

- the factors that most influence a result (in this case, a variable describing a product's feature);
- the weight of each factor;
- the mathematical model that expresses the relationship between factors and result.

Compared to other alternative approaches, in which the effect of any factor is tested individually, in DOE experiments are carried out by varying all factors simultaneously. Operationally, it is necessary to:

- identify the result variable;
- define all the factors that can affect the result variable;
- define a range of variability for each factor (for example, low, medium, or high);
- identify all possible combinations of factors with their relative levels;
- perform as many experiments as there are combinations identified in the previous point;
- analyze the results to develop the mathematical model that describes the relationship between the factors and results.

Figure 17.11 **Quality Loss Function (QLF) vs. Zero Defect (ZD)**

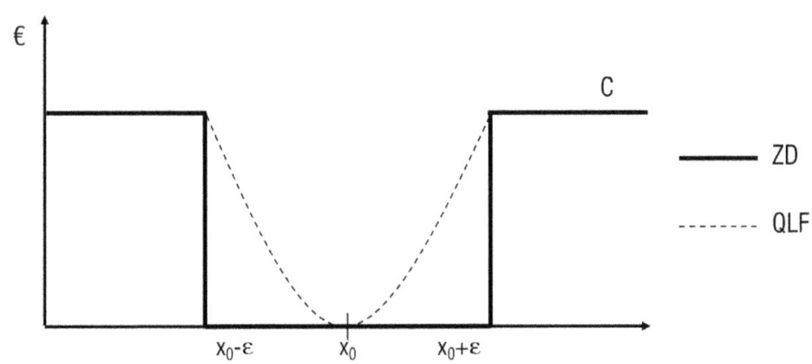

17.3.3 Failure Mode and Effect Analysis

Failure Mode and Effect Analysis (FMEA) is a methodology suitable for developing a list of improvement priorities on which the organization needs to focus. Let us suppose that some product features have been identified with problems of high defects, and that at the same time, the company has a limited budget to devote to improvement projects. In this case it is necessary to use a tool that allows for identification of those characteristics that, more than others, require corrective action.

FMEA is based on the calculation of an indicator called the Risk Priority Number (RPN), obtained as the product of three factors:

- **Occurrence** (O): the probability that a defect (i.e. a *failure mode*) occurs;
- **Severity** (S): the severity of the effects that a certain defect can cause to people or objects;
- **Detection** (D): the probability of not detecting the defect.

After evaluating these factors on a homogeneous scale from 1 (very low) to 10 (very high), RPN is calculated as follows:

$$RPN = O \times S \times D$$

The RPN value associated with each product feature allows for defining an order of priority and thus allocating the available financial resources to the defects characterized by the highest RPN.

17.3.4 Quality Function Deployment

Quality Function Deployment (QFD) is a tool used to ensure alignment between the customer's needs (Voice of the Customer – VOC) and the technical characteristics of the product. The latter, as a rule, are the result of an activity carried out by the Design Department staff, which often has little knowledge of the market and of customer needs, and therefore is likely to develop inadequate designs. This problem can be overcome by using tools and methodologies, such as QFD, which require coordination with the Marketing Department, that on the contrary, knows the customer's needs and is able to guide the Design Department towards technical choices capable of guaranteeing a greater probability of market success.

Operationally, QFD involves the use of the so-called **house of quality**, represented in **Figure 17.12**.

The two key elements of the house of quality are customer needs (**whats**) and technical requirements (**hows**). The former represent the bundle of the relevant product attributes from the customer's perspective (called the Voice of the Customer) and are usually identified by the Marketing Department, which for each of them expresses an evaluation of their degree of importance from the customer's perspective (for

Figure 17.12 House of quality

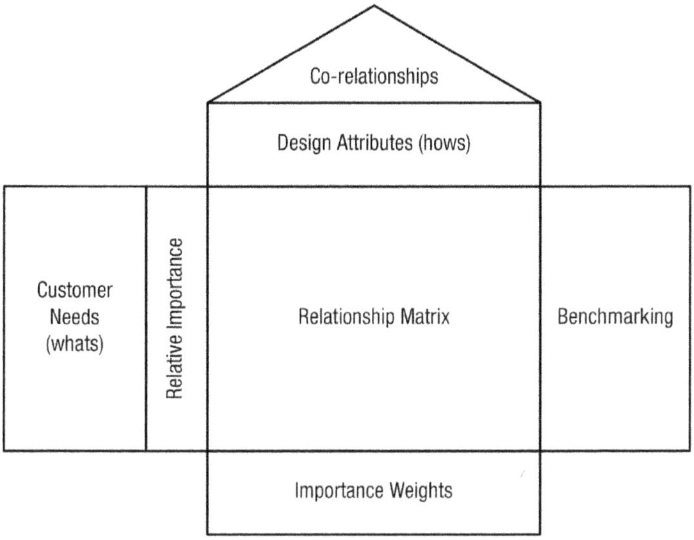

example, a scale of 1 – very low, to 5 – very high). The latter, on the other hand, refer to the technical requirements of the product, which are set by the Design Department. The upper part of the house of quality (**co-relationships**) expresses the intensity of the correlation between pairs of technical requirements.

The crucial element of this tool is the **relationship matrix**, which indicates the correlations between pairs of customer needs and technical requirements. For example, it can be decided that the correlation can be expressed on a scale from 1 (weak), to 3 (medium), or 9 (strong).

Using these pieces of information, it is possible to assign a degree of importance to each technical requirement, considering the customer needs that it can affect and the magnitude of its influence. This measure helps in identifying those technical requirements that deserve more improvement actions.

The house of quality also allows for carrying out benchmarking with competitors, or even with earlier versions of the product. To this end, surveys or focus groups with customers can be used to compare the degree of satisfaction that the market expresses, for each product alternative, in relation to the individual attributes reported in the house of quality (see **Box 17.2**).

Box 17.2 House of equality: an example

Assume that the methodology of the house of quality is adopted to redesign a sushi bar. Building on the evidence of a focus group conducted with some target customers, it has become apparent that their main needs toward this service are the following:

- Quality of the meals, which has scored 5 on a 1 (not important at all) to 5 (very important) scale;
- Cleanliness, scored 4;
- Speed of the service, scored 2;
- Personnel kindness, scored 4.

The design attributes that the company can manage are the following:

- Personnel;
- Equipment;
- Restaurant's layout;
- Quality of the raw materials.

Working with the consulting company in charge of the market research, the management team gains a deeper understanding about the correlation between customer's needs and design attributes, measured on a 1 (weak), 2 (medium), 3 (strong) scale. (See **Figure 17.13**).

Figure 17.13 **Relationship matrix**

		Personnel	Equipment	Layout	Quality of the raw materials
Quality of the meals	5	3	1	–	9
Cleanliness	4	1	–	–	–
Speed of service	2	9	3	3	–
Kindness	4	9	–	–	–

Based on this data, it is possible to compute the level of importance of each design attribute. The procedure requires that, for each combination need-attribute, the score associated with each need must be multiplied by the correlation with the attribute under analysis. The importance of the improvement of a given attribute is measured as the summation of all values previously obtained. Consider the following computation related to "Personnel":

$$Importance_{(Personnel)} = 5 \times 3 + 4 \times 1 + 2 \times 9 + 4 \times 9 = 73$$

Based on the values of the importance of each attribute, it is possible to rank them identifying the most valuable improvement opportunities that deserve a wider budget (See **Figure 17.14**).

Figure 17.14 Importance weights

		Personnel	Equipment	Layout	Quality of the raw materials
Quality of the meals	5	3	1	–	9
Cleanliness	4	1	–	–	–
Speed of service	2	9	3	3	–
Kindness	4	9	–	–	–
		73	11	6	45
		54.1%	8.1%	4.4%	33.3%

18 Physical Distribution & Supply Chain Management

by *Giuseppe Stabilini*

18.1 Physical distribution: balancing service level and logistics cost

Physical distribution logistics deals with all design choices and the management of physical and information flows downstream of the production processes. The main activities concern both **storage** in central, distribution, and local warehouses, and the **transportation and delivery** of products to the customer.

The goal of distribution logistics can be summarized as the ability to optimize the **two main types of performance**:

- **the ability to meet demand in time, with volumes and quality** aligned with the customer's needs, identified as the "service level" provided, namely the quantitative performance of the set of the most important and appropriate measures in the context of the business analyzed;
- **the ability to minimize total logistical costs** through high operational efficiency, represented as a "cost" to serve the market, resulting from the sum of all internal or supply costs that the company incurs to make its products available to customers.

The two elements of performance have a clear **trade-off** between them and require a proper balance in line with the business strategy, customer expectations, and the characteristics of the competitive environment.

The measurement of the service level requires a careful analysis of market requests, often different for each specific segment, to identify the relevant service performance (for example, speed, punctuality, flexibility to volumes or product mix, etc.) and the level of performance expected by customers. For a specific discussion on service measures, please refer to Chapter 4.

Total logistics costs are an expression of all the economic values associated with the distribution activities, whether they are operated directly with their own resources and assets, or indirectly through supply partners. Regardless of the choice on how to manage the activities, the costs can be classified into:

- production costs;
- primary transport cost (trunking);
- secondary transport cost (local delivery);
- facility cost (warehouses and depots);
- inventory holding costs.

Figure 18.1 **Total cost of a distribution network**

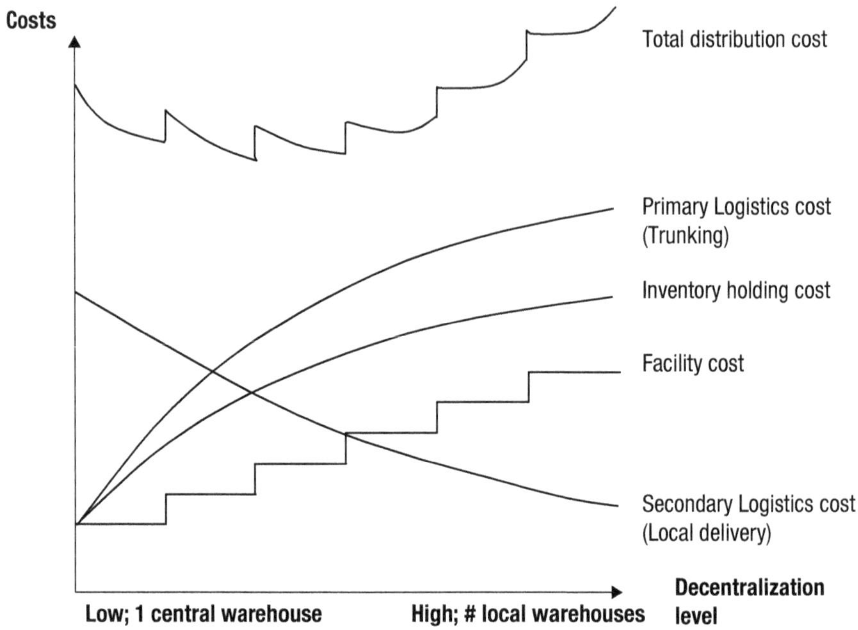

The design choices of the distribution network seek to **balance the trade-off between the different cost items**. In fact, by increasing the number of production sites, the positive effect of economies of scale in production is reduced, and in general, the local delivery costs of distribution are reduced.

For example, with the decentralization of warehouses to bring them closes to demand and customers (and the consequent increase of storage points), the trunking costs of transportation, and warehousing and inventory holding costs will increase. Conversely, local delivery costs will decrease (**Figure 18.1**).

With reference to the inventory holding cost, the increase in the number of warehouses determines an increase in the average stock held in the network. The increase of stock can be determined by employing **Maister's square root law**, namely: average inventory increases proportionally to the square root of the number of locations in which inventory is held.

The formula:

$$Inv\_B = Inv\_A \times \sqrt{\frac{\#\_B}{\#\_A}}$$

where:
Inv_B = future inventory
Inv_A = current inventory
$\#\_B$ = number of future warehouses
$\#\_A$ = number of current warehouses.

Figure 18.2 **Service level and logistics costs: company strategies**

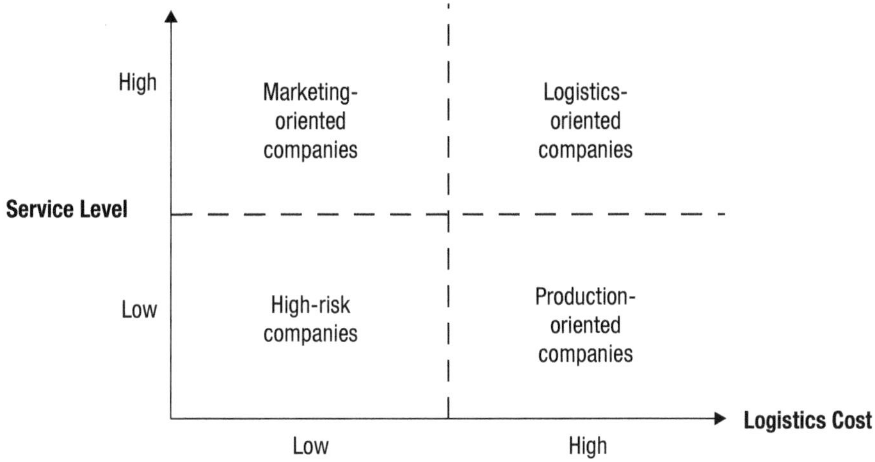

Source: adapted from Ferrozzi et al., 1987.

For example, if the company currently has an inventory of 5,000 units and wants to upgrade from 2 to 8 distribution warehouses, the estimated future inventory level will be equal to

$$5{,}000\ units \times \sqrt{\frac{8}{2}} = 10{,}000\ units$$

From the standpoint of the design of the distribution network, there are sectors in which the sensitivity of customers to the service level is significant, forcing companies to favor the satisfaction of demand even at the expense of the level of operational efficiency (**Marketing-oriented Companies**) (Figure 18.2). Conversely, in contexts of limited profitability and strong attention to prices on the part of customers, distribution logistics will be focused on operational efficiency, even at the expense of the service provided (**Production-oriented Companies**).

The best companies, thanks to a careful analysis of customer requirements and to appropriate logistics network design, manage to **combine both types of performance**, achieving both excellent satisfaction of demand and an efficient cost structure, ensuring high profitability.

18.2 The characteristics of the context and the design of the distribution network

The **design of the distribution network** must be based on a careful analysis of the factors that influence the choices. Although the relationship between single elements, cost structure, and impact on the service is clear, the design must consider that the multiplicity of contextual characteristics (products, customers, geographies)

Figure 18.3 **The factors that influence logistics distribution structure choices**

Product • life cycle of the product • variety of characteristics within the family • product range within the family • product type: functional or innovative • handling characteristics • shelf life • physical size and weight • value • Product Value Density (PVD) **Market** • dispersion and localization of demand • demand level (throughput) • volatility of demand • expected service level	**Source/Production** • economies of scale • demand fulfillment approach • production flexibility • production lead times **Infrastructure and logistics services** • existing infrastructure • transport mode availability • custom/duties/trade areas • legislation

does not allow for the development of ad hoc solutions for each individual distribution flow.

The balance between complexity and segmentation versus simplification and standardization will lead towards **hybrid solutions** that allow for excellent control of flows and of the service provided with a sustainable level of logistics cost.

There are **several factors** that influence the design and management choices of distribution logistics, which can be summarized in the following characteristics (**Figure 18.3**):

- **the product distributed**;
- **the reference markets**;
- **the production system**;
- **infrastructure and logistics services**.

First, we must consider **the characteristics of the distributed product.** The phase of the life cycle affects the need to offer immediate product availability or the company's ability to have reliable forecasts in terms of volumes and mix of products required. The variety of products offered, expressed by the number of configurable variants within the family, determines the need to have a greater or lesser level of flexibility in both production and logistics, generally negatively affecting the level of stock. The shelf life determines the possibility of storing the product, and if the shelf life is limited, the need for rapid transport to the customer. Finally, the characteristics of the product, as well as the weight/volumetric dimension (weight and volume) together with the unit value, determine the level of centralization or decentralization of the warehouses and the need for full efficiency in transport operations.

Secondly, there are **the characteristics of the markets**. The level of dispersion of demand or its location in certain places (such as metropolitan areas) allow for greater or lesser efficiency in designing the production and distribution network. The volume of demand and its volatility influence not only the level of stocks held in the various logistics hubs, but also the level of centralization. Finally, the expected service level determines the need to operate with a local logistics network capable of responding immediately to market demands.

Thirdly, there are **the characteristics of the production system.** Production systems that operate in a "make to stock" approach will allow for different choices of stock polarization at the center or near the customer, due to other factors linked to the market or to the characteristics of the product. Conversely, "assemble to order" or "make to order" production approaches will require logistics choices that operate with distribution connected to customer orders. The possible economies of scale, as well as the high fixed costs of the production system, encourage "focused factories" and global distribution, with high distribution costs more than offset by savings in production.

Finally, we must consider **the characteristics of the infrastructure and logistics services.** Logistics distribution must necessarily consider both the existing infrastructure in the various geographies as well as the storage and transport options available, in terms of speed, capacity, and supervision of qualitative aspects. The presence of trade areas, with different levels of taxation, trade barriers, and volatile exchanges rates between currencies, influences the design of internal or cross-area distribution flows, requiring the design of logistic chains able to optimize both the costs and the lead time.

By limiting this discussion to the **macro design choices** of the distribution network and focusing on purely logistical aspects, and thus excluding the related channel choices governed by marketing evaluations, we can focus the analysis on **two critical decisions**:

1. the choices regarding **polarization of distribution logistics** with the level of centralization or decentralization of inventories and the related flows among warehouses and markets;
2. the choices regarding **speculation or postponement from the logistical and manufacturing standpoint**, defining the position of the decoupling point in the production and distribution network.

18.3 Polarization choices of logistics distribution

The **polarization choices** of the logistics distribution network are mainly influenced by **two variables** (**Figure 18.4**):

- **Product Value Density (PVD)**;
- **the volatility of demand and the reliability of forecasts**.

The **PVD** expresses the value with respect to the most important logistics measure, expressed by the weight or volume per unit. It is calculated as the ratio between the

Figure 18.4 The determinants of logistics polarization choices

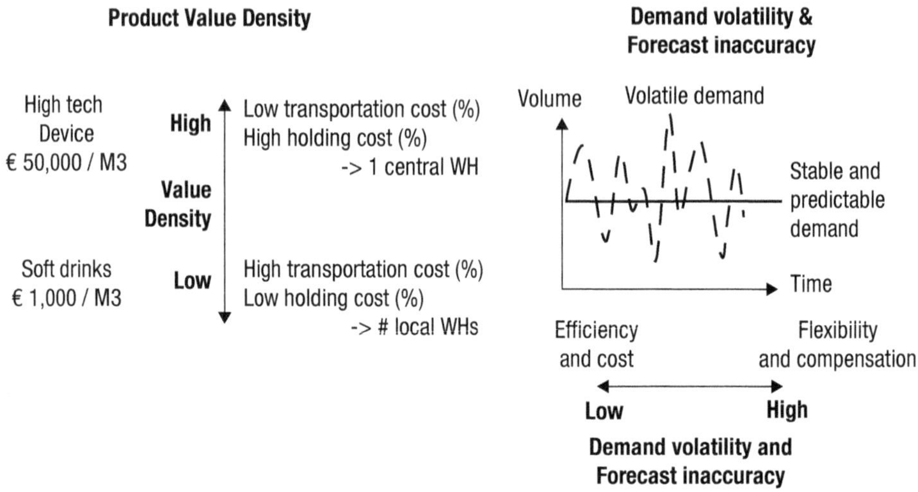

economic value of the single product unit (typically expressed by the sales price in the context of finished product distribution) and the **relevant measure** from the standpoint of logistics costs. This measure makes it possible to relativize logistic costs, that are typically independent of the value of the product but linked to its weight or size.

In situations of **high density of value**, the logistics system will necessarily have to optimize the management of stock in the warehouse, minimizing the value of inventories, which causes a high cost of stock maintenance from both a physical point of view (spaces, management, risks) and a financial one (investments in net working capital). Transport costs, on the other hand, will have a more limited impact on total costs thanks to the high intrinsic value of the product, which makes them marginal.

Contrariwise, in situations of **low value density**, the logistics system will be designed to minimize transport costs. As these are linked to the weight/volumetric dimension, the lower value of the same will make transport more expensive in relative terms. Inventories, on the other hand, will have less importance in terms of net working capital, thus reducing the need for stock optimization. Inventories can be allocated in decentralized warehouses without negatively affecting the total costs.

The **volatility of demand and the reliability of forecasts** influence the ability to plan distribution flows.

Demand with high volatility and unexpected trends of growth or decline determine an inability to be able to plan allocations of products to specific consumption regions or increase the risks of a general imbalance between the products available in stock and the levels of demand. In this context, the search for efficiency often results in numerous compensations that are necessary to manage unexpected events.

Demand with reduced volatility and stable over time in terms of volumes consumed allows for greater certainty in distribution flows, both in terms of transport

Figure 18.5 The polarization systems of distribution logistics

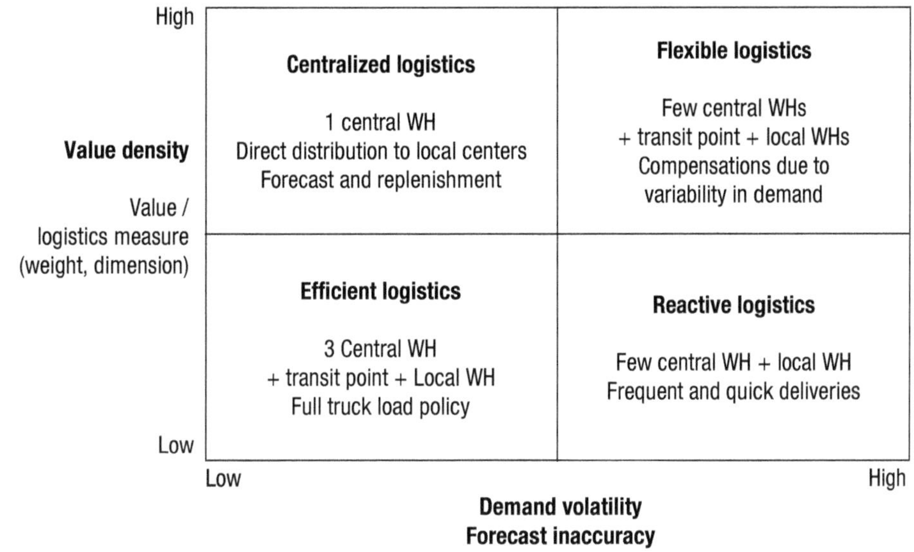

Source: adapted from Lovell et al., 2015.

and stock levels. In this context, there is the possibility of seeking maximum operational efficiency.

The combination of these two variables determines the choice regarding **polarization of the logistics distribution system** (Figure 18.5).

In the case of **demand with low volatility and high reliability of the forecast** there are two alternatives:

- **centralized logistics**. The high value density of the product pushes towards a concentration of stock in a few central warehouses, with consequent direct delivery to customers. Any decentralized stock needed to ensure immediate availability of the product is managed from the center and planned based on reliable forecasting. This structure is adopted by a large multinational in the automotive sector. The spare parts are stored in a central warehouse on the continent, with direct shipments to dealers and retailers based on forecasting and replenishment algorithms.
- **efficient logistics**. The low density of value points towards the search for efficiency in the management of transport, with a logistics network divided into various levels: central warehouses, regional transit points, and local warehouses. Each transport operation is planned to optimize costs, for example by requiring saturation of the vehicles (Full Truck Load, FTL). A multinational company that produces and distributes paper fabrics for hygienic and domestic use operates with 3 central warehouses, 20 transit points, and 80 peripheral warehouses. All transport operations from the center to the markets are planned at FTL, using vehicles sized specifically for the volumes of the section.

In the event of **high demand volatility and low reliability of forecasts**, there are two alternatives:

- **flexible logistics**. The high value density of the product points towards a centralization of stock. However, the volatility dynamics of demand force continuous balancing of stocks aimed at compensating for the imbalances, with transfers both between warehouses at the same level of the network and between center and local markets. These dynamics increase the transport costs, which are however offset by the efficiency achieved in the centralization of the stocks themselves. The main pharmaceutical companies operate this way, with very high value products and small unit sizes.
- **reactive logistics**. The low density of the value of the product points towards a decentralization of stock close to demand. The decentralized logistics network supports efficient transport from the center to the periphery. However, volatile demand forces continuous compensation, with product transfers from one logistics node to another to maintain an adequate service level. This model is used by fashion companies that govern networks of proprietary sales points, where product availability is allocated.

18.4 Logistics and manufacturing speculation or postponement

The design choices of the distribution network also include the design of the **decoupling point (DCP)** from a logistical and production point of view.

The DCP represents an inventory buffer capable of balancing market demands with logistics and production needs (Chapter 6 and 11). The activities downstream of the DCP are directly influenced by demand and customer orders, both in terms of volumes and characteristics of the products requested. The activities upstream of the DCP, on the other hand, are managed with a planning logic based on forecasts, satisfying any need or efficiency from the standpoint of the logistics and production system.

In the distribution field, the choice of **DCP positioning** can be analyzed from two points of view:

- **Logistics**, i.e. positioning of the DCP with respect to the geography of the markets. "**Logistics speculation**" is present when the DCP is positioned close to the customer/demand, and thus with stocks already in local markets and close to customers, available for local demand. "**Logistics postponement**" occurs when the DCP is centrally located with respect to the markets served, far away from customers, and with a few warehouses operating as distribution centers, sometimes close to the production site.
- **Manufacturing**, i.e. positioning of the DCP with respect to operational activities. "**Manufacturing speculation**" is present when the manufacturing process of the product is completed before collecting the orders, with a "make to stock"

approach, planning the activities based on the forecast. "**Manufacturing postponement**" is when the operations process reaches the realization of a work in progress or semi-finished product, keeping them in the buffer of the DCP, and the subsequent activities of completion of the product (for example, assembly) are performed based on a customer order, as happens in the "assembly to order" approach.

Placing the choices in a matrix, four alternatives of speculation/postponement of the distribution network are identified (**Figure 18.6**).

The **full speculation** strategy consists of the creation of the product in a "make to stock" logic with subsequent decentralization of inventories along the distribution network. This logic allows for excellent levels of efficiency from both a production and logistics standpoint, with a high level of customer service in terms of availability and speed. On the other hand, investments in stocks and primary transport costs are high, with the multiplication of storage points in the distribution network. Furthermore, production on forecast does not allow for any customization of the product with respect to any customer needs.

As in the first solution presented above, the strategy of **logistic postponement** consists of the application of a "make to stock" logic, but the product is maintained in a centralized distribution center, to be delivered to customers only after the collection of orders. This solution allows for controlling the production efficiency levels and reducing the investments in stock thanks to the centralization of the inventory. Secondary transport costs are high due to the need to tie distribution to customer orders. Furthermore, customer service is medium-low in relation to the lead time of transfer from the warehouse to the single local customer.

The **production postponement** strategy implies the construction of several production units in the vicinity of the customers, with the realization of the product after the customer's request. The "operations" can cover both simple activities, like packaging or labeling, or complex ones, such as assembly or configuration, up to real production. The warehouse holds the "work in progress" made on plans based on forecasts, often including semi-finished products, but also finished products to be completed. Consistent with the degrees of freedom defined, the production activities allow for the customization of the product and are flexible with respect to the customer's demands. However, the production scale of the individual "mini-factories" does not make it possible to leverage the economies of scale typical of centralized production structures with aggregate volumes. Furthermore, the high costs of primary transport are offset by the efficiency of the planning of replenishments to the peripheral warehouses and by the low costs of secondary transport. The investment in stock is medium-high due to the high number of storage points, only partially limited by the value of the materials themselves, that represent WIP as mentioned. The service level is medium- high, depending on the speed of the operations and logistic proximity.

The last strategy is represented by **full postponement**. Both production and distribution activities are carried out only downstream of the receipt of the customer's request. The structure provides for a central production plant that produces the

Figure 18.6 **The choices of speculation vs. postponement of the distribution network**

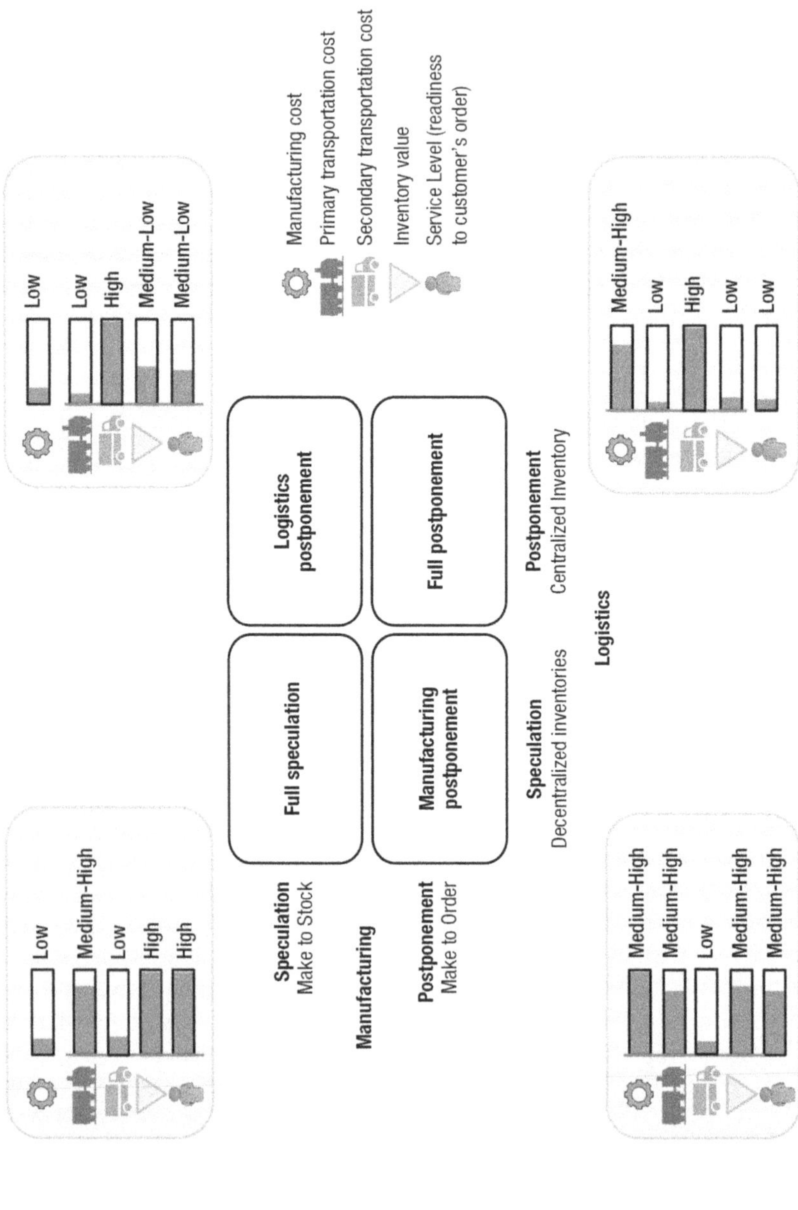

Source: adapted from Pagh and Cooper, 1998.

products requested by customers in an "assemble to order" or "make to order" logic, starting from a centralized WIP warehouse. The advantages are found in the ability to exploit economies of scale in the production of semi-finished products and to allow the customer to customize the product. The negative aspects of this structure are identifiable in the lengthening of the response lead times, due to both the production to order and the subsequent direct distribution to the customer, with a medium-low service level again in terms of response speed and high secondary transport costs.

Each structure highlights how there are positive elements as well as many negative points. The search for an optimal distribution structure is affected both by the volatility of the elements to be considered, be they the costs or characteristics of the demand, and by the co-presence in the portfolio of products with different characteristics. This uncertainty, combined with the need not to excessively multiply the distribution solutions, often leads to the definition of a distribution structure that represents an overall sub-optimal.

By focusing the attention on the two opposite structure choices, that is, full speculation on one side and full postponement on the other, it is possible to identify **different factors** that guide the choice (**Figure 18.7**).

Figure 18.7 The relevant factors in postponement/speculation choices

Some important P/S-decision determinants			Generic P/S-strategies			
			The full speculation strategy	The manufacturing postponement strategy	The logistics postponement strategy	The full postponement strategy
Product	Life cycle	Stage	Introduction	Growth	Maturation	Mat./Decline
		Volume	Low/Medium	←	→	Low/Medium
		Cost/service strategy	Service	←	→	Cost
	Product characteristics	Product type	Standard	←	→	Customized
		Product range	Narrow	←	→	Wide
	Value	Value profile	Initial stages	←	→	Final stages
		Monetary density	Low	Low	High	High
Market and demand		Relative delivery time	Short	←	→	Long
		Delivery frequency	High	←	→	Medium/Low
		Uncertainty of demand	Low	←	→	High
Manufacturing and logistics		Economies of scale	Large	Small	Large	Small
		Special capabilities	Yes	No	Yes	No

Source: adapted from Pagh & Cooper, 1998.

The full speculation strategy is typical of products with high demand volumes, offered as standard and with few variants in the range, that is in any event predefined by the company. They are characterized by a non-volatile and frequent demand, which requires satisfaction immediately or in a very short time. Production processes enjoy large economies of scale and the value entered is high in the initial stages of the processes themselves. **The full postponement strategy**, on the other hand, is typical of products with reduced demand volumes, offered with a fair level of customization that allows for a very wide range able to satisfy the different requests of customers. As a result, demand is on average volatile, with a reduced order frequency and a willingness of the customer to wait between the time of order and delivery, an aspect that allows the execution of customization activities. The production processes are more flexible and are not characterized by significant economies of scale, with high value entered in the final stages of the processes themselves.

The intermediate strategies, of only productive or only logistic postponement, combine and mix the indicated factors.

18.5 Supply Chain Management

Supply Chain Management (SCM) was born as a response to the need for integration of production chains. The interconnection of the various players in the supply chain has significant impacts both on the ability of the same to respond to the needs of final demand, and on the cost-effectiveness and efficiency of the logistic production processes (Romano and Danese, 2010).

SCM thus gathers all of the tools to execute this integration, adopting a "**system**" approach for the management of both physical flows (supplier-producer-customer) and information flows (data and volumes of trade and sales relating to products and services). The two flows, the physical one from upstream to downstream, and information, from downstream to upstream, constitute the backbone of the entire chain, facilitating the sharing of decisions and coordination (**Figure 18.8**). This orientation overcomes the natural dichotomy between customer and supplier, with the traditional pursuit of maximizing value for just one's business.

SCM proposes the optimization of the system with global approaches and solutions, which clashes with the widespread need to preserve the performance of individual players. Furthermore, management approaches of tactical optimization of single actors generate considerable uncertainty in the system, with inability to plan activities and difficulties in making forecasts about the volumes of demand.

Supply chain management thus imposes an integrated vision at two different levels:

- **At the company level**, the supply chain goes beyond the divisions typical of a functional approach, concentrating and coordinating all decisions regarding the management of demand and markets, distribution, production and purchasing from the supply markets. The functions remain from an organizational point of view but report directly to the supply chain manager. In many cases, this inte-

Figure 18.8 Supply Chain management. Material and information flows

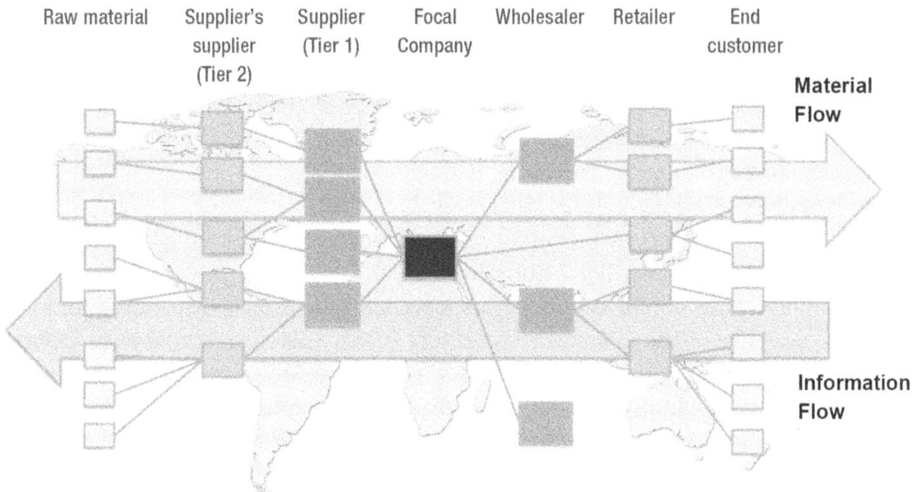

gration also involves innovation and development activities, which determine the performance or constraints of subsequent operations through their choices.

- **At the supply chain level**, the supply chain integrates all the players, starting from the upstream suppliers, passing through all the logistical and production phases (warehouses, transport, factories) to downstream customers, such as distribution channels and end consumers. Integration entails a greater involvement of all the actors than the simple relationships of exchange of products and services, with transparency, sharing of decisions and the achievement of common results.

In SCM, **information is the key element** to be able to build and manage the entire supply chain. The availability of data, its level of detail, and the sharing of the data with all the actors makes it possible to:

- reduce the uncertainty and volatility of phenomena;
- improve the ability of upstream players (suppliers, in general) to make reliable forecasts and to intercept market changes;
- coordinate the activities of the various actors;
- support downstream actors in intercepting constraints and criticalities that may emerge upstream in supply chains.

18.6 The Bullwhip Effect and Supply Chain Management

The performance of the supply chain depends on the quality of the decisions of all the actors involved along the supply chain. An action at one point in the chain can positively or negatively affect all other points.

Although the relationship among all the players is clear from the logistical point of view (supplier-customer relationships), from the economic/financial point of view and from the ability to satisfy the final customers, the orientation of any single company remains focused on the optimization of its own performances.

This attitude has repercussions on the entire supply chain, with evident negative impacts that can manifest themselves both in an increase in costs and a reduction in margins, and with a criticality in the management of stocks, which can lead to an excess of the same or to numerous stock-outs.

The phenomenon that most substantially represents this need for coordination between the actors is the so-called "**Bullwhip Effect**."

In a supply chain with different actors (retailer, wholesaler, distributor, and factory) the **volatility of the orders** that a subject issues towards the upstream supplier (upstream) is greater than the same orders received by the downstream customer. This phenomenon is amplified in **frequency and size** along the supply chain, from downstream to upstream, due to the number of players. It thus appears that an average constant demand from end customers can become, in the upstream part of the supply chain, very volatile in the orders received from the industrial supplier (**Figure 18.9**).

This phenomenon impacts the stocks held by the various players, with fluctuations of the same gradually becoming greater, causing periods of extra stock contrasting with periods of stock-out and backlog (orders from downstream customers not yet satisfied).

The **inefficiencies** caused by this volatility can be seen in the high costs of maintaining stock or associated with stock-outs, as well as the increases in production costs due to the need to continuously adjust the production plan and production capacity to the levels of demand.

Figure 18.10 shows the results, common to several experiments, of a Beer Game simulation conducted with students of the SDA Bocconi MBA program. **The Beer Game** is a business game developed by the Massachusetts Institute of Technology (MIT) that allows you to simulate the effects on orders and stocks of an uncooperative supply chain composed of four actors. The constraints of non-collaboration and the lack of transparency of information always produce very negative results, with the development of the bullwhip effect even in the case of flat and constant end customer demand.

The Bullwhip Effect was identified in the 1990s by **Procter & Gamble**. Although end user consumption of diapers was essentially stable, volumes for their own factories were highly volatile. This volatility, therefore, was not linked to a phenomenon related to consumption or final demand, but was the effect of a series of decisions adopted by all the players along the supply chain.

The Bullwhip Effect is defined as the increase in the volatility of the demand expressed by the volumes ordered from the upstream actor compared to the volatility of the demand expressed by the orders received by the downstream actor.

Since the average demand is a constant throughout the supply chain, i.e. in the period considered (for example a year), and all the actors order a volume of goods equal

Figure 18.9 **The Bullwhip Effect along the supply chain**

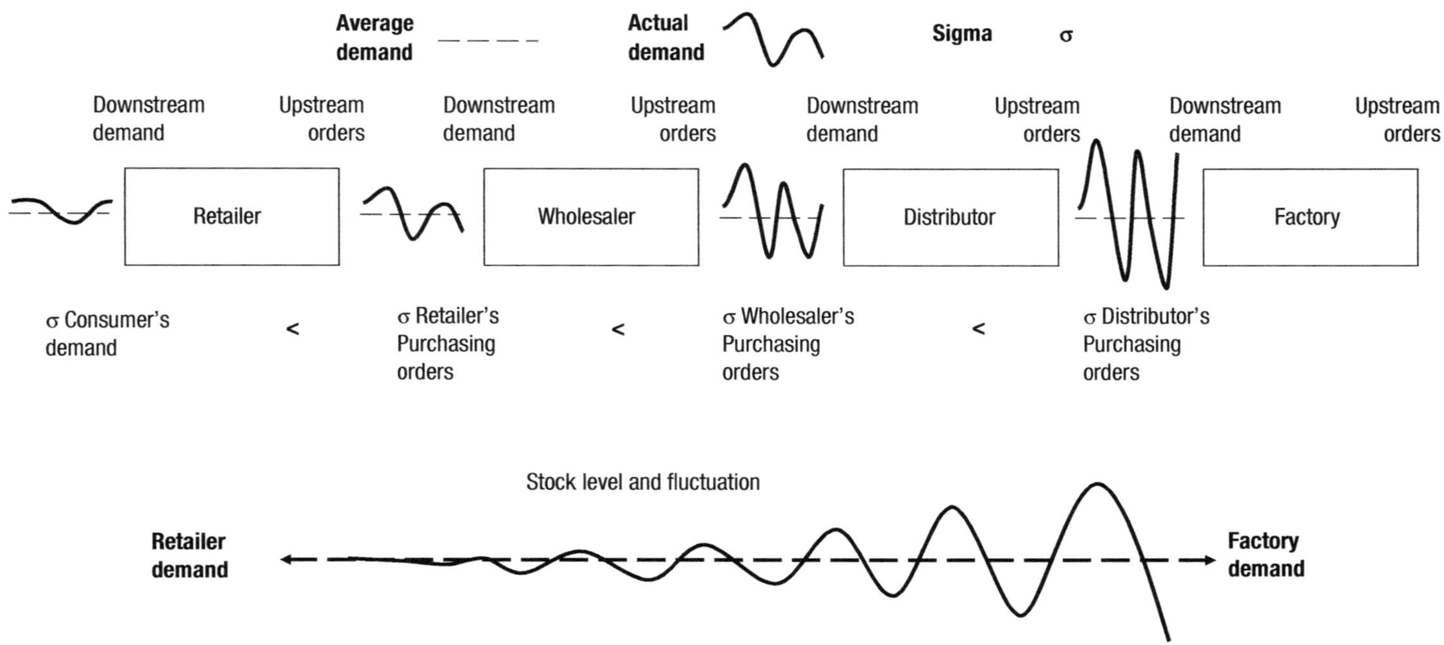

Figure 18.10 Example of a Bullwhip Effect: results of the Beer Game simulation

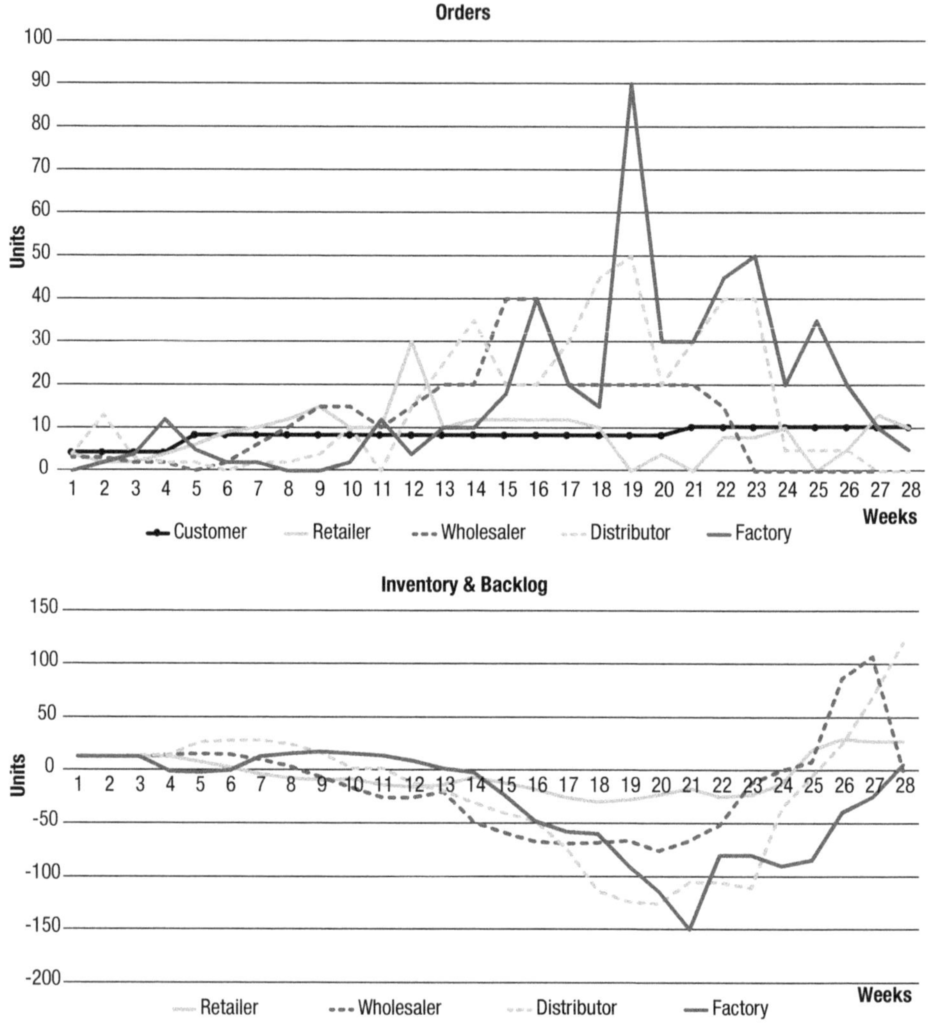

to the demand in order not to increase or decrease the initial stock, the volatility is produced by the choices related to inventory and replenishment orders.

For this reason, the volatility of demand is measured with the **coefficient of variation**, that is the ratio between the standard deviation (sigma) and the mean, which is precisely constant regardless of the position of the subject along the supply chain.

It follows that if in a supply chain the retailer adopts a replenishment policy that always replenishes the warehouse with the same quantity consumed by its customer, the Bullwhip Effect would be neutralized.

However, there are several factors that push each actor to depart from the indicated policy:

- The **frequency of orders** may not be the same as that with which the demand is presented. In fact, a daily demand managed with a monthly replenishment logic, or a longer frequency dictated by the need to schedule deliveries, produces orders that cover long consumption period. In this case, there will be high volume orders interspersed with no demand periods, causing higher induced volatility.
- **Long lead times of ordering**, production and distribution processes. A long lead time often causes the reduction of the order frequency, with the consequences indicated in the previous point.
- **Batch constraints**, i.e. the need to order a volume different than demand in order to optimize both production assets and make purchases or transport efficient. In this case as well, the excess portion of the order will lead to periods with reduced volumes, increasing volatility.
- **Speculative orders**, or volumes dictated by savings opportunities generated by the volatility of the prices of traded goods. In this case also, orders will suffer increases or decreases in volumes due to the behavior of the decision-maker and his willingness to exploit certain advantages.
- **Risk aversion**, resulting in an increase in volumes in order and stocks to avoid product stock-out and consequent disservices. The increase, often also induced by inaccurate forecasts, causes a distortion in the demand generated upstream.
- **Volume discounts and promotions** granted by certain supply chain players. These policies result in an accumulation of stocks in order to exploit favorable price conditions. However, shocks are created in demand, with strong increases in limited time windows, followed by periods of essentially zero demand, precisely due to the consumption of the accumulated stock.
- **Overreacting orders**. In particular situations, an increase in demand, albeit limited to certain periods, can generate the false expectation of a shift in demand towards higher volumes than in the past. This situation results in a higher order volume which generates an increase in volatility. If risk aversion is associated with this situation, there could be a tendency to order volumes greater than those consumed, generating a chain reaction in all the upstream subjects.
- **Risk of dissatisfaction with your orders**. If in the supply chain there is a bottleneck represented by a subject with a finite capacity that cannot be increased, or unavailability of raw materials and components beyond a certain quantity, the situation could lead the subjects downstream to order higher volumes to allow them to satisfy your demand. In fact, if the next order risks not being satisfied in full or suffering delays in delivery, this leads to anticipating the volumes in order. Although in many cases the production capacity or the availability of raw materials and components is balanced with respect to the final demand, this factor causes unjustified volatility.

All of these factors are amplified along the supply chain due to the number of players. The further you move from downstream to upstream, the more the combination of these factors and the decisions associated with them contribute to the generation of the Bullwhip Effect.

To summarize, the main cause of the fluctuations in demand is the **lack of visibility of the real demand from the final consumer** (end user) on the part of the actors in the supply chain. This cause, associated with the factors indicated, generates a volatility of demand that can have negative effects both on the costs of the various subjects and on the service capacity of the supply chain itself.

To eliminate the causes and minimize the distortion effect, various **solutions** can be introduced:

- **Share information** with all players in the supply chain. Giving the upstream actors full visibility of real demand allows for better forecasting and planning of their activities. Furthermore, sharing commercial actions (discounts, promotions, etc.) also prevents an induced change in demand from being identified as a structural trend, misleading suppliers. By the same principle, upstream parties should share information on product capacity and availability with downstream parties, in order to avoid creating stock-outs along the supply chain, to the detriment of customer satisfaction.
- **Synchronize orders** along the supply chain. The simplest way to avoid the development of the Bullwhip Effect is to order from your supplier an amount equivalent to your customer's demand (order-up-to policy). By maintaining an order volume and frequency that replicates customer demand, the upstream subject will enjoy demand that both on average and in variability is exactly equal to the final demand.
- **Reduce lead times**. The ability not to generate a Bullwhip Effect depends on the frequency and speed with which orders and deliveries operate along the supply chain. By reducing the lead times of distribution, production, purchase, and the various administrative cycles, it becomes possible to synchronize all the players and to make the flows more consistent with the dynamics of demand. Technology integration solutions, such as EDI (Electronic Data Interchange) help reduce possible distortion effects.
- **Eliminate order size constraints**. The order volume can be affected by various physical or economic conditions that make it different from demand, causing signs of turbulence in the supply chain and generating the Bullwhip Effect. The elimination of these constraints allows you to align orders with actual consumption. Constraints can be of different kinds. Logistic constraints, such as the saturation of a transport volume (pallet or container), attributable to the FTL, Full-Truck-Load, logic. Economic constraints, such as a lot of volumes higher than real demand that minimizes the total costs of production or purchase, considering the production/order setup costs and the costs of stock maintenance.
- **Production smoothing**. In conditions of high volatility or a peak of demand concentrated in particular periods of the year, sharing information would be sufficient to mitigate the negative effects of the volatility on the supply chain actors. This is especially true for those who have critical issues related to the low flexibility of production capacity. In these cases, a solution is represented by production smoothing. Such an action can be taken both by the subject facing volatile

demand and by the one suffering from production rigidity. Smoothing takes the form of the creation of a buffer stock (inventory) between the volatile demand and the production phase. This stock, comparable to a decoupling point, allows the production phase to plan capacity in a leveled way, saturating the assets and not suffering from the levels associated with peak demand, which would be satisfied by the pre-established stock.

- **Reduce the actors along the supply chain**. The Bullwhip Effect spreads along the supply chain due to the number of actors involved. The redesign of the supply chain with a lower number of subjects, that is, skipping levels and building direct relationships (for example between factory and retailer) contributes to reducing the phenomenon. Not only does the upstream subject acquire better visibility of demand, but the distorting effects deriving from the decisions of the intermediate subjects are avoided. This solution seeks to use the efficiency obtained from a lower volatility of demand to offset an increase in distribution costs given by the lower logistic efficiency.

These solutions are applied in various collaborative practices, illustrated in the next section. The foundation of the practice concerns first of all the ability to establish rules and standardized communication systems between the parties, ensuring the application of the principles that reduce the emergence of a Bullwhip Effect. Secondly, the practice constitutes a collaborative platform that supports synergic dialogue.

18.7 Collaborative practices to integrate the supply chain

The need to integrate the actors along the SC is embodied in a concrete sharing of management logics that go beyond the boundaries of individual companies and create an inter-company infrastructure.

The transition **from a transactional relationship to a collaborative one** requires a different management of some key dimensions:

- **Management of information**, with a greater level of transparency of the data, and sharing it with all members of the SC, upstream or downstream. This information is often represented by sales or consumption data downstream, but it can also be the progress of a production process or the residual capacity of a production asset.
- **Decision-making delegation**, with the transition from decision-making carried out independently by the various subjects, to a process of discussion with numerous actors involved (with consequent sharing of information and resources), up to a full delegation to a single subject. This last step requires not only full visibility of all relevant variables for the subjects, but also rules and principles that allow for overseeing the optimization of performance for all the companies involved.
- **Sharing of resources and technologies**, pooling people, skills and technologies in order to build a platform to support integration. This sharing obviously also involves economic investments, a performance measurement system, and the cre-

ation of principles of equitable attribution of the responsibility, risks, and benefits deriving from it.

The convergence of objectives and the need to face complex phenomena at the supply chain level thus requires the introduction of collaborative logics that allow for "common" actions between the various parties involved.

The planned or achieved **integration can be classified** according to a scheme that proposes three different levels (**Figure 18.11**):

- **Communication**. This first level describes a relationship between companies where there is full sharing of relevant information along the supply chain. This information can be sales data, stock levels, order book or forecasts, as well as certain actions (for example, promotions) carried out by certain subjects. The shared information is sufficient to allow not only for all companies to have full visibility, but to trigger a series of decision-making processes (such as the adjustment of production levels or the preparation of stocks in advance to meet the new demand) without sharing of the same. The ability to better observe the scenario thanks to the wealth of information has a positive impact on decisions and the management of critical issues.
- **Coordination**. The second level envisages, in addition to the communication of relevant information, coordination between the actors. This happens thanks to the definition of the phases and times of sharing, with the subjects involved, formal appointments, structuring of the information path, and communication methods. Furthermore, the process definition also includes feedback loops, with new information/decisions communicated back into the flow, which in turn, transmit a new signal to the supply chain. Structured exchange allows for better decisions in a coordinated and predefined context.

Figure 18.11 The levels of integration along the supply chain

Level of integration	Goal	Key points
Communication	Information sharing	• Continuous and structured sharing of data mainly related to inventory levels, orders, and promotions • Complete autonomy in decision-making Non-binding conditions associated with the exchange • of information
Coordination	Coordination of operating processes	• Coordination through workflow • Coordination of managerial practices applied to the following processes: forecasting, planning, and replenishment. Inter-company business process re-engineering
Collaboration	Strategic processes integration	• Establishment and sharing of common goals through the integration of strategic processes (innovation, new product development, etc.) • Collaboration in workflow and decision-making • Joint decisions and collaborative teams • Long-term goals and Key Performance Indicators (KPI)

- **Cooperation**. The maximum level of integration is achieved with inter-organizational relationships based not only on operational aspects (information and processes) but involving both the strategic perspective and that of evolution and innovation of the relationship itself. At this level, there is often strong integration between companies, including the evaluation of plans for the co-localization of resources (for example, shared planning teams) and sharing not only technological investments but also complex reporting systems on the performance of the entire supply chain.

Although the "collaboration" level represents the maximum expression of integration, requiring the sharing of resources and investments, simple "communication" between the parties allows for better visibility of the variables that influence decisions, contributing at zero cost to improving the performance of the entire supply chain.

The principles of integration are applied in a series of established practices. These practices may provide for different levels of integration depending on the operational rules and organizational choices of the actors involved, increasing their effectiveness and guiding investments. Furthermore, although their application originates in particular contexts, they have now found application in all phases of the supply chain, both downstream (relationships with distributors or retailers) and upstream (B2B industrial contexts and relationships with suppliers).

18.8 Collaborative planning

Collaborative planning is a practice of integration of different actors along the supply chain which involves the sharing by downstream companies of a portfolio of orders and timed prospective needs.

For each item code involved in this program, the supplier can consult the volumes and delivery dates requested on a daily, weekly, or monthly basis according to the characteristics of the business. The temporally closest information is frozen, while the subsequent weeks or months have a randomness level given by a (positive or negative) percentage of variability. The plan is shared in a "rolling" mode, whereby each new period is frozen for the next period and the plan is updated for future months, to always have visibility on the same time horizon.

Given this sharing, the supplier verifies the **feasibility of the delivery plan** and confirms it or proposes changes, both in volumes and in delivery dates (**Figure 18.12**).

Visibility on a medium-term horizon and the possibility to modify that timeframe in terms of volumes and dates allows the **supplier** to ensure a better level of service, especially in terms of punctuality and completeness in deliveries (OTIF, On Time In Full). Furthermore, the information collected allows the supplier to better plan both the production assets and the purchasing activities of its supply chain, sometimes proposing advances or shifts in deliveries that are included in the production plans of the client company.

Figure 18.12 The Collaborative Planning Program

SKU	Family	Material	Type of data	Month 4				Month 5				Month 6	Month 7	Month 8	Month 9	Month 10	Month 11
				Week 13	Week 14	Week 15	Week 16	Week 17	Week 18	Week 19	Week 20						
111	A	Steel	Scheduled	300	300	300	400	300	300	300	400	1,200	1,500	1,000	1,500	2.,500	2,000
			Confirmed	300	300	300	400	...	...	...	...	...	...	...	...	...	...
222	A	Steel	Scheduled	100	200	200	300	100	0	0	400	1,200	2,000	3,500	3,200	3,000	3,500
			Confirmed	100	200	200	300	...	...	...	...	...	...	...	...	...	...
333	B	Steel	Scheduled	200	100	100	100	200	50	200	100	800	600	500	500	400	300
			Confirmed	200	100	100	100	...	...	...	...	...	...	...	...	...	...

Weeks 13–16	Weeks 17–20	Months 6–8	Months 9–11
← Frozen →	← Forecast +/- 20% →	← Forecast +/- 50% →	← Non-binding →

On the other hand, the **client company** has the possibility to collaboratively verify the feasibility of the plan, modifying it according to the availability of its supply chain or reallocating the needs to the available suppliers, thus ensuring its full feasibility and compliance with multiple constraints, both internal and external. The variability declared on the forecasts also makes it possible to maintain a good degree of flexibility with respect to unexpected changes in demand volumes.

This collaborative approach is often embodied both in a **strong integration** between management systems, to make all dialogue efficient, such as the sharing and entering of scheduled and confirmed data, and in significant and direct collaboration between the planning teams of the various companies. involved (sometimes even co-localization of people), in order to support the plan itself with rapid and effective communications.

18.9 Vendor Managed Inventory

Vendor Managed Inventory (VMI) is a method of inventory management that requires the supplier to take responsibility for managing the stock of materials on behalf of the purchasing company. Control, stock replenishment orders, and material delivery times are managed and defined independently by the supplier in compliance with a series of performance parameters shared at the start of the procedure.

Operationally, VMI is based on the definition of **two management parameters**:

- **Minimum stock (S_min)**, which is the minimum stock level in the warehouse. The parameter is defined with a logic similar to the dimensioning of the Re-Order Point (see Chapter 10).
- **Maximum stock (S_max)**, which is the maximum stock level in the warehouse. The parameter is defined as S_min + Economic supply lot, however it is often influenced by logistical evaluations (maximum space to be occupied; optimal logistic lot).

The supplier must ensure a stock level between these two parameters, continuously observing the level of stock and stock discharges in order to organize deliveries independently. The fluctuation of the inventory between S_min and S_max makes it possible both to provide the correct level of service to the consumption processes, and not to exceed capital and material immobilization.

From a logistical point of view, the contract is thus based on the definition of these two parameters, with naturally divergent positions of the parties (**Figure 18.13**):

- The **supplier** is inclined to keep the level of S_min high on average, ensuring the service with the stock (of which it does not bear the maintenance costs) and to increase the S_max in order to have room for freedom in the delivery volumes, both to optimize production processes upstream.
- The **customer**, on the other hand, is inclined to reduce the S_min, without how-

Figure 18.13 **VMI program management parameters and relative positions of the customer and supplier**

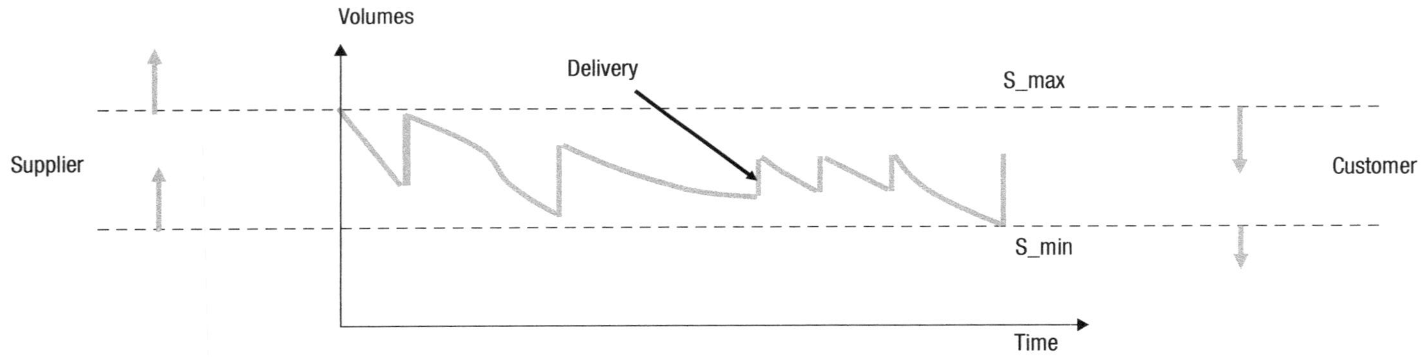

Supplier	Parameter	Customer
The highest possible Goal: increase delivery flexibility (volume/lead time) in order to satisfy delivery efficiency (full truck load) and/or production needs (economic production quantity)	**S_max**	Close to the S_min Goal: reduce the space occupied in the warehouse and reduce the inventory/net working capital
The highest possible Goal: reduce the risk of stock-out	**S_min**	The lowest possible (but coherent with the service level defined) Goal: reduce the net working capital and assure a coherent service level

ever exceeding the level of service, and to keep the S_max close to the S_min, reducing the investment in stock (but consequently increasing the delivery frequency by the supplier).

The **negotiation process** must therefore balance the needs of both parties, with the aim of establishing a collaborative practice that can benefit both the customer and the supplier. The advantages generated by past practice are often precisely what leads to maintaining the exchange prices of materials at historical levels (prior to the introduction of the practice) or market levels.

Operationally, the supplier must have **full visibility** of the following information (**Figure 18.14**):

- stock levels of the material;
- daily stock withdrawals;
- changes and/or replacement of the material due to quality problems or innovation in the technical design;
- forecast of future consumption, or order backlog if the customer is a retailer or operates in "make/assemble to order," or production schedule if the customer operates in "make to stock."

On the basis of the information available and according to the levels of S_min and S_max, the supplier can autonomously organize deliveries, varying both the frequency and the volumes from time to time, trying to optimize both the transport and production processes.

The **advantages for the supplier** are attributable to:

- **better knowledge of the dynamics of demand**, thanks to the full visibility of downstream consumption (stock withdrawals), resulting in a better ability to plan production by seeking economies both at the manufacturing and distribution levels;
- **greater flexibility**, with the possibility of defining volumes and delivery frequency, not subject to the constraints determined by a classic customer order process (volumes and defined delivery date);
- **development of a long-term relationship**, with the certainty of continuity of supply during the life of the contract, and thanks to services in line with the level of service, building a relationship of trust with the customer.

The **advantages for the client company** can be defined as:

- **reduction of administrative and management costs** due to the complete delegation of control, planning, and ordering activities, with a consequent focusing of resources on activities with greater added value or critical for the business;
- **improvement of warehouse rotation** thanks to the definition of appropriate parameters of S_min and S_max;
- **improvement of the service level** with reduction of stock-out phenomena.

Figure 18.14 Information shared in a VMI program

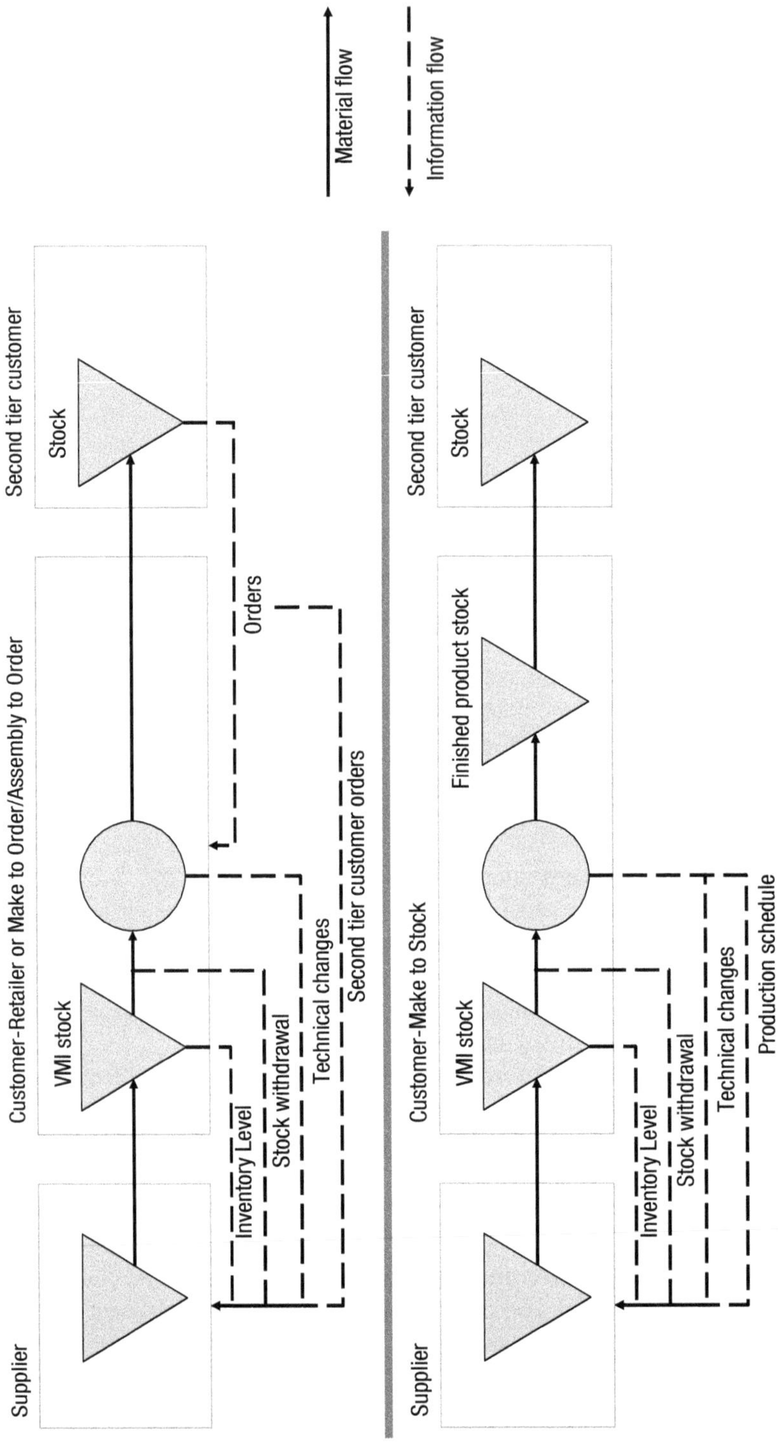

Despite the benefits, VMI management may bring some **risks**. For the supplier, there could be an excessive increase in management costs associated with control and delivery activities or an increase in the costs of maintaining its stocks in order to meet the service levels required by practice. Furthermore, full responsibility for the level of service and stock-out could also lead to penalties.

For the customer, there could be a **lock-in effect** that limits the chances of substitution of the supplier or the excessive disclosure of sensitive information (actual consumption), which could lead to an increase in the supplier's bargaining power.

18.10 Consignment Stock

Consignment Stock (CS) is an evolution of VMI in which the material in storage at the customer's warehouse remains the property of the supplier until the time of sampling and use of the same. Only at this moment does the formal transfer of ownership take place, and the supplier can proceed with the issue of the invoice. Obviously, since the stock is owned by the supplier, the client company does not have to bear the burden of the investment in stock and therefore will be more willing to maintain a high level of stock in the warehouse (considering any physical space constraints, though). Typically, the practice is associated with very short payment terms.

This solution is typically used for **high-value materials** that have **unstable or occasional demand**. The permanence of the material at the supplier's property allows for reducing its impact on NWC (the material is physically available to the customer, but has not yet been paid for), while maintaining a high level of service. The practice is also used in the case of **materials with limited shelf life**. The supplier is required to replace the products that are approaching their expiration date, always allowing the customer to have usable stock without suffering from obsolescence.

As for the VMI method, the S_min and S_max levels that govern inventory management are defined operationally by the parties. However, the transfer of ownership for the actual consumption of the good produces different negotiation behavior (Figure 18.15). The supplier has an interest in lowering the S_min level, having to bear the financial burden of maintaining the stock, while respecting the expected service level. Conversely, as the customer does not have pressure on the NWC, it could request an increase. The S_max, on the other hand, sees trading dynamics like those illustrated in the VMI regime.

This practice is used effectively in several areas:

- **In healthcare contexts** it is often used for rapidly obsolescent medical devices or drugs. The customer recognizes an increased price of the asset (including the value of the service offered) and therefore reduces both the administrative and management burdens, and the costs of inventory depreciation in cases of obsolescence.
- **In industrial contexts**, with CS it is possible to obtain a greater integration of the supply chain, giving full visibility to the supplier regarding consumption and allowing it to recover efficiency in the management of replenishments. The customer reduces the impact on the NWC and gets an improved level of service.

Figure 18.15 CS program management parameters and relative position of the customer and supplier

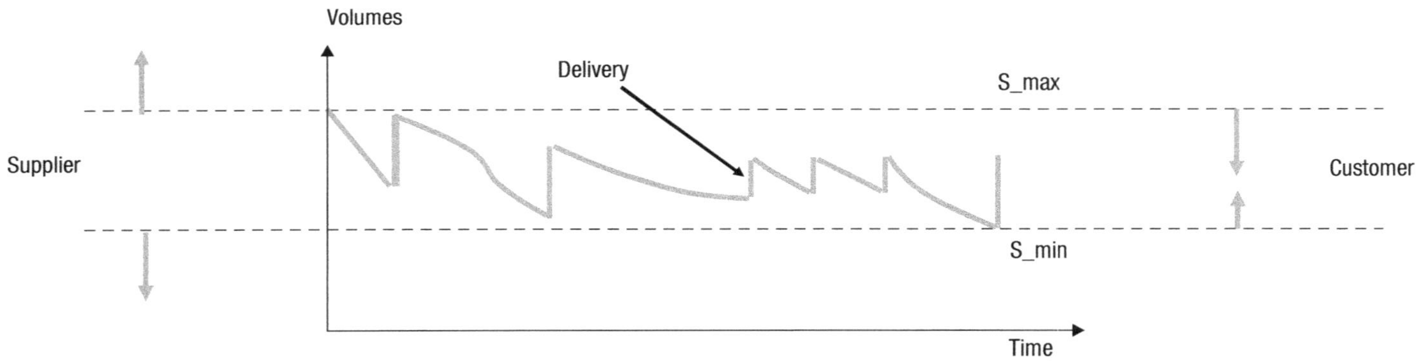

Supplier	Parameter	Customer
The highest possible Goal: increase delivery flexibility (volume/lead time) in order to satisfy delivery efficiency (full truck load) and/or production needs (economic production quantity)	**S_max**	Close to the S_min Goal: reduce the maximum space occupied
The lowest possible, coherent with the service level defined Goal: reduce NWC in its customer warehouse	**S_min**	The highest possible, according to the volume/space occupied Goal: increase the service level. No problem associated with NWC, as the inventory is owned by the supplier

- **In wholesale or retail contexts**, the practice is used by distribution chains by allocating to the supplier (the producer of the goods) both the costs of managing the logistics processes, and the assortment decisions and financial costs of the stock present in the store. This solution is particularly interesting in contexts of high product margins for an industrial company, where the value of the warehouse in store changes radically if valued at cost (cost of goods sold) at the "sell-in" or "sell-out" price." At the same time, the retailer reduces the **risk profile** (higher level of service, lower risk of unsold goods), with a consequent willingness to sustain a higher purchase price than in a traditional trading situation.

18.11 Continuous Replenishment Program

A Continuous Replenishment Program (CRP) is a practice typically adopted in distribution chains. The practice deals with managing the exchanges between the Distribution Centers of the retail chains and the industrial supplier of products. **The supplier collects stock data** on a daily basis, unloads, outstanding orders and forecasts, and proposes to the customer a **delivery plan (items, quantity, delivery date)**, with the aim of remaining within the agreed stock limits and having frequent deliveries for a large number of items (produced or distributed by the same company) in low volumes. This plan, developed by the supplier, makes it possible to optimize various logistical aspects, including the saturation level of the means of transport.

The delivery plan is proposed to the customer who, having verified the contents and approved the times and volumes, issues an executive order (very often without any changes). Against this, the supplier organizes and carries out the shipment (**Figure 18.16**).

The benefits for the customer are similar to those of the VMI, namely:

- improvement of **warehouse rotation**,
- improvement of the **service level**,
- reduction of **stock-outs**,
- reduction of **administrative and management burdens**.

Figure 18.16 The CRP program and the information/decision-making flow

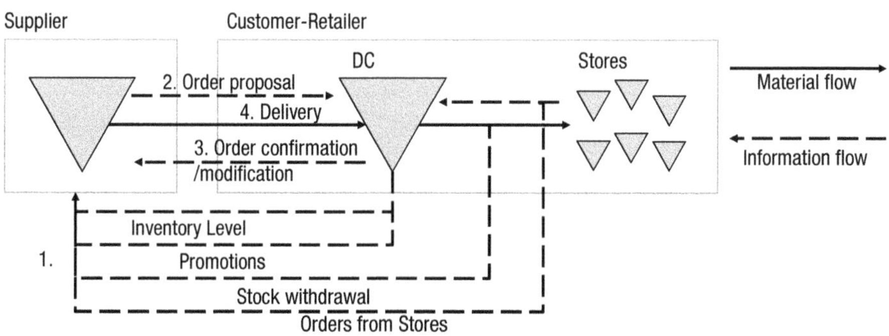

Thanks to this program, the supplier:

- gains **greater visibility** of downstream consumption,
- **optimizes deliveries** and **transport costs**, respecting constraints,
- improve its **service performance**.

The practice is typically used in retail supply chains, where the volatility of demand is not only linked to the dynamics of consumption, but also to any actions by the chain (such as promotions and advertising campaigns). Consequently, the supplier must obtain visibility of the marketing plan and other activities, including confidential ones, planned for future periods.

18.12 Collaborative Planning Forecasting and Replenishment

Collaborative Planning Forecasting and Replenishment (CPFR) constitutes the synthesis of the collaborative practices illustrated above, fully integrating the actors along the supply chain. Typically used in wholesale/retail areas, the CPFR consists of four distinct processes:

- **Strategy and Planning**. The companies involved define the methods and rules of collaboration, developing a joint business plan for the management of individual products. In this phase, the objectives are shared and the KPIs of the collaboration are formalized, such as the strategies for managing the life cycle of the products, their introduction or replacement.
- **Forecasting and demand management**. Each company collaborates in defining the demand forecast, sharing information and developing processing systems for the data used and checking their reliability. As in collaborative planning, the forecast plan entails a short and medium-term vision, with a frozen period; one allowing a certain flexibility and one that is informative regarding the more distant weeks.
- **Replenishment execution**. The forecasts relating to the frozen period become delivery plans, considering the production and transport constraints, the warehouse stocks present in the channels, and the lead times. The document continues to be developed, respecting the constraints of all the actors involved and optimizing the processes in compliance with the objectives.
- **Analysis**. Based on the recorded performance, the companies collaborate in order to improve processes, data collection and analysis systems, and forecast algorithms.

CPFR represents the **maximum level of collaboration along the supply chain**, bringing into practice decisions often taken independently, such as product innovation plans or sales promotion policies. The practice is thus based both on a series of **operational processes** and **structured information exchanges**, typical of any inte-

Figure 18.17 The CPFR program and shared processes

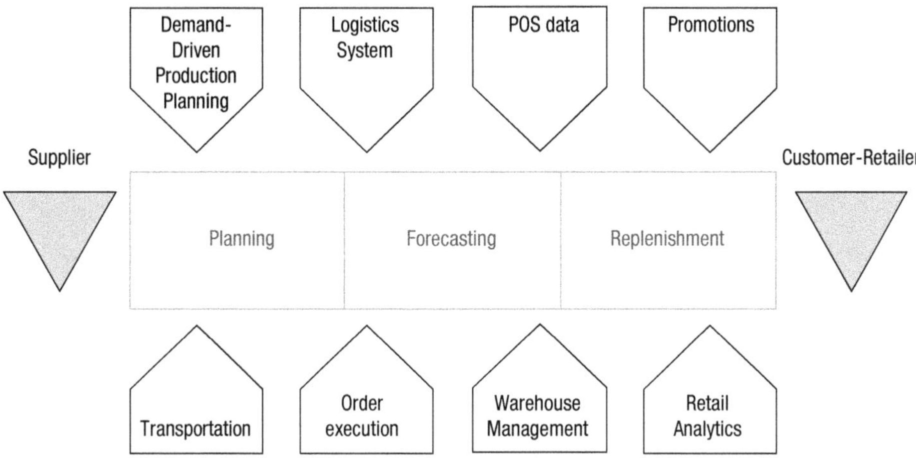

gration practice, and on an **organizational platform** for sharing resources and decisions (**Figure 18.17**).

The benefits of the practice fall on both the customer and the supplier, with an **increase in the level of service**, and a **reduction in stock levels and stock-outs**. The practice is successfully used in distribution/retail areas, where pricing and promotion policies or information asymmetry make it difficult for the industrial manufacturer to follow the demand generated by the wholesaler/retailer.

19 Information Systems and Operations Management

by *Raffaele Secchi*

19.1 Introduction

Most operations managers are aware that fostering effective integration among the company's processes is a precondition to respond to customer requests in a flexible and swift way. In the 1990s, that integration took on a strictly internal perspective, leading to a focus on the coordination of the activities carried out within the company. In the following years, it emerged clearly that concentrating only on internal activities was not enough. That integration needed to be extended also to encompass all the players outside the boundaries of the company.

This **dual integration** – internal and external – requires adequate information flows that allows for monitoring all the activities linked to end-to-end processes, from the generation of the customer request to the final delivery of the product/service.

Operations, as well as other company department, must dedicate particular attention to the selection and implementation of information systems because, as pointed out, their support is essential to ensure an effective and efficient management of business processes. This chapter will illustrate the main software packages that can provide direct and indirect support to operations management.

19.2 Enterprise Resource Planning systems

ERP systems consist of a set of software modules, typically developed by a single vendor, operating on a single database purposely conceived to assure data consistency. They essentially represent the center of gravity of the entire business information system since their integrated solution makes it possible to:

- cover the main information needs of the "core" business processes (administration, accounting and finance, purchasing and sales, production and logistics, human resources);
- manage the master data related to all the resources of the processes (articles, bills of materials, work cycles, production centers, clients, suppliers, etc.);
- manage the data flows underlying the business transactions (warehouse movements, client orders, production orders, supplier orders, etc.).

The broad spread of ERP systems reported in the last few years, both in large corporations and in SMEs, is due to their capability of overcoming two principal limitations of traditional information systems. First, **the lack of uniformity** deriving from the incremental and often customized development of applicative modules covering specific functional needs. Second, the **duplication of the same data**, that is stored in different databases and updated by different actors, causing potential misalignments and inconsistencies.

The new paradigm proposed by ERP systems is based on some distinctive characteristics:

- information univocity, guaranteed by a single database that provides synchronous information to all company processes. With ERP systems, the integration is native, in the sense that the database is unique and common to all the applications that insist on it, the procedures are strictly connected to each other, and the updating of the archives is managed in a unitary and centralized way;
- extension and modularity, that guarantees a wide functional coverage through integrated but independent modules;
- distributed architecture, that emphasizes the spread of the information between the organizational units, according to a horizontal (process) logic, rather than a vertical (department/silos) one.

ERP systems can provide a specific contribution to operations processes from at least two perspectives:

1. direct support for operations management. In particular, we refer to production planning and scheduling, stock management, logistics and warehouse management, shipping, quality management, and maintenance. These activities require purposely designed applications to make their execution efficient and fast. In addition, it is increasingly important for operators to be able to consult reliable and timely reports that allow them to analyze the performance of the activities carried out, and to receive feedbacks to potential risks/criticalities to which the company is exposed (for example, a potential delay in delivery that could give rise to the payment of a significant penalty or the launch of a batch on a production line that should instead be stopped for a preventive maintenance campaign);
2. synchronization of information flows between operations processes and other business processes. Transactions originating from the execution of operations activities must immediately generate the necessary alignments within sales, logistics, administrative, and financial processes. Only in this way it is possible to provide customers with precise and rapid responses that concern the final delivery of the product as well as other information that deeply influence the service level. For example, focusing in particular on B2B contexts, think about the need for the customer care unit to monitor the status of a customer order, or to provide information on the availability of a product or the potential delivery times in the event that some activities have yet to be completed.

In addition to these specific contributions, the introduction of an ERP system usually allows the management to **rethink and reengineer some processes**.

- The main ERP solutions provide process modeling tools to speed up the mapping of the existing physical and information flows. In addition, there is the possibility of accessing a codified set of process models that have been formalized in previous implementation projects. Some of these archives have up to hundreds of different models: some are standard templates representing typical business processes that can be replicated independently from the reference context; others are process models that are valid within specific business areas and that, therefore, can be usefully employed only when there is a reasonable analogy with the reference company. The possibility of accessing these archives, on the one hand, allows companies to limit the commitment, sometimes very expensive, connected to the modeling of their own processes; on the other hand, if there are no high specificities linked to business logic or to the entrepreneurial formula, it allows companies to replicate successful solutions that others have previously developed and tested.
- A close coordination of the flow of activities that affect the design, procurement, production and distribution cycle of the product, both within the company and with the outside partners that make up the supply chain. In this regard, it is interesting to refer to the specific workflow functions that allow you to coordinate the different components of a process. Those functions make it possible to effectively manage:
  - the sequencing of tasks and process activities. In particular, they make it possible to ensure compliance with planned procedures, input/output relationships between pre-established activities and the management of any exceptions in the sequencing of the process flow;
  - the traceability of the activities, in order to keep constantly updated the information on the status (both in terms of progress and location) of the future output of the process;
  - the assignment of the work to the personnel in charge of the single tasks, in order to optimize their use by leveling the loads and as a consequence eliminating potential bottlenecks.

19.3 Advanced planning systems

The ERP systems, even if they offer higher performances compared to those guaranteed by the older IT systems, show some criticalities in completely satisfying the specific information requirements of the operations processes. These systems, in fact, cover in a very satisfactory way the whole transactional area, that is the one related to the accounting records concerning the main flows linked to the company management, but offer much more limited performances with reference to the problems of operations and supply chain planning.

While maintaining a central role within the company's information system, in manufacturing companies the specific functions of ERP systems aimed at planning,

Figure 19.1 **Advanced systems for operations: a reference map**

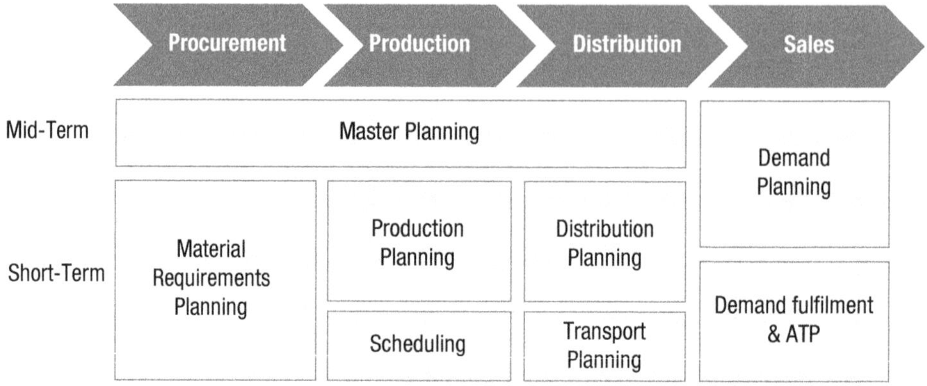

Source: adapted from Fleischmann et al., 2007.

managing and controlling production and logistics are often integrated or even replaced with dedicated application packages.

Figure 19.1 represents the **principal planning activities** classified in relation to the operations and supply chain processes supported and the various time horizons considered. In the remainder of this paragraph, a summary of the main solutions covering the management of information requirements linked to these specific activities will be proposed.

19.3.1 Demand planning

Never as in recent years has **demand forecasting** been such an important lever for increasing company competitiveness. In companies already exposed to forecasting problems as they are organized in a make-to-stock perspective, the reduction of the product life cycle and the contextual expansion of the product ranges have further strengthened the need for tools capable of predicting market dynamics as accurately as possible. In the absence of adequate information about expected demand, companies that are forced to anticipate production expose themselves to increasing risks of stockout and, consequently, to potential demand dissatisfaction. Even in companies that tend to be less exposed to forecasting issues, due to assemble-to-order or make-to-order production organization, the evolution of the competitive scenario has created conditions that make it absolutely important to have reliable projections concerning the qualitative and quantitative profile of market demand.

In particular, if the continuous lead time reduction for the customers' order fulfillment, even for products characterized by a significant degree of customization, is not counterbalanced by adequate lead time contractions, it becomes necessary to shift the **decoupling point** within the production process ever further downstream. This forces companies to expand the extension of the production phases to be managed in a push logic (i.e. based on forecasts) compared to those that can be managed in a pull perspective (i.e. pulled by demand and therefore not affected by forecasting).

This scenario shows how important it is to have specific modules able to support demand forecasting, improving its accuracy and allowing for the calculation of appropriate stock levels to guarantee a certain level of service. The main functionalities of a demand planning module concern:

- the possibility of adopting a **bottom-up approach** for the forecasting of a single item. Due to the scanty computational capacity and the limitations of the models implemented, traditional systems adopt a top-down approach, that is, they carry out the forecasts at the aggregate level (by product families, distribution channels, etc.) and subsequently disaggregate the cumulative data in order to determine the forecast for the elementary item. More advanced systems, thanks to the adoption of more sophisticated models, can work directly on single items, focusing on the peculiarities of their demand pattern;
- the analysis of the demand pattern (for example, trend, seasonality, etc.) in order to choose the most suitable forecasting model;
- the management of promotions, that allows to mitigate their enlarging effect on the time series;
- the management of the product life cycle, in order to properly manage product phase-in and phase-out;
- the management of cannibalizations, that allows for defining to what extent a product can be negatively influenced by the sales of another product. This feature provides a relevant support for the decision-making process related to the launch of new products, simulating the effect on the level of demand for the existing product range.

Since the value of a forecasting method can never be determined a priori, all the forecasting packages are designed to measure the **forecasting error**, the only indicator able to express the effectiveness of the technique adopted. The most sophisticated demand planning modules use forecast errors to automatically recalculate the value of attenuation, trend or seasonality parameters used within the main forecasting methods (so-called self-adaptive methods). They also allow for launching simultaneous demand forecasting sessions in order to assess, through a comparative analysis of the smallest forecasting error, the technique that can guarantee the best result.

19.3.2 Master planning

The main purpose of this module is to synchronize the flow of materials along the entire supply chain, from their procurement to the delivery of the finished product. The **master planning features** support medium-term decisions aimed at managing the trade-offs between production costs, stock maintenance costs, and transportation costs, so as to achieve the best balance between the needs of demand and the constraints of the supply system (internal production and external supply).

The main activities that refer to the master planning module are:

- the aggregate production planning with the goal of defining a master production schedule, i.e. a production plan that allows for satisfying sales demand by product family for each period considered in the planning horizon, minimizing overall costs and respecting production capacity constraints, labor availability, stock levels, and materials to be purchased;
- the make or buy decisions, identifying whether and to what extent it is necessary to activate external production sources to cover the production requirements that cannot be met with internal production resources.

19.3.3 Materials Requirements Planning and Production Planning

These modules aim at supporting decisions regarding production capacity planning and material procurement, defining the quantities to be ordered and determining the safety stocks ensuring a correct supply of production process as defined in the master planning. Since the beginning of the 1980s, these needs have been covered by MRP modules, both in stand-alone software packages and inside ERP systems. For some years now, **advanced planning systems**, known as **APS**, have been developed. They have overcome the main limitations that characterize traditional MRP systems. In order to understand the key differences between MRP and APS systems, it is appropriate to compare some elements that distinguish the two families of applications. The synthesis of this comparison is reported in **Table 19.1**, that requires some additional comments aimed at highlighting the most interesting aspects.

The MRP systems represent vital tools to support the planning activities since they allow for jointly processing hundreds or thousands of data with a high degree of

Table 19.1 A comparison between MRP and APS systems

Basic approach	Transactional	Simulative and interactive
Computation	Batch	Real-time
Expected results	Production plans issued after an iterative check of materials and capacity availability (MRP+CRP)	Feasible production plans thanks to the simultaneous check of constraints on materials and capacities
Capacity constraints	Absent (the capacity of resources is considered infinite)	Present (the capacity of resources is considered finite)
Resources considered	Only production resources (production centers, labor)	Production resources (production centers, labor) and auxiliary resources (tools, equipment, etc.)

precision and efficiency. However, they do not completely replace the human activity, since the order proposals issued by those systems must in any event be screened, in order to heal potential infeasibility. With APS systems, this form of support expands significantly since the data processing, that undoubtedly facilitates the performance of a repetitive task and lightens the workload of the planners, is replaced by a more "intelligent" processing. Such intelligence consists of the possibility guaranteed to the same planners to **interact on multiple levels with the APS system**: before the processing, through the configuration of a series of parameters (for example, the resolution technique to be adopted, the objective function to be maximized, etc.) that can influence the effectiveness of the plan; during the process, through the launch of parallel simulation sessions with different settings to verify the possibility of improving the overall performance of the planning.

Another significant difference regards the processing methods, that are substantially affected by the innovations related to IT technologies. In MRP systems, data is processed in batch mode, within specific time windows (usually at night) considering the enormous use of computing capacity associated with the launch of an MRP run. In APS systems, the generation of a plan, even a complex one, takes place in a **very short time** (even a few seconds). This enormous increase in response speed is essentially due to the fact that APS systems load all of the data necessary for the processing directly in the RAM memory. The greater amount of memory required (up to the level of gigabytes) does not represent an obstacle since the cost of hardware has dropped sharply in recent years.

These foundational differences are accompanied by additional differences linked specifically to the way in which production plans are generated. The traditional solutions adopted for the calculations of the requirements lead to a determination of a series of planned orders relating to materials, both manufactured internally and purchased externally. These plans must be appropriately screened to verify that there are no temporal infeasibilities (orders that should have been issued in the past due to anticipation of lead times) or infeasibilities linked to the unavailability of production capacity. About the first, using consolidated MRP solutions the planner intervenes to remove the constraints or manage their causes (for example, putting pressure on the supplier with an intense expediting activity, or defining a higher priority, in agreement with production, in order to accelerate the order fulfillment). For the potential lack of availability of production resources, the planner proceeds to launch a **CRP (Capacity Requirement Planning)** run, that makes it possible to surface any over- or under-absorption of capacity. Since MRP and CRP process the data in succession, any potential conflict between materials and capacity is managed independently of the impact it can have on decisions already taken; in this way, it is possible that a **vicious circle** is created whereby the need for materials determines an infeasibility in terms of capacity which, when removed, in turn makes the plan infeasible from the point of view of materials. This aspect probably represents the **primary limit of the MRP procedure**: since it is not able to simultaneously consider the constraints on materials and capacity, it does not automatically resolve situations of infeasibility. At the most, the MRP system merely reports criticalities, assuming that the planner can find the appropriate corrective actions. **APS systems** completely solve these limitations since they are able to **simultaneously plan materials and production resources**, the latter considered in their finite dimension. The enormous processing capacity required is partly linked to the need to jointly check all the possible constraints and to resolve them in a single process step. It is also necessary to specify that, unlike their predecessors, APS systems consider not just production resources (work centers and labor), but also constraints associated with auxiliary resources (tools, equipment, etc.) that in many contexts can assume an often underestimated importance.

19.3.4 Scheduling

The purpose of this module is to generate a detailed production schedule over a relatively short time interval. The formulation of an operating plan makes it possible to determine, for each order to be executed within the scheduling interval, the start and end time of the resources required to perform the planned work and specifies the sequence of orders to be processed on a given resource.

The scheduling module is able to develop **finite capacity production plans** using heuristic algorithms. It takes into account:

- the priorities of production orders related to the delivery dates of customer orders;
- the setup times depending on product attributes and production campaigns;
- the capacity constraints in terms of labor, machine resources, and auxiliary resources (such as molds and/or specific equipment).

The main scheduling solutions available on the market entail:

- the creation of a **GANTT diagram** in electronic form that illustrates the allocation and sequence of the orders on production resources. According to this approach, a level of priority is assigned to all work orders waiting to be scheduled, the order with the highest priority is selected and all phases of the work cycle are scheduled. During this process, each order is loaded on a specific resource when two conditions occur together: there is a significantly long time interval to complete the operation foreseen by the cycle (unless a split of the order over multiple concomitant production resources is allowed) and any sequencing constraints present in the production cycle are respected;
- the use of **dispatching rules** to assign priorities to various competing orders. The level of priority is determined according to a specific rule that is driven by a particular managerial objective (for example, the order allocated is the one with the shortest work time or the shortest setup time on the specific production resource, or the order with the closest delivery time). It is not possible to state that one rule is better than another in absolute terms, since each of them can guarantee ade-

quate solutions with respect to the goal it sets: for example, the adoption of a rule that assigns priority to the orders so as to respect the delivery dates, could perform poorly in terms of setup times reduction;
- the interaction between the scheduling package and the operator, who can intervene to modify the scheduling proposed by the system by simply **dragging and dropping an order** from one resource to another and verifying immediately, within just a few seconds, the impact of that choice on the overall results of the scheduling. For example, the service level could improve by anticipating the delivery of an order, but at the same time the process time of another order could worsen because it has been shifted to a non-optimal production resource (with a lower production speed).

These applications are generally called "**interactive scheduling systems**," due to the possibility they offer to the operator to interact with the tool to which is entrusted the task to propose a first feasible solution. The interactive scheduling systems are based on heuristic models, that is, they find solutions that respect the assigned constraints, but do not necessarily identify the optimal solution with respect to the objective function to be maximized. If there are particularly critical operating conditions that require the identification of an optimal solution (think of the minimization of setup times within a plant where tooling costs can cause a positive or negative marginality of the order), the most advanced scheduling systems have libraries of optimization algorithms that can be called directly by the user according to specific needs; in some cases, are also provided simple programming languages that allow you to create, combining different optimization rules, routines tailored to solve particular cases of planning.

19.3.5 Distribution planning

The purpose of this module is to determine the best possible **distribution plan** considering the demand of the various players along the supply chain (for example, branches, wholesalers, and final customers). The parameters that are taken into consideration to obtain an optimal distribution plan are the current stock at the various nodes of the supply chain and the service level to be guaranteed.

The management activities that refer to the distribution planning model regard:

- the planning of stock along the entire distribution chain in order to guarantee an adequate service level at the various nodes of the network. These modules use sophisticated mathematical models that suggest for each item the stock level (operating stock and safety stock) to be held in the various nodes based on the planned service levels, the demand characteristics (variability, order frequency), the production and supply constraints (minimum production and shipping lots, production frequency, procurement lead times, etc.), and production uncertainties (lead time variability);

- the determination of the needs for finished products in the various nodes of the distribution chain in relation to demand forecasts, planned stock, and service levels;
- the control of stock availability in the various points of the network and its potential reallocation, through transfer orders, to other logistics nodes where stockout situations are likely or to strategic clients to avoid stockout situations.

19.3.6 Transportation planning

The main purpose of these modules is to **optimize the logistics flows** in order to minimize the transportation costs and/or maximize the use of means of transport. The calculations performed are based on information relating to the service levels agreed with the clients, the specific delivery settings, the loading capacity of the means of transport, the geographical delivery routes, and the costs of the various means and methods of transport.

The main activities supported by the transportation planning modules regard:

- the planning of any type of transportation request (inbound or outbound, mono or multi-modal) through dashboards and maps that allow for identifying the most efficient and effective solution based on the availability and performance of the entire fleet, individual carriers, or specialized logistics operators (3PL);
- the search for the minimization of transportation costs, through an automatic comparison of the performance of all the available operators with respect to the routes to be organized;
- the optimization of delivery processes, taking into account planning constraints (loading/unloading points, orders, and vehicles) in order to contain transportation costs;
- the evaluation of the overall KPIs through the monitoring of transportation performance in terms of service level, costs, delivery times, and environmental impacts.

19.3.7 Demand fulfillment

Speed and on time delivery are service profiles by now considered indispensable for competing in the current competitive environment. Alongside these two fundamental dimensions, a third way of providing service to the client is gradually emerging, especially in business to business relationships: the ability to promptly communicate the confirmation of the delivery date requested by the customer or the presumed order fulfillment date, estimated on the basis of the availability of the item requested. This is an important piece of information, because it allows the client, properly advised of any inability to receive goods on a certain date, to revise its production plans or seek out supply sources capable of meeting those needs.

This particular requirement can be effectively supported through the use of appropriate functionalities commonly known as **Available to Promise (ATP)**. They allow for verifying the possibility of promising the customer a precise delivery date based on the actual availability of the finished products. The availability is influenced, within a given time horizon, by the initial level of the stock and by the dynamics generated by the replenishment processes (planned and confirmed production orders) and consumption processes (client orders).

19.4 Manufacturing Execution Systems (MES)

To complete the picture about information systems supporting operations management, it is necessary to illustrate the **MES systems**. These are focused on monitoring and control of operational processes and connect the layer of planning with the layer of execution, i.e. where planned actions are actually implemented.

These systems are essential for two reasons: on the one hand, they allow for detecting any criticalities or anomalies in manufacturing and logistics processes; on the other, they feed the planning systems with constantly updated data, a fundamental aspect for the reliability of the plans.

The most advanced MES systems support the following functions:

- *dispatching*: issuing of documentation to operators and machines (operating instructions, drawings, part-programs) and indication of the priorities relating to the activities to be carried out (with visualization on a PC or display of the lists of jobs sequenced according to the plans issued);
- *data collection*: acquisition through suitable interfaces of the data collected and generated by the operators and/or by the equipment making them usable by other applications that make up the MES system;
- *monitoring*: instantaneous and real-time verification of the operating status of the production process, with particular focus on the principal resources involved (labor, equipment) and the progress of production;
- *performance analysis*: elaboration of synthetic performance indicators, both at single and aggregate resource level, in order to keep under control the overall performance of the production system;
- *product/process quality*: analysis of product and process quality through the detection of non-conformities with respect to the target parameters of quality and tolerance set;
- *tracking and tracing*: search and visualization of information relating to materials that are going through internal and external production phases or have already completed their production process.

MES systems play an increasingly important role in the panorama of IT solutions supporting operations processes. The growing spread of **demand-driven strategies**, in which the service level to the end customer represents the main competitive advantage, requires companies to adopt tools that allow them to respond in real-time to continuous and sudden changes in operating conditions. The traditional **Plan-then-Execute** approach is gradually substituted with a more flexible **Plan-while-Executing** approach. For this reason, MES systems, from merely operative solutions for the control of production processes, are becoming a fundamental element to create that closed-loop planning and control cycle that places operations managers in the position to have full control over what is happening and to be able to anticipate, rather than be subject to, situations that can cause difficulties for the operations system.

19.5 Adoption of ERP/APS/MES systems: some possible configurations

After the in-depth analysis of the functional coverage of the various solutions illustrated in the previous paragraphs, the discussion should be completed with a reflection on the possible configurations for the adoption of ERP, APS and MES systems.

By cross-examining two dimensions of analysis – the level of specialization of the solutions adopted and its degree of pervasiveness within the company – **four reference models** can be derived with respect to the possible articulation of information systems for the planning, management, and control of manufacturing and logistics processes (Candussio, 2013) (See **Figure 19.2**).

Figure 19.2 ERP/APS/MES integration: reference models

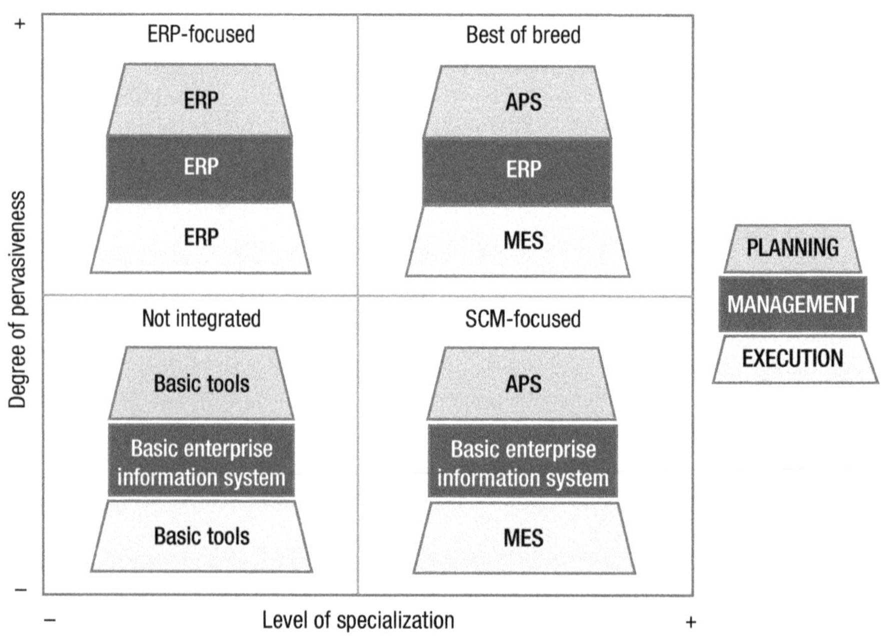

Source: adapted from Candussio, 2013.

Not integrated

Core processes are managed through a basic enterprise information system, implementing a few essential modules that are often customized to fit the company's operating model. Production control is not supported by dedicated IT tools and data coming from the shop floor is manually gathered and then asynchronously loaded to the enterprise information system. Planning is often entrusted to personal initiative and carried out through basic tools (especially Microsoft Excel and Access), that are not structurally integrated with the company's database.

ERP-focused

The path to maturity starts with a review of organizational processes and supporting tools, resulting in the adoption of a modern ERP system. The awareness of the importance of working within a natively integrated solution leads to implementing production planning and control modules embedded into the ERP suite. While on the one hand this solution guarantees native integration of the database and information flows, on the other, it can surface some gaps in the coverage of specific needs related to the operations processes.

SCM-focused

The process of evolution within the company leads to the awareness that manufacturing and logistics requirements must be addressed with targeted tools. The central role of the enterprise information system in managing the operational flows (issue of sales and purchase orders, master data management, etc.) is not questioned, mainly for budget and organizational issues. However, application suites for production control (MES) and resource and material planning (APS) are introduced and structurally integrated with the enterprise information system.

Best of breed

The maturity stage requires a twofold awareness: on the one hand, the importance of adopting consolidated and structured systems such as ERPs for the integrated management of the company processes; on the other, the recognition of the strategic nature of dedicated solutions – such as APS and MES – for a truly efficient monitoring of manufacturing and logistics flows. This is clearly the most advanced configuration, that is typically adopted by medium to large-sized companies that are exposed to a significant competitive pressure.

20 Industry 4.0: The Digital Evolution of Operations

by *Raffaele Secchi*

20.1 The fourth industrial revolution

The economic history of Europe and North America is characterized by three major industrial revolutions. **The first**, which occurred between the 18th and 19th centuries, essentially saw the **development of mechanics** and the **use of water and steam power** within factories. The motion generated by steam engines was transferred to machines with articulated mechanical systems consisting of pulleys and belts, thus giving rise to the first industrial production sites. The **second revolution** used **electricity to develop mass production**. The possibility of physically decoupling the power units from the machines dedicated to production processes, multiplying the production resources operating within the factories, allowed for a significant increase in production volumes. The **third revolution** used **electronics and information technology** to automate production. As from the 1970s, the first numerical control machines appeared, that could be programmed using predefined codes (originally stored on paper punch cards and subsequently replaced by electronic files). The new ways of programming machines made it possible to gradually free the use of equipment from the operators' control, and made it easier to set up the operating machines so as to run multiple production cycles in the available time.

Today we are experiencing a **fourth industrial revolution**, which is commonly referred to as Industry 4.0. The new paradigm entails the **digital transformation of the industrial system**, thanks to a combination of technologies that enable the creation of an extended connection between machines and smart objects, capable not only of communicating with each other, but also with the surrounding environment and operators. Creating this interconnection presents a much greater level of complexity compared to the three previous industrial revolutions. While the formers were based on a specific technology, Industry 4.0 is based on a combination of nine technological pillars.

20.2 The enabling technologies

20.2.1 Advanced manufacturing systems

Advanced manufacturing systems represent a rather broad and diversified set of technological solutions. Focusing our attention on the most innovative and consistent proposals related to the constituent principles of the 4.0 paradigm, we will explore two specific applications: collaborative robots and mobile robots.

Collaborative robots are anthropomorphic robots[1] that can operate in close symbiosis with the human operator, sharing the same work area in fully safe conditions. In fact, they are equipped with capacitive sensors that, by detecting the proximity of a human body, send signals to the robot's control system to decrease its speed or immediately stop its operation. The advantage of using collaborative robots lies in the possibility of dividing the production activities of a work cycle between the operator and the robot to exploit their respective capabilities. For example, very repetitive or hard tasks, or tasks requiring a very high degree of precision over long periods of time, can be more effectively assigned to a collaborative robot. Conversely, tasks that are subject to a high degree of variability (due to different positioning of pieces to be processed or their characteristics) can be performed more quickly and efficiently by a human operator. Collaborative robots are also easier to program and reprogram than traditional ones. Instead of developing special programs requiring knowledge of specific languages, the user can record the movements and actions that the collaborative robot must perform through very user-friendly graphic interfaces, or even directly maneuvering its "limbs." However, it should be noted that although collaborative robots represent the latest evolution of industrial robotics, they do not represent the panacea for any automation needs in factories. Traditional anthropomorphic robots, at their current stage of technological evolution, are more efficient to perform tasks that require extremely high levels of precision and speed. Moreover, in production contexts where the objects to be handled have weights greater than a few kilograms, the use of traditional robots cannot be ignored.

A second set of advanced solutions for production is represented by **mobile robots** that, thanks to laser scanners, cameras, and CPUs holding maps and algorithms for localization and navigation, can move autonomously and reach a destination avoiding even mobile obstacles, such as operators moving around inside the shop floor. Robots are thus able to optimize routes by defining, like the navigators that equip cars, the most efficient paths to reach the intended destination.

20.2.2 Additive manufacturing

Additive manufacturing (or 3D printing) is a production method that, by using different technologies, allows the creation of objects (components, semi-finished or finished products) through depositing layers of liquid, solid, or powder material. In additive manufacturing, the traditional approach that entails the production of a piece through a process of subtraction from the solid (such as lathing or milling) is reversed.

[1] Anthropomorphic robots are so called because they can reproduce the appearance and movements of a human arm. They are part of the broader category of industrial robots, i.e., devices equipped with servo mechanisms that can move in the surrounding space, as well as manipulate tools, pieces to be processed or other instruments.

Additive manufacturing is not an absolute novelty in industry. Its use began at the end of the 1980s, but for many years it remained limited to the realization of prototypes. The possibility of rapidly producing what was designed provided an interesting opportunity both for marketing representatives, that could present a solid mock-up to customers so as to immediately get any feedback, and for R&D/Engineering representatives, that could more precisely assess any interferences between the various components of a sub-assembly or any criticalities in their assembly. Subsequently, the possibility of obtaining more stringent manufacturing tolerances and using materials with better mechanical characteristics made it possible to build products or components that could be used in a real environment.

Over the last ten years, a robust technological evolution has taken place, causing a progressive reduction in costs, an increase (albeit marginal) in the speed of deposition, and an improvement in the precision with which the product is made. Therefore, additive manufacturing is no longer considered as merely a **technology for rapid prototyping**, but as a **true production technology** for the realization of components and finished products.

However, it is worth noting a few points of caution regarding the use of additive manufacturing. Although the materials are expanding, they are still limited with respect to those that can be used in traditional processes. The depositing speed of additive machines, although increasing, is still not comparable with the speed of traditional equipment. Therefore, the former cannot replace the latter in repetitive large volumes production, while they represent a valid option to produce small and medium quantities and items requiring a high level of customization. The quality of the objects obtained through additive manufacturing, especially in contexts where the aesthetic dimension is an important issue for the customer, is still far from that reached through traditional technologies.

20.2.3 Augmented and virtual reality

Augmented reality allows for enriching sensorial perception by making available to the operator, through appropriate devices (for example, a pair of glasses, a tablet, a smartphone, a smartwatch) information that would not be perceivable with the five senses. The information is provided pro-actively, appearing to the operator while he is performing his activities and thus facilitating the decision-making process. Augmented reality consists of a set of technologies that belong to the ecosystem of context computing (algorithms that recognize objects, geolocalization algorithms, artificial intelligence algorithms, wearable devices, and augmented reality interfaces).

Like additive manufacturing, augmented reality does not represent an absolute novelty in the industrial field. The first applications of augmented reality, mainly related to maintenance and manual assembly activities, were used in Boeing at the beginning of the 1990s. Subsequently, also thanks to the continuous performance improvement, augmented reality has found interesting applications in supporting the execution of activities characterized by a low degree of repetitiveness and/or a high degree of complexity, such as picking, tooling, and maintenance activities.

When carrying out picking activities in the warehouse, an operator wearing a specific visor can be guided in the path to be followed and can read bar-code or QR-code labels placed on objects to be picked without necessarily using his/her hands. This obviously leads to significant improvements in both effectiveness and efficiency of picking activities. The adoption of augmented reality in tooling and maintenance activities makes it possible to lead the operator in recognizing the target part and to show the proper mode of replacement. The most advanced applications have spread to contexts where the effectiveness of maintenance is critical, such as in the aerospace and automotive industry and in process-based manufacturing systems.

Augmented reality should not be confused with **virtual reality**, that entails the creation of an entirely virtual environment, separate from the actual environment. The use of this technology to train operators to perform specific production activities is increasingly widespread thanks to technological improvements that provide a realistic experience in the physical interaction with the virtual object to manipulate and the equipment to use. The virtual reality can also be used for other purposes, in particular to support the design of a production/logistics system. It in fact allows for virtually moving inside a plant before it is completed, thus making it possible to evaluate in advance alternative layouts with reference to loading docks for forklifts, space available for storage and materials handling, and so on.

20.2.4 Simulation

Thanks to some sophisticated software packages available today, it is possible to develop a **computer-based model of a manufacturing and logistics system and simulate its behavior** in order to evaluate its performance with reference to various conditions (such as the number of resources, layout, materials management policies, etc.). These simulations allow the decision-maker to identify which configuration of layout and flow management methods could guarantee the best performance of the physical system and, therefore, to make an extremely informed choice. Generally, the simulation is mainly used to support strategic decisions, for example the evaluation of different layout alternatives of a production facility, or tactical decisions, that is the appraisal of different production flow options (such as an intermediate buffer to decouple two sequential phases of a process).

Although the benefits are rather evident, the use of simulation for **process design** purposes is still fairly limited due to the scarce availability of the necessary skills, the costs of the software, and the considerable time required for the development of the simulation models. On the other hand, the adoption of simulation packages is quite consolidated in **product design** processes. The three-dimensional model of an object can be studied in a virtual environment – the production process in which the object will subsequently be produced or the context in which it will be used – with the goal of evaluating any criticalities in production and/or functioning before the design is definitely released. Furthermore, it is possible to reduce engineering lead times, improve product quality, and reduce potential waste of materials.

20.2.5 Integration

The integration promoted within Industry 4.0 is characterized by two dimensions. **External, or horizontal, integration** is meant as integration with suppliers and customers in order to foster decision-making flows more efficient and faster within the supply chain. This is not a new challenge, either in terms of practices or technologies. Over the years, numerous solutions have been developed to allow upstream and downstream integration with respect to the focal company, such as electronic kanban, web portals to increase supply chain visibility, and collaborative planning systems. Despite this, valuable cases of horizontal integration are rather limited. To pursue true integration the actors in the supply chain must be willing to collaborate. Technology is a facilitator, a means to make the desired collaboration concrete; but first trust must be developed among the various actors in the supply chain, aligning objectives and promoting the joint adoption of collaborative practices.

With reference to the **internal**, or **vertical integration**, it consists in the possibility of activating mutual communication between machines and devices operating inside the factory. The key element for integration is the interconnection between the different devices, such as the PLCs that equip systems and robots, supervision, control, and data acquisition systems (SCADA), shop floor data collection systems (MES), and enterprise information systems (typically ERP). Although some important progress has been made towards making these elements "speak to each other," no single standard is currently able to fully cover all the needs for integration. This constraint postpones to the future the concrete possibility that the production resources can autonomously exchange information on their status as well as the status of the work in process, so as to promote a greater autonomy and a gradual decentralization of decision-making processes.

20.2.6 Big data and analytics

Big data commonly refers to the large mass of data that is generated within manufacturing and logistics processes operating in a 4.0 environment. Compared to the recent past, datasets are characterized not only by the **volume** of data stored, but above all by the **variety** and the **speed** with which they are generated and become obsolete. Focusing on the variety, in addition to purely transactional data (all those that are actually created and stored within the enterprise information system), we are witnessing a substantial proliferation of sources: sensors that equip production systems or products going through shop floor processes, videos, images, and texts.

With reference to this enormous amount of data that is not necessarily structured, it is essential to encourage and support an increasingly **data analytics-oriented approach** that:

Figure 20.1 **Different approaches to data analytics**

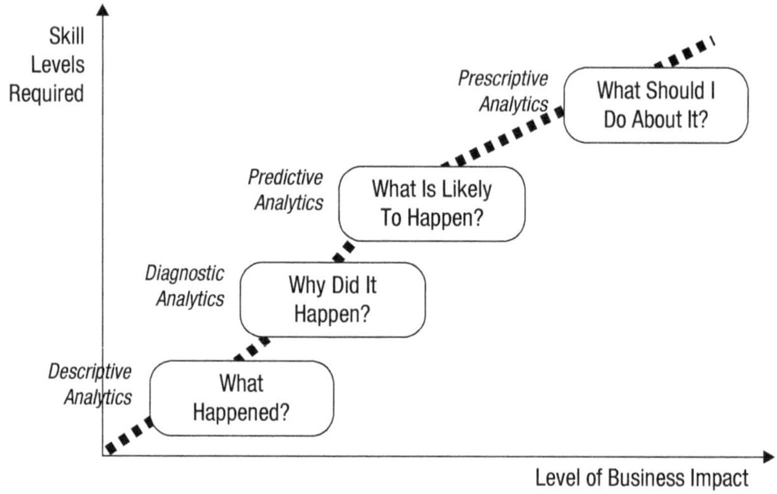

Source: Gartner, 2013.

- uses the available data to produce measures with strong informative value;
- is based on a systematic application of appropriate mathematical and statistical techniques in order to identify significant relationships between data;
- makes use of effective, interactive, and user-friendly modes of visualization and interrogation.

As shown in Figure 20.1, data analysis is usually conducted following four approaches: descriptive, diagnostic, predictive, and prescriptive. The **descriptive** approach is used to understand what is happening or has just happened. It usually exploits classification and clustering algorithms to represent the available data and provides historical reporting, statistical analysis, comparison of data sets, and dashboards for real-time monitoring of information coming from "the field." The **diagnostic** approach focuses on the investigation of the causes that led to a certain phenomenon, in order to steer the improvement activities. The **predictive** approach takes into account historical data to which statistical algorithms, such as regression, and machine learning are applied, in order to predict trends and future behavior. Lastly, the **prescriptive** approach, that is typically based on optimization algorithms, goes beyond the prediction of future behavior, providing the decision-maker indications on the best course of action to be taken.

20.2.7 Other enabling technologies: IoT, cloud, and cybersecurity

The Industry 4.0 paradigm contemplates three other pillars that will be illustrated together since they represent **the layer of hardware and software technologies** that enable the real world to be joined with the digital world.

The need to connect the different components of a production system – such as equipment, loading and handling units, people who operate within the process through wearable devices – is enabled by a layer of technologies commonly identified with the acronym IoT – **Internet of Things**. This refers to various technologies, since it is needed to cover a very wide range of applications, from wearable devices, designed to communicate over distances of a few centimeters, to asset management and monitoring applications, to sensors that can communicate over distances of several kilometers. Various solutions are currently available, such as short-range and long-range IoT technologies, that are differentiated by speed of data transmission, distances covered, and energy consumption. The proliferation of different technological solutions highlights the criticalities linked to the co-existence of multiple protocols, that could find a possible solution with the development of 5G.

A second technological pillar is **cloud computing**, i.e. the set of technologies that enable the use of hardware and software resources (such as mass data storage) offered in a service logic by an outside provider. At the shop floor level, the cloud is essential for the integration of connected devices, manufacturing execution and control systems (MES), and specific enterprise data repositories. If expectations of a significant improvement of data transfer/access times, on the magnitude of milliseconds, are maintained, in the near future we could see a greater propensity for **cloud-based production data management**, allowing for the birth of "data-driven" services for production systems.

The third pillar regards **cybersecurity**. The widespread adoption of some of the technologies illustrated above, in particular IoT and the cloud, significantly increases what experts call the "attack surface," that is, the possibility of being the target of attacks by cyber criminals. These malicious actions can compromise both the availability of production resources, causing potential losses of productivity, and their integrity, which is essential to prevent malfunctions that could cause harm to people, products, or other resources operating in the production environment. In this direction we must mention the attacks recently suffered by several companies, even large ones and therefore potentially more prepared to face similar situations, which have led to production breakdowns in several sites for not negligible periods of time.

20.3 Expected impact on industrial processes

After having presented the technological layer at the base of the 4.0 paradigm, we illustrate the benefits that can be achieved through the application of those technologies to the principal industrial processes. The goal is to provide some ideas about the opportunities that can be seized, with particular reference to the improvement of the competitive position of the company.

20.3.1 Use of production assets

Focusing on internal processes, the adoption of a 4.0 approach can **maximize the use of production assets**, such as critical equipment or tools. This is achieved by being able to monitor the operating conditions of these resources in real time, highlighting any downtime states that can be made evident to operators (perhaps through devices of immediate use such as tablets or smartwatches) and allow immediate intervention to restore the operating conditions. A more sophisticated approach is grounded on predictive maintenance models. Based on correlations between the values of the operating parameters and the values of the operational performance such as efficiency or conformity, these models are able to predict potential anomalies. This makes it possible to implement a series of actions aimed at avoiding asset downtime, cutting costs for urgent intervention of the maintenance staff and avoiding lost margins due to lack of production.

20.3.2 Productivity of human resources

Analogous considerations can be made regarding human resources involved in production. In this area, too, it is possible to monitor operators' activities with greater precision and timeliness and, therefore, create the conditions for increasing their **productivity**. For example, the failure to achieve a certain volume of production within the assigned standard time due to the unavailability of equipment or materials can immediately trigger a review of the process aimed at resolving the situation encountered. The increase in productivity could also be achieved by improving the working conditions of the operators. On the one hand, through the identification of the operator with a wearable transponder, it is possible to configure the workplace by creating the best ergonomic conditions for each person. Incidentally, we should consider the potential offered by this dynamic mode of **configuring the workplace**, thinking both of disabled people, who could thus find themselves in work situations that are more consistent with their deficits, and older people, who could require different settings of the work area to best express their potential. On the other hand, improvements of productivity can be obtained through better **training of operators** on the work cycles. Also in this case, digitization can contribute in various ways, from a simple system of graphic presentation of the cycles through user-friendly devices (typically tablets or screens positioned near the workstation), to more advanced systems that use augmented reality tools to allow the operator to receive on a wearable device (typically glasses) only the information considered relevant to perform the operation according to the expected standards. The use of these solutions goes beyond just workstations in a factory environment, to potentially extend to other activities:

- picking, within traditional warehouses in order to indicate to the operator the items to be withdrawn or deposited;

- service, when the operator is in the field and has to carry out a maintenance intervention that requires the use of specific equipment or the execution of certain sequences to be strictly observed.

20.3.3 Synchronization of production and logistics activities

The expected benefits that can be achieved through better **synchronization of production and logistics activities** do not regard only the efficiency, but also the speed of the processes. Examples in such sense regard the potential reduction of setup times, and consequently, of production activities due to the possibility of verifying the status and position of the tools (that actually become nodes of a network receiving and sending data). Analogous considerations can be made in relation to internal logistics, since operators can directly receive instructions on the routes to follow in order to optimize the picking activities. Remaining focused on process synchronization, the possibility of exchanging and sharing increasingly rich information at decidedly limited costs can provide an additional impulse for the development of collaborative planning practices among different companies within a supply chain.

20.3.4 Reorganization of product development processes

Customer expectations regarding high level of customization and short delivery times is one of the areas where greater competitive pressure is likely to emerge in the near future. Significant improvements have already been achieved in the past years by companies that have radically rethought the organization of their product development process. They adopted new approaches commonly known as concurrent engineering or simultaneous engineering, which push to organize and manage the process in a simultaneous and integrated logic. While these management practices have greatly contributed to streamlining and accelerating product development activities, additional improvements can be achieved through the adoption of new 4.0 technologies that tend to make this process increasingly virtual. It is no longer necessary to develop a physical prototype and put it through leak or stress tests under particular conditions. Thanks to advanced **simulation software packages** it is possible to simulate the functioning of the object in the different conditions in which it could operate and get a fast and consistent feedback on the expected performance and overall reliability.

One of the fundamental elements to run these simulations is the creation of a digital model, a perfect virtual replica of the physical object, that is currently indicated as **digital twin**. In addition to the simulations already mentioned, the availability of a digital twin of the physical product enables a series of interesting opportunities. In the sales area, for example, it provides the possibility of making the object of the potential sale immediately visible to the client. Furthermore, thinking to sectors where very complex products are manufactured (such as packaging lines, assembly lines, etc.), the digital model associated with a virtual reality solution allows the potential client, navigating in a virtual context, to be able to examine some critical aspects of the product that would be difficult to understand starting from traditional paper charts. In the presence of a digital twin of a factory layout, it is even possible to verify how the production line fits into the available space and thus evaluate the positioning of feeding paths, service areas, safety lanes, and so on.

As indicated, the adoption of these technologies has numerous effects, but they can be summarized as follows:

- the impact on product development lead time is evident. Therefore, it is desirable their wide diffusion in all the industrial contexts that have been and still are subject to competitive pressure on time to market;
- these technologies not only lead to a progressive reduction of resources (man hours), but they also favor greater efficiency of processes, cutting the cost of the product development process. This reduction is mainly due to the drastic reduction of costs for the engineering change. In fact, by increasing visibility of what has been decided during the upstream steps, simulation can reduce the likelihood of making significant changes to products or production processes in very advanced stages of development.

20.4 Operations 4.0 and new business models

The digital transformation promoted by the Industry 4.0 paradigm also offers enormous potential in reconfiguring the business models.

A first area regards the opportunities fostered by new additive manufacturing technologies. First of all, through additive manufacturing, costs and time – in many cases particularly relevant – to build molds and tools can be significantly reduced. Although the ideal state of producing one piece at a time may not necessarily be reached, additive technologies create the conditions for an effective **reduction of the production scale**. The potential implications are easy to conceive: first, the possibility of customizing products without negatively affecting production costs and, consequently, sales prices; second, the review of make or buy choices that in the past have led to outsource the production of small batches not considered economically viable.

Second, the **lower costs of prototyping activities**, together with the greater speed with which it is possible to move from the design phases to the industrialization and production phases, unlock additional opportunities in terms of product customization, with an intensification of customer involvement from the earliest stages of the process. The practices of co-design and early supplier involvement, that many companies have adopted in recent years with their suppliers, may and must be extended towards their customers, who will be more and more decisive in orienting companies' decisions.

Third, additive manufacturing systems can produce parts and components with very complex geometries in a single production phase. On the one hand, this allows for introducing a strong simplification of internal production processes, eliminating the division between manufacturing and assembly of finished products. This separation is one of the main reasons for long manufacturing lead times. On the other hand, more integrated production processes lead to a reduction in the number of actors involved, making coordination within a supply chain simpler and less costly. One can, therefore, configure **lean supply chains**, with important impacts in terms of reduction of lead times, stock levels, and logistics costs.

For some manufacturing companies that intend to focus on additive manufacturing, another strategic path is represented by the development of service platforms that can aggregate customers and suppliers, in an ecosystem logic. Being owners of a platform capable of bringing together supply and demand for value added services could have a precise strategic value, especially if we consider that in the near future the mere ownership of production equipment may not represent the most important asset. It is very likely that the production capacity to manufacture products based on customer specifications will become a "commodity" concentrated in specific "**printing farms**." In this regard, some platforms already exist on which it is possible to upload the production specifications (drawings, quality requirements, volumes, cost targets, etc.) in order to obtain specific supply proposals, in a logic of virtualization of the manufacturing service as it happens in IT services. It is also difficult to imagine that the 3D models necessary for printing products, being totally digital and easy to share, could constitute a distinctive competence. 3D scanner systems are in fact already capable of generating sophisticated digital models of products scanning their physical shape.

20.5 The three pillars of the digital transformation of operations

Although technologies are the driving force behind the digital transformation that companies have launched in recent years, it is not appropriate to give them the supremacy over other elements. It is undoubtedly necessary to find the proper balance between the three pillars of the digital transformation: the **technological pillar**, the **organizational pillar**, linked to processes and change management, and the **human pillar** represented by available skills.

A technology-driven digital transformation process is doomed to fail, or at least not to produce the expected results, until the underlying processes have been appropriately reviewed and the necessary skills have been set up to lead the change. As already mentioned, the digital transformation must be approached by properly assessing the competences related to process reengineering and change management. Methods and tools relating to process mapping as well as physical and information flows redesign must be adopted to reengineer the process before, or at least in parallel with, the choice of enabling technologies. Not least, it is essential to adopt the appropriate change management approaches to support the changes emerging within the existing organization. What is needed, then, are organizational structures prepared not only to absorb change, but also to manage it proactively, leveraging human resources ready to take up and lead this challenge.

The issue of human resources cannot be dissociated from that of available skills, the third key dimension of any digital transformation process. On this front, multiple reflections arise, but one question seems crucial: how is work changing inside factories, with reference to both operational and managerial levels?

As regards operational aspects, the implications of these technological innovations on factory work are important, complex, and in part contradictory. On the one hand, we are witnessing the emergence of new, very specialized professions, with particularly in-depth technical skills. On the other hand, the first effects of the so-called "technological unemployment" phenomenon are beginning to emerge, with the disappearance of some jobs and the decline of others. Blue collars must be accompanied along the development of new skills that are necessary to manage the enormous amount of data generated by increasingly intelligent and interconnected production resources. This raises the issue of both the quantity/quality of data to be analyzed and the ability to transform data into information that is needed to make decisions quickly, sometimes even in real time, without requiring the support of higher-level organizational managers. To summarize, shop floor workers will need to possess more management than execution skills. Carrying out an in-depth analysis of the information provided by the production assets, they will be more responsible for taking operational decisions.

With reference to management levels, there are several fronts that must be considered. First, specific skills are needed to steer the difficult process of evaluating and selecting technologies and suppliers. Secondly, a strategic perspective must be assumed because it is vital to understand the impact of the digital transformation on the company's business model and business processes. Third, it is critical that adequate skills are available to lead, oversee and manage the digital transformation project. Moreover, the ability to conduct such a pervasive project within the organization cannot but require strong leadership skills among the top managers. On the one hand, they must build and convey a clear vision of change and innovation linked to the digital transformation, and on the other hand, they must concretely drive change to achieve the expected goals.

20.6 The key elements to implement the Industry 4.0 paradigm

First of all, any Industry 4.0 project must be situated within a definite path of strategic evolution of the business. It should be as clear as possible how the project can concretely improve the company's competitiveness with reference to both the main value drivers (cost, quality, service, customization), and the possibility of developing innovative business models. These strategic reflections must be placed within a structured roadmap that allows for understanding the steps – in terms of technologies to be introduced, resources to be involved, and reorganizations to carry out – and the time required.

In this strategic perspective, it is essential for the IT Function to evolve with the aim of supporting the business. It should be capable of orienting the choices towards the enabling technologies that are coherent with the strategic vision.

The "combinatory" nature of the technologies that constitute the Industry 4.0 paradigm limits for most companies, and in particular for smaller ones, the possibility of managing the entire innovation process, from the identification of the technological solutions to be implemented, to the search and selection of the most suitable suppliers, to the integration between internal and external actors involved in the project.

Therefore, the ability to activate and consolidate relationships with all the players who are potentially able to provide a differential contribution for the development of an innovation regarding products, processes or business models becomes fundamental. On the one hand, the traditional relationships with suppliers and customers must be enhanced with a view to **open innovation**, so as to generate a continuous stimulus towards innovation processes. On the other hand, it is necessary to embrace a broader vision of the **ecosystem**, where companies that have produced concrete results in completely different business contexts can be carriers of innovative spurs. In this perspective, opportunities for relations with start-ups, universities and research centers must be strengthened. The need to have adequate skills in **scouting for innovative suppliers**, which can represent the source of innovation that cannot be developed internally, should also be highlighted.

On the organizational side, it is advisable to adopt **agile approaches**. The project should be based on numerous "check & go" stages in order to quickly validate the solution and immediately report the need for any changes or additions to what has already been developed. Instead of delivering a solution that perfectly complies with the specifications but in a very long time, it is preferable to release partial solutions in a shorter time, but still working, which will be gradually consolidated and improved, also according to the feedback received, until the release of the final solution.

The multiple perspectives and competences needed impose strongly **cross-functional organizational solutions**, capable of expressing different sensibilities as a mix of technological and organizational skills. It would be appropriate to involve competences related to research and development, processes industrialization, manufacturing, logistics and planning, organization, IT, management control, and safety.

The responsibility of leading an Industry 4.0 project must be entrusted to a manager with the ability to: coordinate complex projects and cross-functional teams; manage change; analyze/map processes according to lean management principles; understand the technological scenario (without necessarily being an expert on the single Industry 4.0 technologies); identify, assess and select technology providers (technology scouting).

Industry 4.0 projects must fit into an organizational context that has already been exposed to a wide-ranging and in-depth reengineering aimed at eliminating the main situations generating waste. Beyond the specific techniques and tools that lean management suggests, it seems necessary to adopt this "obsession" with all forms of waste as a precondition for any investment in automation related to Industry 4.0.

Lastly, but paradoxically just to reinforce the concept, a transition path towards the digital factory cannot but see as protagonists the people who work there daily and who represent the very engine of change. As already pointed out, it is not conceivable that the digital revolution should be achieved exclusively through investment in new technologies, without human capital being minimally involved in the review of the processes that will be reshaped by those technologies. It is necessary to accompany the change with adequate training courses that, at the various levels of the organization, can provide methodologies and tools to activate, lead and implement the digital transformation of the factory.

References

Akmal, A., Greatbanks, R. and Foote, J. (2020). "Lean thinking in healthcare – findings from a systematic literature network and bibliometric analysis." *Health Policy*, 124(6), 615-627.

Baglieri, E. (ed.) (2004). *La gestione strategica degli approvvigionamenti*. Milan: ETAS.

Baldwin, C.Y. and Clark, K.B. (2000). *Design Rules: The Power of Modularity* (vol. 1). Cambridge, MA: The MIT Press.

Beamon, B.M. (1999). "Measuring supply chain performance." *International Journal of Operations & Production Management*, 19(3), 275-292.

Belvedere, V. (2015). *La misura delle prestazioni produttive e logistiche nelle aziende industriali*. Milan: EGEA.

Belvedere, V., and Grando, A. (2017). *Sustainable Operations and Supply Chain Management*. Chichester: John Wiley & Sons.

Bi, Z. (2021). *Practical Guide to Digital Manufacturing: First-Time-Right for Design of Products, Machines, Processes and System Integration*. Cham: Springer.

Bicheno, J. and Holweg, M. (2016). *The Lean Toolbox: The Essential Guide to Lean Transformation* (5th ed.). Buckingham: PICSIE Books.

Bortolotti, T. and Romano, P. (2012). "'Lean first, then automate': a framework for process improvement in pure service companies. A case study." *Production Planning & Control*, 23(7), 513-522.

Bozarth, C.C. and Handfield, R.B. (2019). *Introduction to Operations and Supply Chain Management*. New York: Pearson.

Brandolese, A., Brugger, G., Garetti, M. and Misul, E. (1985). "Analisi dei sistemi di produzione manifatturieri." *Finanza, Marketing e Produzione*, 1, 65-97.

Brundtland, G.B. (edited by) (1987). *Our Common Future: World Commission on Environment and Development*. Oxford: Oxford University Press.

Cachon, G. and Terwiesch, C. (2013). *Matching Supply with Demand* (3rd ed.). New York: McGraw-Hill Education.

Candussio F. (2013). "I sistemi informativi a supporto della produzione" in De Toni, A.F., Panizzolo, R. and Villa, A., *Gestione della produzione*. Turin: Isedi.

Carrol T.C. (2013). *Six Sigma for Powerful Improvement*. Boca Raton: CRC Press, Taylor & Francis Group.

Chase, C.W. (2016). *Next Generation Demand Management: People, Process, Analytics, and Technology*. Hoboken, NJ: John Wiley & Sons.

Chase, R.B., Jacobs, F.R. and Aquilano, N.J. (2006). *Operations Management for Competitive Advantage with Global Cases* (11th ed.). Boston: McGraw-Hill-Irwin.

Chase, R.B., Jacobs, R.F., Aquilano, N.J., Grando, A. and Sianesi, A. (2008). *Operations Management nella produzione e nei servizi*. Milan: McGraw-Hill.

Chatfield, C. (1978). "The Holt-winters forecasting procedure." *Journal of the Royal Statistical Society: Series C (Applied Statistics)*, 27(3), 264-279.

Chopra, S. and Meindl, P. (2007). *Supply Chain Management: Strategy, Planning, and Operation* (3rd ed.). Upper Saddle River, NJ: Pearson Prentice-Hall.
Coman, A. and Ronen, B. (2010). "Icarus' predicament: managing the pathologies of overspecification and overdesign." *International Journal of Project Management*, 28(3), 237-244.
Corbett, L.M. (2009). "Sustainable operations management: a typological approach." *Journal of Industrial Engineering and Management*, 2(1), 10-30.
Crosby, P.B. (1984). *Quality Without Tears: The Art of the Hassle-Free Management.* New York: McGraw-Hill.
D'Aveni, R. (2015). "The 3-D printing revolution." *Harvard Business Review*, 93(5), 40-48.
Da Villa, F. (1991). *Logistica manifatturiera: organizzazione e gestione dei sistemi produttivi.* Milan: ETAS.
Da Villa, F. (2000). *La logistica dei sistemi manifatturieri.* Milan: ETAS.
Dale, B.G., van Der Wiele, T. and Van Iwaarden, J. (2013). *Managing Quality.* Hoboken, NJ: John Wiley & Sons.
De Toni, A.F. and Panizzolo, R. (2018). *Sistemi di gestione della produzione.* Turin: Isedi.
Deming W.E. (1982). *Out of the Crisis: Quality, Productivity and Competitive Position.* Cambridge: Cambridge University Press.
Deming, W.E. (1989). *L'impresa di qualità.* Turin: Petrini.
EFQM (2019). *The EFQM Model.* Brussels: EFQM.
Elkington, J. (1994). "Towards the sustainable corporation: win-win-win business strategies for sustainable development." *California Management Review*, 36(2), 90-100.
Elkington, J. (1997). *Cannibals with Forks: The Triple Bottom Line of 21st Century Business.* Oxford: Capstone.
Ellen MacArthur Foundation (2012). *Towards the circular economy: economic and business rationale for an accelerated transition* (vol. 1).
Ellen MacArthur Foundation (2013). *Towards the circular economy: opportunities for the consumer goods sector* (vol. 2).
European Commission (2011). *Communication from the Commission to the European Parliament, the Council, the European Economic and Social Committee and the Committee of the Regions: A renewed EU strategy 2011-14 for Corporate Social Responsibility.* Available at: http://eur-lex.europa.eu/legal-content/EN/TXT/?uri=CELEX:52011DC0681
Feigenbaum, A.V. (1991). *Total Quality Control* (3rd ed.). New York: McGraw-Hill.
Ferrozzi, C., Shapiro, R.D. and Heskett, J.L. (1987). *Logistica e strategia*, Milan, Isedi.
Fleischmann, B. and Meyr, H. (2001). "Supply Chain Planning" in Sebastian, H.J. and Grünert, T. (eds.), *Logistik Management.* Wiesbaden: Vieweg+Teubner Verlag.
Gelderman, C.J. and Semeijn, J. (2006). "Managing the global supply base through purchasing portfolio management." *Journal of Purchasing & Supply Management*, 12(4), 209-217.
Gelderman, C.J. and Van Weele, A.J. (2003). "Handling measurement issues and strategic directions in Kraljic's purchasing portfolio model." *Journal of Purchasing and Supply Management*, 9(5-6), 207-216.
Genc, T.S. and De Giovanni, P. (2020). "Closed-loop supply chain games with innovation-led lean programs and sustainability." *International Journal of Production Economics*, 219, 440-456.
Goldratt, E.M. and Cox, J. (1984). *The Goal: A Process of Ongoing Improvement.* Great Barrington, MA: North River Press.
Goodwin, P. and Wright, G. (2014). *Decision Analysis for Management Judgment.* Chichester: John Wiley & Sons.
Grando, A. (1995). *Organizzazione e gestione della produzione industriale.* Milan: EGEA.

Grando, A. (2018). *Produrre con successo.* Milan: Università Bocconi Editore.
Grando, A. and Turco, F. (2005). "Plant capacity and productivity measurement: conceptual framework and a single-machine case." *Production Planning & Control*, 16(3), 309-322.
Grando, A. and Vicari, S. (2019). *Gestione della tecnologia, dell'innovazione e delle operations.* Milan: EGEA.
Grasley, A. (2006). *Operations Management.* Chichester: John Wiley & Sons.
Harris, F.W. (2013). "How many parts to make at once." *Factory, The Magazine of Management*, 10(2), 135-152.
Hax, A.C. and Candea, A. (1984). *Production and Inventory Management.* Englewood Cliffs, NJ: Prentice-Hall.
Hayes, R.H. and Wheelwright, S.C. (1984). *Restoring our Competitive Edge. Competing Through Manufacturing.* New York: John Wiley & Sons.
Hill, T. (2005). *Operations Management* (2nd ed.). New York: Palgrave Macmillan.
Holweg, M. (2007). "The genealogy of lean production." *Journal of Operations Management*, 25(2), 420-437.
Hübner, R. (2007). *Strategic Supply Chain Management in Process Industries: An Application to Specialty Chemicals Production Network Design.* Berlin, Heidelberg: Springer-Verlag.
Ishikawa, K. (1972). *Guide to Quality Control.* Tokyo: Asian Productivity Organization.
Jacobs, F.R. and Chase, R.B. (2018). *Operations and Supply Chain Management* (15th ed.). New York: McGraw-Hill Education.
Johnson, G., Scholes, K. and Whittington, R. (2005). *Exploring Corporate Strategy* (7th ed.). Harlow: Financial Times/Prentice Hall.
Juran, J.M. (1988). *Juran's Quality Control Handbook.* New York: McGraw-Hill.
Kraljic, P. (1983). "Purchasing must become supply management." *Harvard Business Review*, 61(5), 109-117.
Lee, H., Padmanabhan, V. and Whang, S. (1997). "The bullwhip effect in supply chains." *MIT Sloan Management Review*, 38(3), 93-102.
Lee, H.L., Padmanabhan, V. and Whang, S. (1997). "Information distortion in a supply chain: the bullwhip effect." *Management Science*, 43(4), 546-558.
Leenders, M.R., Fearon, H.E., Flynn, A.E. and Johnson, P.F. (2002). *Purchasing and Supply Management* (11th ed.). Boston, MA: McGraw-Hill Education.
Levitt, J. (2011). *Preventive and Predictive Maintenance* (2nd ed.). New York: Industrial Press.
Lovell, A., Saw, R. and Stimson, J. (2005). "Product value-density: managing diversity through supply chain segmentation." *The International Journal of Logistics Management*, 16(1), 142-158.
Magee, J.F. and Boodman, D.M. (1967). *Production Planning and Inventory Control* (2nd ed.). New York: McGraw-Hill.
Mather, H. (1987). *Bills of Materials.* Homewood, IL: Dow Jones-Irwin.
Mintzberg, H. and Waters, J.A. (1985). "Of strategies: deliberate and emergent." *Strategic Management Journal*, 6(3), 257-272.
Morelli, M., Stabilini, G. and Zerbini, F. (2018). "Spending Analysis @ Barilla." The Case Centre. Available at: http://casecent.re/p/158506
Nakajima, S. (1988). *Introduction to TPM: Total Productive Maintenance.* Cambridge, MA: Productivity Press.
Oakland, J.S. (2008). *Statistical Process Control* (6th ed.). Oxford: Butterworth-Heinemann.
Orlicky, J. (1975). *Materials Requirement Planning.* New York: McGraw-Hill.

Pagh, J.D. and Cooper, M.C. (1998). "Supply chain postponement and speculation strategies: how to choose the right strategy." *Journal of Business Logistics*, 19(2), 13-33.

Pande, P.S., Neuman, R.P. and Cavanagh, R. (2000). *The Six Sigma Way: How GE, Motorola, and Their Top Companies Are Honing Their Performance*. New York: McGraw-Hill.

Perrini, F., Pogutz, S. and Tencati, A. (2006). *Developing Corporate Social Responsibility: A European Perspective*. Northampton, MA: Edward Elgar.

Plossl, G.W. (1994). *Orlicky's Material Requirements Planning* (2nd ed.). New York: McGraw-Hill.

Puschmann, T. and Rainer, A. (2005). "Successful use of e-procurement in supply chains." *Supply Chain Management: An International Journal*, 10(2), 122-133.

Reid, R.D. and Sanders, N.R. (2005). *Operations Management: An Integrated Approach*. Hoboken, NJ: John Wiley & Sons.

Romano, P. and Danese, P. (2010). *Supply Chain Management* (2nd ed.). Milan: McGraw-Hill.

Rother, M. and Shook, J. (2003). *Learning to See: Value Stream Mapping to Add Value and Eliminate Muda*. Cambridge, MA: Lean Enterprise Institute.

Russel, R.S. and Taylor III, B.W. (2009). *Operations Management along the Supply Chain* (6th ed.). Hoboken, NJ: John Wiley & Sons.

Rüßmann, M., Lorenz, M., Gerbert, P., Waldner, M., Engel, P., Harnisch, M. and Justus, J. (2015). "Industry 4.0: the future of productivity and growth in manufacturing industries." Boston Consulting Group [online]. Available at: bcg.com.

Sanders, N.R. (2017). *Forecasting Fundamentals*. New York: Business Expert Press.

Schmenner, R.W. (1981). *Production/Operations Management: Concepts and Situations*. Chicago: Science Research Associates.

Secchi, R. (2000). *Produrre e gestire informazioni per integrare la supply chain*. Milan: Guerini.

Secchi, R. and Rossi, T. (2018). *Fabbriche 4.0: percorsi di trasformazione digitale della manifattura italiana*. Milan: Guerini Next.

Shankar, R. (2009). *Process Improvement Using Six Sigma: A DMAIC Guide*. Milwaukee, WI: ASQ Quality Press.

Shewart, W.A. (1931). *Economic Control of Quality of Manufactured Product*. New York: D. Van Nostrand.

Sianesi, A. (2011). *La gestione del sistema di produzione: pianificazione, programmazione, controllo, misura e miglioramento*. Milan: ETAS.

Siemsen, E. (2015). *Demand Forecasting for Managers*. New York: Business Expert Press.

Skinner, W. (1974). "The focused factory." *Harvard Business Review*, 52(3), 113-121.

Slack, N., Brandon-Jones, A. and Johnston, R. (2016). *Operations Management* (8th ed.). Harlow: Pearson.

Slack, N., Chambers, S. and Johnston, R. (2007). *Operations Management* (5th ed.). Harlow: Financial Times/Prentice Hall.

Stabilini, G. (ed.) (2005). *Acquistare prodotti e servizi: processi, logiche e soluzioni gestionali*. Milan: ETAS.

Taguchi, G. and Clausing, D. (1990). "Robust Quality." *Harvard Business Review*, 68(1), 65-75.

Trent, R.J. (2010). "Creating the Ideal Supplier Measurement System." *Supply Chain Management Review*, March/April.

van Weele, A.J. (2014). *Purchasing and Supply Chain Management* (6th ed.). London: Thomson Learning.

van Weele, A.J. (2018). *Purchasing and Supply Chain Management* (7th ed.). London: Cengage Learning EMEA.

Vollmann, T.E., Berry, W.L. and Whybark, D.C. (1984). *Manufacturing Planning and Control Systems*. Homewood, IL: Dow Jones-Irwin.

Wallace, T.F. and Stahl, R.A. (2008). *Sales & Operations Planning: The How-To Handbook* (3rd ed.). Cincinnati, OH: T.F. Wallace & Co.

Wilson R.H. (1934). "A scientific routine for stock control." *Harvard Business Review,* 13(1), 116-128.

Womack, J.P., Jones, D.T. and Roos, D. (1990). *The Machine that Changed the World*. New York: Rawson Associates.

Wortmann J.C. (1983). "A Classification Scheme for Master Production Schedule" in Berg, C., French, D. and Wilson, B. (ed.), *Efficiency of Manufacturing Systems*. New York: Plenum Press.

Zijm, H., Klumpp, M., Regattieri, A. and Heragu, S. (2019). *Operations, Logistics and Supply Chain Management*. Cham: Springer.

About the Authors

Alberto Grando is Professor of Operations Management at the Management and Technology Department of Bocconi University, Milan (Italy), where he teaches Operations Management, Sustainable Operations Management, and Innovation and Technology Management. He is also Professor of Production & Supply Chain Management at Operations and Technology Management Knowledge Unit of SDA Bocconi School of Management. His research interests focus on manufacturing performances measurement, supply chain and operations management.

Valeria Belvedere is Associate Professor of Operations and Supply Chain Management at the Catholic University of the Sacred Hearth, Milan, where she is the Director of the MSc Innovation and Technology Management. She teaches courses in the field of Operations and Supply Chain Management both in undergraduate and graduate courses. Her main fields of research are manufacturing strategies, manufacturing and logistics performance measurement systems, technology-adoption in operations, sustainable operations management.

Raffaele Secchi is Professor of Operations Management at LIUC – Università Cattaneo, Castellanza (Italy), where he teaches Operations and Supply Chain Management. Since January 2017, he is the Dean of LIUC Business School. At Bocconi University he teaches Supply Chain Management and Technology, Innovation and Operations Management. He is also visiting professor at IÉSEG School of Management. His research interests focus on lean management, Industry 4.0 and the digital transformation of operations.

Giuseppe Stabilini is Associate Professor of Practice of Procurement & Supply Management at SDA Bocconi School of Management, where he is the Operations and Technology Management Faculty Deputy. He is the Director of the Executive Program in Procurement & Supply Management and head of the Supply Chain Management module in the Full Time MBA. At Bocconi University he teaches Global Operations & Supply Chain Management. His research activities focus on chief procurement officer responsibilities and decisions in industrial or service companies.

www.ingramcontent.com/pod-product-compliance
Ingram Content Group UK Ltd.
Pitfield, Milton Keynes, MK11 3LW, UK
UKHW051448290926
15471UKWH00008B/1085

9 788831 322416